Structural Performance

Structural Performance

Probability-based Assessment

Christian Cremona

First published 2011 in Great Britain and the United States by ISTE Ltd and John Wiley & Sons, Inc.

ISTE Ltd
27-37 St George's Road
London SW19 4EU
UK

www.iste.co.uk

John Wiley & Sons, Inc.
111 River Street
Hoboken, NJ 07030
USA

www.wiley.com

© ISTE Ltd 2011

Library of Congress Cataloging-in-Publication Data

Cremona, Christian.
  Structural performance : probability-based assessment / Christian Cremona.
    p. cm.
  Includes bibliographical references and index.
  ISBN 978-1-84821-236-7 (hardback)
  1. Structural analysis (Engineering)--Statistical methods. 2. Structural failures--Risk assessment--Statistical methods. I. Title.
  TA640.2.C84 2011
  624.1'71--dc22

2011003204

British Library Cataloguing-in-Publication Data
A CIP record for this book is available from the British Library
ISBN 978-1-84821-236-7

Printed and bound in Great Britain by CPI Antony Rowe, Chippenham and Eastbourne.

Table of Contents

Preface

"When one admits that nothing is certain one must, I think, also admit that some things are much more nearly certain than others."

Bertrand Russell, Am I An Atheist Or An Agnostic?

The development of useful and relevant methods for assessing and managing structures has become an important challenge today. The economic, societal, and environmental stakes attached to this development are a growing concern for public or private owners and stakeholders. The need for relevant and efficient approaches, taking into account the uncertainties to do with loading, geometry, material properties, design and construction, and their operating conditions, is widely felt. The reliability theory, which is based on a probabilistic formulation of structural performances, conceptually answers these questions, in an adapted way. Nevertheless, it often raises difficulties, on the theoretical front as well as the numerical and practical front. However, the theory provides an original alternative for requalifying structures.

This book has been designed to provide students, engineers, and researchers with a panorama of methods in order to implement a probabilistic approach to structural performance in their respective frameworks. This book is the product of much teaching in engineering schools, or in continuing education. Also, it presents different concepts by giving examples which, mostly, can be carried out *by hand*. Based on measuring the possible on concrete examples, these concepts try both to illustrate theoretical approaches and to demonstrate their interest and practical implementation. This book is then, neither a textbook on reliability theory, nor a technical set of guidelines for assessing structure's performances. It is something in between, passing through the field of theoretical concepts, as well as exploring practical problems found in structural assessments.

This book is divided into six chapters. The reader may be surprised to find a long chapter (Chapter 1), referring back to probabilities and statistics. This cannot be overlooked as the reader needs to have a minimum theoretical knowledge in this domain, in order to approach the following chapters peacefully. Chapter 2 presents a whole series of concepts on hazards and performances. It summarizes the basic concepts used in the other chapters. Chapters 1 and 2 form the backbone of this book, with the other chapters develop or branching off from previously presented concepts. Chapter 3 introduces the concepts of reliability theory. It does not try to present its theoretical elements as exhaustively as possible, but to highlight its practical implementation. It is, then, intended for students, engineers and researchers, who – across the range of problems that they are studying – want to translate their studied problems into the principles of reliability theory. Chapter 4 defines performance ratings, and their practical implementation. To a certain extent, it draws the link between probabilistic concepts and the reliability theory of structures, as well as the problems of structural assessment. As this type of study requires a probabilistic description of variables, Chapter 5 therefore proposes various models and data, allowing an *a priori* performance study that could be updated as soon as knowledge is gradually improved. The book ends on a chapter (Chapter 6) devoted to decision theory. A series of concepts for maintenance and inspection are presented within this chapter, along with their contribution, their integration into performance assessment and their help for the management of structures.

This book is the fruit of many exchanges and contributions, both in France and abroad. The list would be too long to name them all, and some names risk being forgotten: let them all be assured that they have my most sincere gratitude. Nonetheless, I would like to directly thank C. Bois from the *Conseil général des Ponts et Chaussées*, T. Kretz and J. Raoul at the *Service d'études sur les transports, les routes et leurs aménagements,* B. Godart, B. Mahut, A. Orcesi, A. Patron (today *Consultora Mexicana de Ingenieria*) and M. Thiery from the *Laboratoire central des Ponts et Chaussées* (today *Institut Français des Sciences et Technologies des Transports, de l'Aménagement et des Réseaux*) for their encouragement and advice for the introduction and development of a probabilistic approach to the assessment and management of existing bridges. Many ex-students have also greatly contributed to making this book: G. Byrne, M. Lukic, S. Mohammadkhani, O. Rasson, B. Richard, C. Santerelli, H. Sempere. Their contribution has been considerable in the creation of this book. In terms of producing this manuscript, I am grateful to C. Ménascé and A. Toulze from Hermes-ISTE for their effort to publish this book. Also K. Docwra did a valiant job translating a mass of French texts. The book would certainly not exist in this form without the effort of these people.

I would finally like to thank my wife, Françoise, for her patience and encouragement during the weekends and holidays spent dedicated to writing this book.

Christian CREMONA

Chapter 1

Concepts from Probability Theory and Statistics

1.1. The role of probability in civil engineering

Today, the civil engineering domain has many quantitative modeling and analysis methods at its disposal. According to their degree of sophistication, these methods rely on idealized hypotheses, meaning that the information retrieved from these models may more or less reflect reality. In the development of structural design, but also in its management (which is the framework for this book), decisions are made independently of the information quality. In this sense, the consequences of a decision may be determined with perfect confidence. Not counting the fact that this information is deduced by modeling processes or imperfect knowledge, many problems in civil engineering involve phenomena which are, essentially, random. The effect of these uncertainties on the performance of structures must be approached with appropriate mathematical methods, and the theory of probability is one of these methods.

The theory of probability is an essential element for apprehending structural safety. Yet, if many engineers in the civil engineering domain recognize the rationality and importance of probabilistic modeling and certain phenomena (studies in the behavior of bridges towards wind have familiarized the technical community with the random nature of certain physical phenomena), the number of engineers familiar with probability theory is still low.

The importance of probabilistic and statistical concepts gives us reason to approach them in the first chapter of this book, rather than in an appendix which is often the case. These concepts are an invaluable prerequisite for fully understanding the principles of structural safety. It is not, however, the objective of this chapter to propose a lesson on the subject, but to present a sufficiently exhaustive panorama. For this, the chapter is strewn with application examples taken from practical examples that the engineer may encounter.

1.2. Physical and statistical uncertainties

When a particular phenomenon is studied (load variability, for example), it is therefore necessary to distinguish two types of uncertainty: *intrinsic uncertainty* and *statistical uncertainty*. Intrinsic or physical uncertainty concerns the random nature of physical phenomena. Physical variables are described by random variables or stochastic processes. Statistical uncertainty is used for assessing the parameters which describe the probabilistic models taken from sampling.

Statistical analysis is the most traditional method of characterizing physical uncertainties. To study a random phenomenon, firstly we must specify the conditions necessary to observe it. When these conditions are fixed by the observer, they define a *randomized experiment*. In the opposite case, they characterize a random test during which the phenomenon takes place (wind, rain, etc.).

When defining an experiment known as $\mathcal{E}$, the experimenter must list the set Ω of all the possible *outcomes* or *observations*, called sample set. Here, there is a real difficultly which we must be wary of. Another phenomenon of the same nature may help to define many randomized experiments. In addition, the set of possible results from a randomized experiment may be an *infinite countable* or an *uncountable* set. Very often, the envisaged experiment will be comprised of many stages: for example, measuring the temperature each day. Each stage, by itself, is an experiment with a corresponding sample set. The sample set Ω_i is associated with the i-th stage; it is, then, clear that the sample set of the complete experiment is Ω, the Cartesian product of Ω_i. When each stage is indexed (per each day of measurement, as for temperature for example), the succession of experiments is called the *process*.

EXAMPLE 1.1.– Let us consider fifteen reinforced concrete beams, made from the same concrete. These beams are subject to a three points bending test. For each test, the load corresponding to the appearance of the first crack is recorded. Each loading test constitutes a randomized test $\mathcal{E}$. The set of possible results, or sample set Ω is *a priori* the set of real possible outcomes (uncountable). A test provides a result or a sample point. Carrying out 15 tests constitutes an experiment whose the set for possible outcomes is made up of 15 values, $(F_1, \cdots, F_n)$ where F_i is the outcome or observation (load making the first crack appear, rounded to the nearest kN) from the i-th test.

47	38	32	23	29
48	27	27	43	29
42	27	27	26	29

□

Often, in a randomized experiment, the experimenter considers the studied phenomenon from different angles, translating the appearance of an event as a realization of one of many outcomes. Of course, the experimenter may also be interested in an event that does not occur. The immediate consequence of this definition is that an *event* is a sub-set of the sample set Ω.

1.3. Axiomatics

1.3.1. *Probabilities*

The probability of an event may be considered as a number, assigned to a statement about an outcome of an experiment, that expresses the chance that the statement is true. This number, called the *probability measure*, can be established by:

– assumptions on the mechanisms controlling the appearance of the events;

– frequencies of past observations;

– intuitive or subjective assumptions.

The probability of an event A is written $P(A)$ where $P(.)$ is *the measure of probability*.

1.3.2. *Axioms*

Measures of probability must, however, verify a certain number of axioms:

– that the probability of an event A is between 0 and 1;

– that the probability $P(\Omega)$ of the sample set is equal to 1 (the event Ω is called the *certain event*). The probability $P(\varnothing)$ of the empty set is equal to 0 (the event $\varnothing$ is called the *impossible event*);

– the probability of the sum of two events A_1, A_2 allowing an empty intersection $A_1 \cap A_2 = \varnothing$ (these events are known as *mutually exclusive*, i.e. the two events cannot be true simultaneously) is equal to the sum of the probability for each of the events:

$$P(A_1 \cup A_2) = P(A_1) + P(A_2)$$
[1.1]

There are many ways of defining a *probability* $\mathcal{P}$. The first approach is called *classical probability*:

$$P(A) = \frac{\text{number of sample points belonging to } A}{\text{number of possible outcomes}}$$

This definition is based on the principle of equal probability: in fact, all the possible outcomes are equally likely.

EXAMPLE 1.2.– An engineering office is responsible for building a structure requiring an energy and water provider. Let us presume that the need for water is around 10 or 20 units, and the need for energy is around 30 or 60 units, thus the possible set of outcomes is $\{E_{10}P_{30}, E_{10}P_{60}, E_{20}P_{30}, E_{20}P_{60}\}$, where E_iP_j denotes the need for i water units and j energy units. When there is no assumptions for these needs, equal probability will be presumed, which leads us to assign the probabilities $\mathcal{P}(E_iP_j) = 1/4$. □

The second approach is called *empirical probabilities*. We estimate that an experiment can be reproduced many times, and we will count the number of times that the event A – for which we are trying to find the probability – occurs. Its probability is then assessed by the ratio:

$$P(A) = \frac{\text{number of times } A \text{ occurs in } n \text{ test}}{\text{number of tests}}$$

EXAMPLE 1.3.– Let us take the data from example 1.1 and consider the event $A = \{F_i \leq 27\}$. The empirical probability for this event can be defined by $\mathcal{P}(A) = 6/15 = 0.4$. □

EXAMPLE 1.4.– Let us use example 1.2. The events E_iP_j are obviously incompatible. In this case, the probability of i water units is:

$$\mathcal{P}(E_iP_{30} \cup E_iP_{60}) = \mathcal{P}(E_iP_{30}) + \mathcal{P}(E_iP_{60}) = 0.25 + 0.25 = 0.5 \qquad □$$

Finally, a third approach is called *subjective probability*. This is often used for everyday situations; for example, "there is a 90% chance of there being wind tomorrow". Formally, the event of "there being wind tomorrow" has a probability of 90%, or 0.9. Here we are dealing with an opinion or a degree of subjective confidence on the event.

1.3.3. *Consequences*

Various results can be deduced immediately for the definition of a probability measure:

$$\forall \ A, \ P(A) = 1 - P(\overline{A})$$
$$\forall \ A, B, \ A \subset B, \ P(A) \leq P(B) \qquad\qquad [1.2]$$
$$\forall \ A, B, \ P(A \cup B) = P(A) + P(B) - P(A \cap B)$$

EXAMPLE 1.5.– For example 1.2, the probability of the demand for 10 water units, or the need for 60 energy units can be calculated as the following:

$$P\left(\left(E_{10}P_{30} \cup E_{10}P_{60}\right) \cup \left(E_{10}P_{60} \cup E_{20}P_{60}\right)\right)$$
$$= P\left(E_{10}P_{30} \cup E_{10}P_{60}\right) + P\left(E_{10}P_{60} \cup E_{20}P_{60}\right) - P\left(\left(E_{10}P_{30} \cup E_{10}P_{60}\right) \cap \left(E_{10}P_{60} \cup E_{20}P_{60}\right)\right)$$
$$= (0.25 + 0.25) + (0.25 + 0.25) - P\left(E_{10}P_{60}\right) = 1 - 0.25 = 0.75$$

□

1.3.4. *Conditional probabilities*

Let us consider the probability of the intersection of two events $P(A \cap B)$; we define the *conditional probability* $P(A/B)$ of an event A given B, by:

$$P(A/B) = \frac{P(A \cap B)}{P(B)} \qquad\qquad [1.3]$$

In an identical way, it is possible to construct the probability B given A, $P(B/A)$, so we can write:

$$P(A \cap B) = P(B)\,P(A/B) = P(A)\,P(B/A) \qquad\qquad [1.4]$$

If (B_n) is a collection of events verifying $P\left(\bigcup_{n \in N} B_n\right) = 1$, $P(B_i) > 0$, then the following result may be expressed:

$$\forall A \in T, \quad P(A) = \sum_{n} P(B_i)\,P(A/B_i) \qquad\qquad [1.5]$$

This result is known as the *law of total probability*.

If we take the same hypotheses as for the previous theorem, moreover by assuming that B_n are incompatible in pairs, thus, the following rule (known as the *Bayes theorem*) can be stated:

$$\forall A \in T, \quad \mathcal{P}(B_i \,/\, A) = \frac{\mathcal{P}(B_i)\,\mathcal{P}(A\,/\,B_i)}{\sum_n \mathcal{P}(B_n)\,\mathcal{P}(A\,/\,B_n)} \qquad [1.6]$$

The two events A and B are said to be *independent*, if the following condition is verified:

$$\begin{cases} \mathcal{P}(A\,/\,B) = \mathcal{P}(A) \\ \text{or} \\ \mathcal{P}(A \cap B) = \mathcal{P}(A)\,\mathcal{P}(B) \end{cases} \qquad [1.7]$$

EXAMPLE 1.6.– (adapted from [BEN 70]) Let us consider that an engineer has assessed the compressive strength of a concrete from an existing structure. By studying the structure's documents, he proposes to assign the strength into three possible clusters: C1, C2 and C3, by associating them with *a priori* probabilities (0.3; 0.5; 0.2). Core samples are taken from the structure to better help specify the structure's concrete cluster. The engineer knows that the information coming from a test is reasonably reliable, but not, however, conclusive. Consequentially, it attributes a confidence probability to the results coming from the test on a core sample:

– if the concrete really belongs to cluster C1, test E will indicate cluster C1 in 70% of cases, cluster C2 in 20% of cases, and cluster C3 in 0% of cases;

– if the concrete really belongs to cluster C2, test E will indicate cluster C1 in 30% of cases, cluster C2 in 60% of cases, and cluster C3 in 30% of cases;

– if the concrete really belongs to cluster C3, test E will indicate cluster C1 in 0% of cases, cluster C2 in 20% of cases, and cluster C3 in 70% of cases.

Let us assume that a core sample is taken from the structure and that it indicates an E1 outcome (C1 concrete from the test). The conditional probabilities of the concrete belonging to each of the clusters knowing this outcome, and the associated uncertainties, are given by the Bayes theorem:

$$P(C1/E1) = \frac{P(E1/C1)\,P(C1)}{P(E1/C1)\,P(C1)+P(E1/C2)\,P(C2)+P(E1/C3)\,P(C3)}$$

$$= \frac{0.7\times0.3}{0.7\times0.3+0.3\times0.5+0.0\times0.2} \approx 0.583$$

$$P(C2/E1) = \frac{0.3\times0.5}{0.7\times0.3+0.3\times0.5+0.0\times0.2} \approx 0.417$$

$$P(C3/E1) = \frac{0.0\times0.2}{0.7\times0.3+0.3\times0.5+0.0\times0.2} = 0.0$$

Sample E1 may exclude the C3 cluster and favor the C1 cluster. To reduce the uncertainty of determining the concrete cluster, the engineer may take an additional core sample which, this time, will indicate a C2 concrete (outcome E2). If we presume that this new outcome is independent from the first one, we obtain:

$$P(C1/E1\cap E2) = \frac{P(E1\cap E2/C1)\,P(C1)}{P(E1\cap E2/C1)\,P(C1)+P(E1\cap E2/C2)\,P(C2)+P(E1\cap E2/C3)\,P(C3)}$$

$$= \frac{P(E1/C1)P(E2/C1)\,P(C1)}{P(E1/C1)P(E2/C1)\,P(C1)+P(E1/C2)\,P(E2/C2)P(C2)+P(E1/C3)P(E2/C3)\,P(C3)}$$

$$= \frac{0.7\times0.2\times0.3}{0.7\times0.2\times0.3+0.3\times0.6\times0.5+0.0\times0.3\times0.2} \approx \frac{0.042}{0.132} \approx 0.318$$

$$P(C2/E1\cap E2) = \frac{0.3\times0.6\times0.5}{0.132} \approx 0.682 \text{ and } P(C3/E1\cap E2) = 0.0$$

Let us have a look closer on the first equation. It can be rewritten as follows thanks to the Bayes theorem:

$$P(C1/E1\cap E2) = \frac{P(E1/C1)P(E2/C1)\,P(C1)}{P(E1/C1)\,P(C1)+P(E1/C2)\,P(C2)+P(E1/C3)\,P(C3)}$$

$$\times \frac{P(E1/C1)\,P(C1)+P(E1/C2)\,P(C2)+P(E1/C3)\,P(C3)}{P(E1/C1)P(E2/C1)\,P(C1)+P(E1/C2)\,P(E2/C2)P(C2)+P(E1/C3)P(E2/C3)\,P(C3)}$$

$$= \frac{P(E2/C1)P(C1/E1)}{P(E2/C1)\,P(C1/E1)+P(E2/C2)\,P(C2/E1)+P(E2/C3)\,P(C3/E1)}$$

$$= \frac{0.2\times0.583}{0.2\times0.583+0.6\times0.417+0.2\times0.0} \approx \frac{0.175}{0.367} \approx 0.318$$

The second calculation consists, then, of replacing the *a priori* probabilities (0.3; 0.5; 0.2) by *a posteriori* conditional probabilities which were obtained from E1

(0.583; 0.417; 0.000). These *a posteriori* probabilities will become new *a priori* probabilities for calculating the conditional probabilities in relation to E2. Thanks to the Bayes theorem, updating these probabilities can be achieved when new information becomes available. We will note on this example that the second test largely favors cluster C2, which is coherent with the engineer's *a priori* idea. A third test will help to decide between clusters C2 and C3.   □

1.4. Random variables – distributions

1.4.1. *Definitions*

A probability function, however it is defined, tries to describe the random nature of a phenomenon through an analysis of its possible outcomes. Thus, wind is a random phenomenon which can be described by its speed. Wind speed is therefore also a randomized measurement: we are, then, referring to *random variables*.

A *random variable X* is therefore an application of the set of outcomes Ω in the set of real numbers $\mathbb{R}$:

$$
\begin{aligned}
X: \quad & \Omega \;\rightarrow\; \mathbb{R} \\
& \omega \;\rightarrow\; X(\omega)
\end{aligned}
\qquad [1.8]
$$

The probability measure $\mathcal{P}$ of the random phenomenon represented by the variable X is called the *probability distribution* of X and is written $\mathcal{P}_X(.)$.

EXAMPLE 1.7.– Let us use example 1.1. The beam test is a random experiment; the force value corresponding to the appearance of a crack is its outcome. The random nature of the test can be seen by this force value which, then, defines a random variable.   □

1.4.2. *Sampling*

Sampling is a random experiment which consists of randomly generating one or many sample values from the set of possible outcomes. This sub-set of sample values is called a *sample*.

EXAMPLE 1.8.– Example 1.1 is an example of sampling. The 15 obtained values constitute a sample.   □

A sample can be any size, but we can see intuitively that the bigger the sample, the more relevant its information drawn from the analysis will be. A traditional representation of a sample $(x_1,\cdots,x_n)$ $(x_1,\cdots,x_n)$ is the *histogram*. This can be rough or standardized. It is generated by dividing the interval of the sample's variation $\left[\min_i x_i, \max_i x_i\right]$ by a number of smaller intervals, known as the bins of the histogram. For each given bin, we count the number of times that a outcome from the sample belongs to this bin. If the number of occurrences in a bin is divided by n, the total number of outcomes, the value for each bin interval is then called the frequency and the histogram is called the frequency distribution (Figure 1.1). The sum of the class frequencies must be equal to 1.

Let us write these different frequencies as f_i. Let us associate to each bin, the successive partial sum of frequencies up to this interval, and connect the points between them. The created function is a cumulative function (Figure 1.1). The distribution for a random variable is entirely determined by knowing this function. This function is called the *probability distribution function* or *cumulative distribution function* (CDF) of the random variable X.

$$F_X(x) \underset{\text{notation}}{=} P(X \leq x) \underset{\text{notation}}{=} P(\omega \in \Omega / X(\omega) \leq x) \qquad [1.9]$$

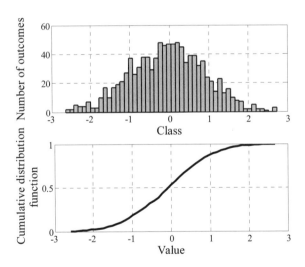

Figure 1.1. *Frequency histogram and cumulative distribution function*

EXAMPLE 1.9.– Thirty wire segments are sampled from a suspension cable. The rupture resistance F_i is given in kN to the nearest decimal.

24.4	25.5	25.0	24.1	23.2	25.9	24.7	24.8	24.4	23.1
22.6	22.4	24.9	23.9	24.7	23.4	23.7	25.4	23.0	21.7
24.9	22.5	25.5	22.4	24.3	25.0	24.9	22.8	25.7	24.0

The interval $\left[\min_i F_i, \max_i F_i\right]$ is $[21.7; 25.9]$. The histogram is based on 0.5 kN intervals starting from 20 kN. This is illustrated in the figure below, which also gives the cumulative distribution function.

$$F_F\left(20 + 0.5\left(i - 1\right)\right) = \frac{1}{30}\sum_{k=1}^{i-1} N_i \quad i = 2, \cdots, 14$$

$$F_F\left(20\right) = 0.0; \quad F_F\left(27\right) = 1.0$$

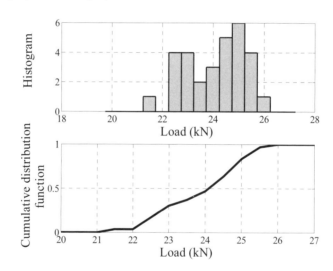

1.4.3. *Probability density function*

We say that the distribution $\mathcal{P}_X\left(.\right)$ of X has a density function, if an $f\left(.\right)$ function exists, verifying:

$$\int_{\mathbb{R}} f_X(x)dx = 1; \quad F_X(x) = \int_{-\infty}^{x} f_X(u)\,du \qquad [1.10]$$

The function $f(.)$ is the *probability density function* (PDF) of the variable X. According to [1.10], we then have: $f(x) = F'(x)$. The density function is therefore a continuous form of the histogram in Figure 1.1. In particular, we obtain:

$$F_X(x) = \mathcal{P}(X \leq x) = \int_{-\infty}^{x} f_X(u)\, du \qquad [1.11]$$

$$\mathcal{P}(x_1 \leq X \leq x_2) = F_X(x_2) - F_X(x_1) = \int_{x_1}^{x_2} f_X(u)\, du \qquad [1.12]$$

EXAMPLE 1.10.– Let us consider a concrete slab carrying a load. The load can be localized at any point. The probability that this load will be located in a particular domain is assumed to depend on its distance D in relation to the sides of the slab. If this slab has the following dimensions $2L \times 2L$, this hypothesis leads us to write that the probability of D being lower than a value d is:

$$\mathcal{P}(D \leq d) = F_D(d) = \frac{\text{shaded area}}{\text{total area}}$$

$$= \frac{4L^2 - (2L - 2d)^2}{4L^2} \qquad 0 \leq d \leq L$$

$$= 1 - \left(\frac{L-d}{L}\right)^2$$

The density function is, then, easily obtained by deriving this distribution function:

$$f_D(d) = F_D'(d) = 2\frac{L-d}{L^2} \qquad \square$$

1.4.4. *Main descriptors of a random variable*

An important and practical summary of the properties of a random variable may be obtained using a certain number of parameters that we can classify, primarily into three categories.

1.4.4.1. *Location measures*

These parameters give us an idea of the amplitude of the variable. The *mean* of a random variable X is given by:

$$\mathbb{E}(X) = \int_{-\infty}^{+\infty} x \, f(x) dx \qquad [1.13]$$

The mean is also called the first order *moment* of the variable X.

A *quantile* $\mathbb{Q}_{i,p}(X)$ is a solution of the equation:

$$F_X\left(\mathbb{Q}_{i,p}(X)\right) = \frac{i}{p} \quad 1 \le i \le p-1 \qquad [1.14]$$

When $p = 4$, the quantiles are called *quartiles*, first, second, and third respectively $(i = 1, 2, 3)$. The second quantile is also called the *median* of X.

In general, the fractiles $\mathbb{F}_p(X)$ are defined as solutions for the equations:

$$F_X\left(\mathbb{F}_p(X)\right) = p \qquad [1.15]$$

where p is a probability. The fractile concept is often used for characterizing material properties and loads. In fact, a fractile of order p of a random variable X, means that the probability of exceeding this value is $1 - p$:

$$P\left(X \ge \mathbb{F}_p(X)\right) = 1 - F_X\left(\mathbb{F}_p(X)\right) = 1 - p \qquad [1.16]$$

EXAMPLE 1.11.– The lifespan of a structural component Q is a random variable described by a density function $f_Q(q) = a \, e^{-aq} \ q \ge 0$. The mean of this variable is therefore given by:

$$\mathbb{E}(Q) = \int_0^\infty q \, f_Q(q) \, dq = \int_0^\infty aq \, e^{-aq} \, dq$$

$$= \left[aq \frac{e^{-aq}}{-a} \right]_0^\infty + \int_0^\infty e^{-aq} \, dq = \left[\frac{e^{-aq}}{-a} \right]_0^\infty = \frac{1}{a}$$

The pth order fractile is given by:

$$F_Q(\mathbb{F}_p(Q)) = 1 - \exp\left(-a \, \mathbb{F}_p(Q)\right) = p \Rightarrow \mathbb{F}_p(Q) = \frac{1}{a} \ln \frac{1}{1-p} \qquad \square$$

1.4.4.2. *Dispersion measures*

The *variance* of X is defined by:

$$\mathbb{V}(X) = \mathbb{E}\left(\left(X - E(X)\right)^2\right) = \int_{-\infty}^{+\infty} (x - \mathbb{E}(X))^2 \, f(x) \, dx \qquad [1.17]$$

Variance is also called the *second order moment* of the variable X. *The standard deviation* and the *coefficient of variation* of the variable X are defined by:

$$\sigma(X) = \sqrt{\mathbb{V}(X)}; \; \mathbb{COV}(X) = \frac{\sigma(X)}{\mathbb{E}(X)} \qquad [1.18]$$

These parameters can be used to measure the variable's dispersion around its mean value.

EXAMPLE 1.12.– Let us go back to example 1.11. The standard deviation of the equipment's lifespan is calculated in an identical way to the mean:

$$\sigma(Q) = \int_0^\infty \left(q - \frac{1}{a}\right)^2 f_Q(q) \, dq = \left[-\left(q - \frac{1}{a}\right)^2 e^{-aq}\right]_0^\infty + \int_0^\infty 2\left(q - \frac{1}{a}\right) e^{-aq} \, dq$$

$$= \frac{1}{a^2} + \frac{2}{a} \int_0^\infty \left(q - \frac{1}{a}\right) a \, e^{-aq} \, dq$$

$$= \frac{1}{a^2} + \frac{2}{a} \left(\left[-\left(q - \frac{1}{a}\right) e^{-aq}\right]_0^\infty + \frac{1}{a} \int_0^\infty a \, e^{-aq} \, dq\right)$$

$$= \frac{1}{a^2} + \frac{2}{a} \left(-\frac{1}{a} + \frac{1}{a}\left[-e^{-aq}\right]_0^\infty\right) = \frac{1}{a^2} + \frac{2}{a}\left(-\frac{1}{a} + \frac{1}{a}\right) = \frac{1}{a^2}$$

$\square$

1.4.4.3. *Shape measures*

We can also define k-th *order moments*:

$$\mathbb{E}\left[(X - \mathbb{E}[X])^k\right] = \int_{-\infty}^{+\infty} (x - \mathbb{E}[X])^k \, f(x) dx \qquad [1.19]$$

The third order moment divided by $\sigma(X)^3$ is called the skewness coefficient and can be used to measure the symmetry of variable X in relation to the mean:

$$\gamma_1 = \frac{\mathbb{E}\left[\left(X - \mathbb{E}[X]\right)^3\right]}{\sigma(X)^3} \qquad\qquad [1.20]$$

The fourth order moment divided by $\sigma(X)^4$ is called the Kurtosis coefficient and assesses the flatness of the distribution:

$$\gamma_2 = \frac{\mathbb{E}\left[\left(X - \mathbb{E}[X]\right)^4\right]}{\sigma(X)^4} \qquad\qquad [1.21]$$

The Kurtosis coefficient measures the concentration intensity of the variable around its mean. It is usually compared to the value 3 (which is the Kurtosis coefficient value for a normal variable). If $\gamma_2 - 3 > 0$, the variable is said to be more peaked than the normal distribution, otherwise it is said to be flatter. The coefficients γ_1 and $\gamma_2 - 3$ are sometimes called the first and second *Fisher coefficients*.

EXAMPLE 1.13.– For the variable defined by the density function in example 1.11, the first and second Fisher coefficients for the component's lifetime are given by:

$$\gamma_1(Q) = \frac{1}{\sigma(Q)^3} \int_0^\infty \left(q - \frac{1}{a}\right)^3 f_Q(q)\, dq$$

$$= \frac{1}{\sigma(Q)^3} \left[-\left(q - \frac{1}{a}\right)^3 e^{-aq}\right]_0^\infty + \frac{1}{\sigma(Q)^3} \frac{3}{a} \underbrace{\int_0^\infty a\left(q - \frac{1}{a}\right)^2 e^{-aq}\, dq}_{\sigma(Q)^2}$$

$$= \frac{1}{\sigma(Q)^3} \left(-\frac{1}{a^3} + \frac{3}{a^3}\right) = \frac{1}{\frac{1}{a^3}} \frac{2}{a^3} = 2 > 0$$

$$\gamma_2(Q) = \frac{1}{\sigma(Q)^4} \int_0^\infty \left(q - \frac{1}{a}\right)^4 f_Q(q)\, dq$$

$$= \frac{1}{\sigma(Q)^4} \left[-\left(q - \frac{1}{a}\right)^4 e^{-aq}\right]_0^\infty + \frac{1}{\sigma(Q)} \frac{4}{a} \frac{1}{\sigma(Q)^3} \underbrace{\int_0^\infty a\left(q - \frac{1}{a}\right)^3 e^{-aq}\, dq}_{\gamma_1(Q)}$$

$$= \frac{1}{\sigma(Q)^4} \left(\frac{1}{a^4}\right) + \frac{1}{\sigma(Q)} \frac{4}{a} 2 = \frac{1}{\frac{1}{a^4}} \frac{1}{a^4} + \frac{1}{\frac{1}{a}} \frac{8}{a} = 9 > 3$$

The distribution therefore has a long right tail and is less flattened than the normal distribution.   □

1.4.5. *Joint variables*

When two or many random variables are considered simultaneously, their joint behavior is, then, determined by a *joint probability distribution*. In an identical way to the case of a single variable, a *joint cumulative distribution function* may be expressed by[1]:

$$F_{X,Y}(x,y) = \mathcal{P}\left(X \le x \cap Y \le y\right) = \int_{-\infty}^{x} \int_{-\infty}^{y} f_{X,Y}(u,v)\, du\, dv \qquad [1.22]$$

where $f_{X,Y}(.,.)$ is the *joint probability density function* subject to whether the joint distribution function is derivable in relation to x and y: in this case, we obtain:

$$f_{X,Y}(x,y) = \frac{\partial^2}{\partial x \partial y} F_{X,Y}(x,y) \qquad [1.23]$$

It is sometimes interesting to have *marginal density functions* for the variables X and Y. For variable X for example, we must integrate the joint density on the set of values taken by Y:

$$f_X(x) = \int_{-\infty}^{+\infty} f_{X,Y}(x,y)\, dy \qquad [1.24]$$

By extending the concept of conditional probability, the concept of the *conditional density function* can be defined as:

$$f_{X/Y}(x,y) = \frac{f_{X,Y}(x,y)}{f_Y(y)} \qquad [1.25]$$

The *conditional distribution function* is, then, given by:

$$F_{X/Y}(x,y) = \int_{-\infty}^{x} f_{X/Y}(u,y)\, du \qquad [1.26]$$

---

1 These definitions are generalized for *n* variables.

As for the random variables, it is possible to define *means* and *conditional variances*:

$$\mathbb{E}(Y/X) = \mathbb{E}(Y/X = x) = \int_{-\infty}^{+\infty} y \, f_{Y/X}(y,x) \, dy \qquad [1.27]$$

$$\mathbb{V}(Y/X) = \mathbb{V}(Y/X = x) = \int_{-\infty}^{+\infty} (y - \mathbb{E}(Y/X))^2 \, f_{Y/X}(y,xx) \, dy \qquad [1.28]$$

1.4.6. *Independent variables*

When the conditional density function $f_{X/Y}(x,y)$ is identical to $f_X(x)$, X and Y are said to be *independent*. This means that the events related to X are independent of those related to Y. We thus obtain:

$$f_{X/Y}(x,y) = f_X(x); \, f_{Y/X}(x,y) = f_Y(y)$$
$$F_{X/Y}(x,y) = F_X(x); \, F_{Y/X}(x,y) = F_Y(y) \qquad [1.29]$$
$$f_{X,Y}(x,y) = f_X(x) \, f_Y(y); \, F_{Y,X}(x,y) = F_X(x) \, F_Y(y)$$

1.4.7. *Correlation coefficient*

The *correlation coefficient* is defined by:

$$\rho(X,Y) = \frac{\int_{-\infty}^{+\infty} \int_{-\infty}^{+\infty} [x - \mathbb{E}(X)][y - \mathbb{E}(Y)] f_{X,Y}(x,y) \, dx \, dy}{\sigma(X) \, \sigma(Y)} \qquad [1.30]$$

This coefficient is always understood to be between −1 and +1. The correlation coefficient is a measurement of the *linear dependence* between two variables. If $\rho(X,Y) = 0$, the two variables X and Y are not related linearly, but may be related in another way (non-linear relationship). If the two variables are independent, then $\rho(X,Y) = 0$. The reciprocal proposition is generally false. When $\rho(X,Y) = 0$, the two variables X and Y are said to be *uncorrelated*. Figure 1.2 gives an outline of the meaning of a variation coefficient value.

The term $\displaystyle\int_{-\infty}^{+\infty}\int_{-\infty}^{+\infty}\big[x-\mathbb{E}(X)\big]\big[y-\mathbb{E}(Y)\big]f_{X,Y}(x,y)\,dx\,dy$ is called the *covariance*,

$\mathbb{COV}\big[X,Y\big]$ between X and Y. The means of X and Y are deduced, either by marginal distributions, or by the joint distribution:

$$\begin{aligned}
\mathbb{E}(X) &= \int_{-\infty}^{+\infty} x\, f_X(x)\,dx \\
&= \int_{-\infty}^{+\infty}\int_{-\infty}^{+\infty} x\, f_{X,Y}(x,y)\,dx\,dy
\end{aligned}$$

[1.31]

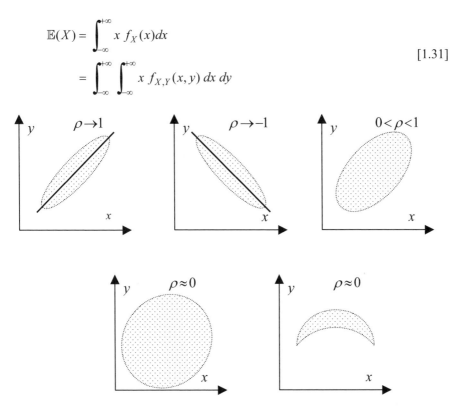

Figure 1.2. *Correlation coefficient*

EXAMPLE   1.14.–   Let   us   consider   the   joint   density   function
$f_{X,Y}(x,y)=C\dfrac{2a-x}{2a},0\le x\le 2a,0\le y\le a$. The marginal density functions are given by:

$$f_X(x)=\int_0^a C\frac{2a-x}{2a}\,dy=C\frac{2a-x}{2a}\,a,\quad 0\le x\le 2a$$

$$f_Y(y) = \int_0^{2a} C \frac{2a-x}{2a} \, dx = C \, a, \quad 0 \le y \le a$$

The joint density is given by:

$$f_{X/Y}(x,y) = \frac{f_{X,Y}(x,y)}{f_Y(y)} = \frac{C \dfrac{2a-x}{2a}}{C \, a} = \frac{1}{a} \frac{2a-x}{2a}$$

$\square$

1.4.8. *Functions of random variables*

In most applications, it is normal to deal with functions having many random variables. For this, it is necessary to introduce the joint probability distribution.

The average (or mean) of the variable $Z = g(X_1, \cdots, X_p)$ is given by the expression:

$$\mathbb{E}(Z) = \int_{-\infty}^{+\infty} \cdots \int_{-\infty}^{+\infty} g(x_1, \cdots, x_p) \, f_{X_1, \cdots, X_p}(x_1, \cdots, x_p) \, dx_1 \cdots dx_p \qquad [1.32]$$

with the variance being expressed in an identical way:

$$\mathbb{E}\left(\left(Z - \mathbb{E}(Z) \right)^2 \right)$$
$$= \int_{-\infty}^{+\infty} \int_{-\infty}^{+\infty} \left(g(x_1, \cdots, x_p) - \mathbb{E}(Z) \right)^2 f_{X_1, \cdots, X_p}(x_1, \cdots, x_p) \, dx_1 \cdots dx_p \qquad [1.33]$$

EXAMPLE 1.15.– Let us consider the variable $Z = g_1(X_1, \cdots, X_p) + g_2(X_1, \cdots, X_p)$. By applying equation [1.32], we directly obtain the following $\mathbb{E}(Z) = \mathbb{E}\left(g_1(X_1, \cdots, X_p) \right) + \mathbb{E}\left(g_2(X_1, \cdots, X_p) \right)$. From this, we can deduce:

If $Z = X + Y$, then $\mathbb{E}(X + Y) = \mathbb{E}(X) + \mathbb{E}(Y)$

If $Z = (X - \mathbb{E}(X))(Y - \mathbb{E}(Y)) = XY - \mathbb{E}(X)Y - \mathbb{E}(Y)X + \mathbb{E}(X)\mathbb{E}(Y)$
$$\text{then } \mathbb{E}(Z) = \mathbb{COV}(X,Y) = \mathbb{E}(XY) - \mathbb{E}(X)\mathbb{E}(Y)$$

If $Z = \sum_k a_k X_k$ then $\mathbb{E}(Z) = \sum_k a_k \mathbb{E}(X_k)$

If $Z = \sum_k a_k X_k$ then $\mathbb{V}(Z) = \sum_k a_k^2\, \mathbb{V}(X_k) + 2\sum_k \sum_{l=k+1} a_k\, a_l\, \mathbb{COV}(X_k, X_l)$

and therefore, if the variables are not correlated: $\mathbb{V}(Z) = \sum_k a_k^2\, \mathbb{V}(X_k)$.   □

It is also possible to generate the distribution of variable Z by the joint cumulative distribution function:

$$F_Z(z) = \mathcal{P}(Z \le z) = \int_{\mathcal{D}_z} f_{X_1, \cdots, X_p}(x_1, \cdots, x_p)\, dx_1 \cdots dx_p \qquad [1.34]$$

where $\mathcal{D}_z = \{(x_1, \cdots, x_p)/g(x_1, \cdots, x_p) \le z\}$. This expression is particularly interesting when it is applied to a sum of two random, independent variables:

$$F_Z(z) = \mathcal{P}(Z \le z) = \int_{\mathcal{D}_z} f_{X,Y}(x,y)\, dx\, dy = \int_{\mathcal{D}_z} f_X(x)\, f_Y(y)\, dx\, dy$$

$$= \int_{-\infty}^{+\infty} f_X(x) \left(\int_{-\infty}^{z-x} f_Y(y)\, dy \right) dx \qquad [1.35]$$

$$= \int_{-\infty}^{+\infty} f_X(x)\, F_Y(z-x)\, dx$$

This equation enables us to establish the density function of variable Z:

$$f_Z(z) = \frac{d}{dz} \int_{-\infty}^{+\infty} f_X(x)\, F_Y(z-x)\, dx = \int_{-\infty}^{+\infty} f_X(x)\, \frac{d}{dz} F_Y(z-x)\, dx$$

$$= \int_{-\infty}^{+\infty} f_X(x)\, f_Y(z-x)\, dx \qquad [1.36]$$

EXAMPLE 1.16.– Let us consider the sum of the two following independent variables:

$$f_X(x) = \alpha\, e^{-\alpha x}; f_Y(y) = \beta\, e^{-\beta y}$$

By applying the previous equations to the interval $[0, +\infty[$ (because the density function is only defined on this interval), we obtain:

$$f_Z(z) = \int_{-\infty}^{+\infty} f_X(x) f_Y(z-x)\, dx = \int_0^z \alpha\, e^{-\alpha x} \beta\, e^{-\beta(z-x)}\, dx$$

$$= \alpha\, \beta\, e^{-\beta z} \int_0^z e^{(\beta-\alpha)x}\, dx$$

$$= \alpha\, \beta\, \frac{e^{-\alpha z} - e^{-\beta z}}{\beta - \alpha}$$

Replacing $-\infty$ by 0 can be explained by the fact that $f_X(x)$ is zero for negative x. Replacing $+\infty$ by z can be explained by the fact that $f_Y(y)$ is zero for negative values, and therefore zero when $y = z - x < 0$.   □

1.4.9. *Approximate moments*

The previous sections have given the exact expressions of means or variances of variables as a function of other correlated or uncorrelated random variables. These means or variances are expressed according to the means and covariances associated with different variables.

For non-linear functions, such expressions are difficult to access analytically, and using approximations may be of great use. Thus, by using a multi-dimensional Taylor series development, a *second order approximation* of the mean according to random variables can be given by:

$$\mathbb{E}(Z) = \mathbb{E}(g(X_1, \cdots, X_p))$$

$$\approx g\left(\mathbb{E}(X_1), \cdots, \mathbb{E}(X_p)\right) + \frac{1}{2} \sum_{i=1}^{p} \sum_{j=1}^{p} \left.\frac{\partial^2 g}{\partial x_i \partial x_j}\right|_{\mathbb{E}(X_1),\cdots,\mathbb{E}(X_p)} \mathbb{COV}(X_i, X_j) \qquad [1.37]$$

Obtaining a second order approximation is more difficult for the variance, so it is common practice to use a first order approximation:

$$\mathbb{V}(Z) \approx \sum_{i=1}^{p} \sum_{j=1}^{p} \left.\frac{\partial g}{\partial x_i}\right|_{\mathbb{E}(X_1),\cdots,\mathbb{E}(X_p)} \left.\frac{\partial g}{\partial x_j}\right|_{\mathbb{E}(X_1),\cdots,\mathbb{E}(X_p)} \mathbb{COV}(X_i, X_j) \qquad [1.38]$$

EXAMPLE 1.17.– Expressing the strength bending moment of a reinforced concrete rectangular section is sometimes given in standards as an expression of the following type:

$$\mathfrak{M} = A\,f_y\left(d - k\,\frac{A\,f_y}{\sigma_b\,b}\right)$$

where A, d, b, f_y, σ_b respectively represent: the steel section, the beam section height, the section width, the ultimate strength of the reinforcement steel and the concrete compressive strength. k is a coefficient which depends on the shape of the stress distribution over the compressed concrete section.

Let us assume these variables to be independent. An application of the expressions discussed previously gives:

$$\mathbb{E}(\mathfrak{M}) \approx \mathbb{E}(A)\,\mathbb{E}(f_y)\left(\mathbb{E}(d) - \mathbb{E}(k)\,\frac{\mathbb{E}(A)\,\mathbb{E}(f_y)}{\mathbb{E}(\sigma_b)\,\mathbb{E}(b)}\right)$$

$$-\mathbb{E}(k)\,\frac{\mathbb{E}(f_y)^2\,\mathbb{V}(A) + \mathbb{E}(A)^2\,\mathbb{V}(f_y)}{\mathbb{E}(\sigma_b)\,\mathbb{E}(b)}$$

$$-\mathbb{E}(k)\,\frac{\mathbb{E}(A)^2\,\mathbb{E}(f_y)^2}{\mathbb{E}(\sigma_b)\,\mathbb{E}(b)}\left(\frac{\mathbb{V}(\sigma_b)}{\mathbb{E}(\sigma_b)^2} + \frac{\mathbb{V}(b)}{\mathbb{E}(b)^2}\right)$$

$$\mathbb{V}(\mathfrak{M}) \approx \mathbb{E}(A)^2\,\mathbb{E}(f_y)^2\,\mathbb{V}(d) + \left(\mathbb{E}(A)\,\mathbb{E}(d) - 2\mathbb{E}(k)\,\frac{\mathbb{E}(A)^2\,\mathbb{E}(f_y)}{\mathbb{E}(\sigma_b)\,\mathbb{E}(b)}\right)^2\mathbb{V}(f_y)$$

$$+\left(\mathbb{E}(f_y)\,\mathbb{E}(d) - 2\,\mathbb{E}(k)\,\frac{\mathbb{E}(f_y)^2\,\mathbb{E}(A)}{\mathbb{E}(\sigma_b)\,\mathbb{E}(b)}\right)^2\mathbb{V}(A)$$

$$+\left(\mathbb{E}(k)\,\frac{\mathbb{E}(f_y)^2\,\mathbb{E}(A)^2}{\mathbb{E}(\sigma_b)\,\mathbb{E}(b)}\right)^2\left(\frac{\mathbb{V}(b)}{\mathbb{E}(b)^2} + \frac{\mathbb{V}(\sigma_b)}{\mathbb{E}(\sigma_b)^2}\right)$$

□

1.5. Useful random variables

1.5.1. Discrete variables

The expressions from the previous sections are only applicable to continuous random variables (i.e. defined in the set of real numbers). When the variables are *discrete*, i.e. taking outcomes within a set of countable values such as natural numbers, then similar definitions can be given, where the integrals are replaced by sums. Some of the most frequently used discrete variables are given further on. These are what are known as the *Bernoulli sequences*. Here, it is a question of studying the occurrence of an event in a series of random trials which are based on the following hypotheses:

   – each trial has two possible events, known as success or failure;

   – the probability of occurrence of each event is constant for each trial;

   – the trials are statistically independent.

1.5.1.1. *Binomial distribution*

The *binomial distribution* $B(n, p)$ gives the probability of k successes in n trials:

$$P(X = k) = \binom{n}{k} p^k (1-p)^{n-k}$$ [1.39]

where $\binom{n}{k} = \dfrac{n!}{k!(n-k)!}$. The parameters of the distribution are the number of independent trials and the probability of success for a trial. The first and second order moments are given by:

$$\mathbb{E}(X) = n\,p;\ \mathbb{V}(X) = n\ p\ (1-p)$$ [1.40]

EXAMPLE 1.18.– A suspension cable is inspected at a given cross-section. Out of 10,976 wires inspected, 127 broken cables were detected. If a wire failure in another cross-section of the cable is assumed to be distributed in a binomial way, the probability of k wire failures is given by:

$$P(X = k) = \binom{10976}{k} \left(\frac{127}{10976}\right)^k \left(1 - \frac{127}{10976}\right)^{10976-k} = \binom{10976}{k} 0.01^k\ 0.99^{10976-k}$$

□

1.5.1.2. *Geometric distribution*

The *geometric distribution* $G(p)$ gives the possibility of success on the n-th trial:

$$P(X = n) = p_X(n) = p(1-p)^{n-1}$$ [1.41]

Its cumulative distribution function is given by:

$$F_X(n) = \sum_{i=1}^{n} P(X = i) = p \sum_{i=1}^{n} (1-p)^{i-1} = 1 - (1-p)^n \tag{1.42}$$

The parameters for this distribution are given by:

$$\mathbb{E}(N) = \frac{1}{p}; \ \sigma(N) = \frac{(1-p)}{p^2} \tag{1.43}$$

EXAMPLE 1.19.– In the design or assessment of civil engineering structures which must be resistant to flood or strong winds, it is necessary to check their resistance in relation to load effects with low probabilities, but with high amplitudes. Let us suppose that an engineer, using previous data, estimates that the design value X_d for a given load effect has a probability of $p = 0.02$ of being exceeded over a one year period. The engineer also presumes that the occurrence of such an effect is independent from year to year. In this case, if exceeding this design value over a one year period is considered a "success", the probability of this design value being exceeded over the structure's working lifetime T (in years) is given by a binomial distribution $B(T, p)$. If $T = 100$ years, this probability is then:

$$P(X \geq X_d) = 1 - \binom{100}{0}(0.02)^0 (0.98)^{100} = 1 - 0.98^{100} \approx 0.87$$

The number of years N, at the end of which the design value occurs, is a geometric variable of probability $p = 0.02$. The probability of this value being exceeded after the first ten years is then:

$$P(N > 10) = 1 - F_N(10) = (1 - 0.02)^{10} \approx 0.82$$

The mean value of this distribution is therefore $1/p = 1/0.02 = 50$ years. It represents the average period between two exceedances of the design value, or in other words, the average time between two occurrences of this value. This period is called the *return period*. □

The *return period* is an important concept in the study of certain random phenomena, such as wind, traffic, earthquakes, etc. It is, then, essential to fully understand it.

Let us point out, in addition, that for a Bernoulli stream of events where the tests are independent, the average time interval between two occurrences of the same event is also the first occurrence time of this event.

We presume that discrete time is a multiple of a reference period τ (a week, year, month, etc.). In this case, the average time interval between the two occurrences of an event, expressed according to the reference period τ is given by:

$$\mathbb{E}(T) = \lim_{t \to \infty} \sum_{n=1}^{t} n\tau \, p(1-p)^{n-1}$$
$$= \lim_{t \to \infty} \tau \, p \left(1 + 2(1-p) + 3(1-p)^2 + \cdots + t(1-p)^{t-1}\right) \qquad [1.44]$$
$$\approx -\frac{\tau}{\ln(1-p))}$$

If $p \ll 1$, we obtain:

$$\mathbb{E}(T) \approx \frac{\tau}{p} \qquad [1.45]$$

Equation [1.45] expresses the return period of an event. It is the inverse of the probability of the event over the reference period.

1.5.1.3. Poisson distribution

The *Poisson distribution* provides the probability of k occurrences of an event over a time (or space) interval t, knowing that the average occurrence rate per time (or space) unit ν (known also as intensity) is given by:

$$\mathcal{P}(X = k) = \frac{(\nu t)^k}{k!} \exp(-\nu t) \qquad [1.46]$$

The parameters of this distribution are the average occurrence rate per time unit ν, and the interval t. The first and second order moments are:

$$\mathbb{E}(X) = \nu t; \ \sigma(X) = \sqrt{\nu t} \qquad [1.47]$$

EXAMPLE 1.20.– Inspecting reinforced concrete elements (3 m long) of a structure has led us to note that the average number of cracks along an element is two per meter. By presuming that the spatial distribution of these cracks is represented by a

Poisson distribution, the probability of there being less than five cracks on an element is:

$$P(N<5) = \sum_{i=0}^{4} \frac{(2\times3)^i}{i!} \exp(-2\times3) = e^{-6}\left(1+\frac{6}{1}+\frac{6^2}{2}+\frac{6^3}{6}+\frac{6^4}{24}\right) \approx 0.28$$

$\square$

1.5.2. *Normal distribution*

A very useful continuous distribution in practice is the *normal distribution*. Its cumulative distribution function is given by:

$$F(x) = \int_{-\infty}^{x} \frac{1}{\sqrt{2\pi}\sigma} \exp\left(-\frac{(y-\mu)^2}{\sigma^2}\right) dy \qquad\qquad [1.48]$$

where $\mu = \mathbb{E}(X)$ is the mean and σ is the standard deviation. The normal variable is sometimes known as the *Gaussian variable* (Figure 1.3).

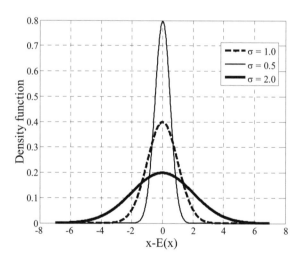

Figure 1.3. *Probability density function of a normal distribution*

When the mean is zero and the standard deviation is equal to 1 ($\mu = 0, \sigma = 1$), the variable is called the *standard normal variable*. Its distribution function and density are generally written $\Phi(.)$ and as $\varphi(.)$ respectively.

The normal distribution naturally appears in random phenomena whose physics are microscopic but are observed on a macroscopic scale. In other terms, the Gaussian distribution is the distribution for any variable whose values are the result of a multitude of independent factors. This is justified by one of the versions of the *central limit theorem* (CLT) which will be fully explained in section 1.6. The normal distribution is often adopted as an approximation of other distributions. This is the case for the Poisson distribution when $v\,t$ is large.

Here we will note a very important result: the sum of many normal independent variables is also a normal variable. The expressions from example 1.15 are, then, applicable for calculating the mean and the variance of this new random variable. It is, finally, possible to demonstrate that the kth order moments are given by the following expressions:

$$\mathbb{E}\left[\left(X-\mathbb{E}(X)\right)^k\right]=\frac{k!}{2^{\frac{k}{2}}\left(\dfrac{k}{2}\right)!}\sigma^k \quad k=2,4,\cdots$$

$$\mathbb{E}\left[\left(X-\mathbb{E}(X)\right)^k\right]=0 \quad k=1,3,\cdots$$

[1.49]

EXAMPLE 1.21.– Let us consider a construction supported by three bearings A, B, and C. The bearing conditions are not perfectly known and the settlements are presumed to follow normal independent distributions (which is rarely the case, but which is presumed here for simplicity) with respective means and coefficients of variations 2, 2.5 and 3 cm, and 20%, 20% and 25%. The probability that the maximum settlement d is higher than 4 cm is given by:

$$\mathcal{P}(d>4\,\text{cm})=1-\mathcal{P}(d\leq4)=1-\mathcal{P}(d_A\leq4\cap d_B\leq4\cap d_C\leq4)$$

$$=1-\Phi\left(\frac{4-2}{2\times0.2}\right)\Phi\left(\frac{4-2.5}{2.5\times0.2}\right)\Phi\left(\frac{4-3}{3\times0.25}\right)$$

$$=1-\Phi(5)\,\Phi(3)\,\Phi(1.333)\approx1-1\times0.9986\times0.9088\approx0.0925$$

□

1.5.3. *Lognormal distribution*

Another very useful distribution is the *lognormal distribution*. A variable X is said to be lognormal if $\ln(X)$ is a normal variable. The cumulative distribution function is given by:

$$F(x) = \int_{-\infty}^{\ln(x)} \frac{1}{\sqrt{2\pi}s} \exp\left(-\frac{(y-\mu)^2}{s^2}\right) dy = \Phi\left(\frac{\ln(x/\mu)}{s}\right) \qquad [1.50]$$

where μ is the average and s is the standard deviation for variable $\ln(X)$. The mean and standard deviation of $\ln X$ is obtained by the following expressions:

$$\mathbb{E}(\ln X) = \ln\left(\frac{\mathbb{E}(X)}{\sqrt{1+\frac{\mathbb{V}(X)}{\mathbb{E}(X)^2}}}\right); \quad \mathbb{V}(X) = \ln\left(\mathbb{CDV}(X)^2 + 1\right) \qquad [1.51]$$

As for the normal distribution, the lognormal distribution (Figure 1.4) appears in the phenomena issued from a multitude of factors. It is used frequently in modeling hydraulic data, but also in building models linking earthquake amplitude to their occurrence intervals.

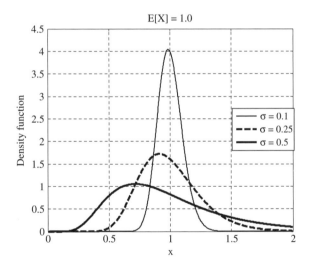

Figure 1.4. *Probability density function of a lognormal distribution*

EXAMPLE 1.22.– By using tests on welded joints, the number of rupture cycles N_r is identified as following a lognormal distribution with a mean of 166,000 cycles and a standard deviation of 52,000 cycles (tests performed under a constant stress variation of 202 N/mm^2). According to equation [1.51], it is possible to write:

$$s^2 = \ln\left(\mathbb{CDV}(N_r)^2 + 1\right) \approx 0.09; \quad \mu = \ln\left(\frac{\mathbb{E}(N_r)}{\sqrt{1 + \dfrac{\mathbb{V}(N_r)}{\mathbb{E}(N_r)^2}}}\right) \approx 12.01$$

When the variation coefficient is lower than 0.2, a good approximation of s is given by:

$$s^2 = \ln\left(\mathbb{CDV}(N_r)^2 + 1\right) \approx \mathbb{CDV}(N_r)^2$$

The cumulative distribution function of the number of rupture cycles is therefore given by:

$$F_{N_r}(n) = \Phi\left(\frac{\ln(n/\mu)}{s}\right) = \Phi\left(\frac{\ln(n/12.1)}{0.3}\right) \quad n \geq 0 \qquad\qquad \Box$$

1.5.4. *Beta distribution*

The *beta distribution* is defined by the following cumulative distribution function:

$$F(x) = \frac{1}{B(a,b)} \int_0^x \frac{(t-c)^{a-1}(d-t)^{b-1}}{(d-c)^{a+b-1}} dy \quad x \in [c,d] \tag{1.52}$$

where a and b are two (positive) shape parameters and $B(a,b)$ is the incomplete beta function:

$$B(a,b) = \int_0^1 t^{a-1}(1-t)^{b-1} dy = \frac{\Gamma(a)\,\Gamma(b)}{\Gamma(a+b)} \tag{1.53}$$

with $\Gamma(x) = \int_0^{+\infty} t^{x-1} e^{-t}\, dt$. The mean and standard deviation of X are obtained by the following expressions:

$$\mathbb{E}(X) = c + \frac{a}{a+b}(d-c); \quad \mathbb{V}(X) = (d-c)^2 \frac{1}{(a+b)^2} \frac{a\,b}{a+b+1} \tag{1.54}$$

The beta distribution is interesting because, according to the choice of parameters, it may cover a wide variety of shapes. It contains, in particular, rectangular ($a = 1$, $b = 1$) and triangular ($a = 1$ or 2, $b = 2$ or 1) distributions.

EXAMPLE 1.23.– In the study of a structure's behavior to wind effects, the direction of the wind is taken to be described by a beta distribution. The mean and variance are given by *in situ* daily measurements and have respective values of 144° and 24°. As we are dealing with wind direction (limited between 0° and 360°), we obtain $c = 0°$, $d = 360°$. Let us estimate, according to equations [1.54], the parameters of the corresponding beta distribution:

$$\mathbb{E}(X) = 144 = \frac{a}{a+b} 360 \Rightarrow \frac{a}{a+b} = 0.4$$

$$\mathbb{V}(X) = 360^2 \frac{1}{(a+b)^2} \frac{a\,b}{a+b+1} = 24^2 \Rightarrow \frac{1}{(a+b)^2} \frac{a\,b}{a+b+1} = 0.004$$

From this we can obtain: $a \approx 21.2$ and $b \approx 31.8$.   □

1.5.5. *Exponential distribution*

The exponential distribution is built upon the Poisson distribution and characterizes the moment of the first occurrence of an event described by the Poisson distribution:

$$F_T(t) = \mathcal{P}(T \leq t) = 1 - \mathcal{P}_{poisson}(X = 0) = 1 - \frac{(v\,t)^0}{0!} \exp(-v\,t) = 1 - e^{-v\,t} \qquad [1.55]$$

The mean and the variance can be obtained very easily:

$$\mathbb{E}(X) = \frac{1}{v}; \quad \mathbb{V}(X) = \frac{1}{v^2} \qquad [1.56]$$

The exponential distribution is used in a process, without memory or ageing, where it represents the time law separating the successive outcomes of a given event. The average value for this variable $\mathbb{E}(X) = \frac{1}{v}$ must be brought closer to the return period of the discrete Poisson variable.

EXAMPLE 1.24.– Analyzing seismic data in a region of the world shows that over a period of 125 years, 16 earthquakes, reaching more than 6 on the Richter scale have been recorded. If the occurrence of such earthquakes is presumed to follow an exponential distribution, the probability of having such an earthquake in the next two years is given by:

$$P(T \leq 2 \text{ years}) = 1 - \exp(-v\,2) = 1 - \exp\left(-\frac{16}{125} \times 2\right) = 0.226$$

$v = 16/125 \approx 0.128$ represents the annual number of earthquakes. The probability that no earthquakes of this amplitude will appear in the next ten years is given by:

$$P(T > 10 \text{ years}) = \exp(-0.128 \times 10) = 0.278$$

The return period for an earthquake reaching 6 on the Richter scale is $1/0.128 \approx 7.8$ years.   □

1.5.6. *Gamma distribution*

The *gamma distribution* $\Gamma(a,b)$ appears when random variable X is considered to be the result of certain operations carried out on normal or exponential variables. Its density function is given by:

$$f(x) = \frac{b^a}{\Gamma(a)} x^{a-1} e^{-b\,x} \quad x \in \mathbb{R}^+ \tag{1.57}$$

where $a, b, \Gamma(.)$ are respectively two strictly positive parameters and the Gamma function is defined by:

$$\Gamma(a) = \int_0^{+\infty} x^{a-1} e^{-x}\, dx \tag{1.58}$$

The mean and standard deviation are obtained by the following expressions:

$$\mathbb{E}(X) = \frac{a}{b}; \quad \mathbb{V}(X) = \frac{a}{b^2} \tag{1.59}$$

For $a = 1$, the Gamma distribution is identical to the exponential distribution.

An interesting case deserves to be mentioned here: the *chi-square distribution*. It corresponds to the Gamma distribution with parameters $a = n/2$ and $b = 1/2$ where n is a natural integer.

$$f(x) = \frac{1}{2^{\frac{n}{2}} \Gamma(n)} x^{\frac{n}{2}-1} e^{-\frac{x}{2}} \quad x \in \mathbb{R}^+ \tag{1.60}$$

This distribution is called the chi-square distribution with n degrees of freedom and is frequently written χ_n^2. It is often used in statistical tests, but it is also the sum of the square of n independent random normal variables. This result is particularly interesting when studying the effects of quadratic equations.

1.5.7. Students t-distribution

Another distribution is also widely used in statistics: *Student's t-distribution* with n degrees of freedom. Its density function is given by the following expression:

$$f(x) = \frac{1}{\sqrt{n\,\pi}} \frac{\Gamma\left(\dfrac{n+1}{2}\right)}{2^{\frac{n}{2}} \Gamma\left(\dfrac{n2}{}\right)} \left(1 + \frac{x^2}{n}\right)^{-\frac{n+1}{2}} \tag{1.61}$$

This distribution is written T_n, with the mean and the standard deviation given by:

$$\mathbb{E}(X) = 0 \ (n > 1); \quad \mathbb{V}(X) = \frac{n}{n-2} \ (n > 2) \tag{1.62}$$

If $n = 1$, the Student t-distribution is called the *Cauchy distribution*; it has neither mean, nor variance!

1.6. Limit theorems

The probability theory states important convergence theorems. They rigorously demonstrate results confirmed by intuition and experience. In return, this agreement with reality justifies the theory's axiomatics. This being so, developing this theory provides useful tools which can effectively approach real problems. These theorems are linked to questions of converging series of random variables (X_k) according to different modes, which would be useful to specify here.

The series (X_k) *will converge in p-th order mean* to the variable X if and only if:

$$\lim_{k \to \infty} \mathbb{E}\left[(X_k - X)^p\right] = 0 \tag{1.63}$$

When $p = 1$, we are dealing with a convergence in mean; if $p = 2$, we are referring to quadratic mean convergence.

The series (X_k) will be *convergent in probability* to the variable X if and only if:

$$\lim_{k \to \infty} \mathcal{P}\left(|X_k - X| > \varepsilon\right) = \lim_{k \to \infty} \mathcal{P}\left(\omega \in \Omega / |X_k(\omega) - X(\omega)| > \varepsilon\right) = 0 \tag{1.64}$$

The series (X_k) *converges in distribution* to the variable X when:

$$\lim_{k \to \infty} F_{X_k}(x) = F_X(x) \tag{1.65}$$

where $F_{X_k}(.), F_x(.)$ are the cumulative distribution functions of the variables X_k and X.

It can also be easily proved that any linear combinations of convergent series (whatever the mode) constitute a convergent series towards the linear combination of limit random variables.

$$\alpha X_k + \beta Y_k \xrightarrow[\substack{\text{mean} \\ \text{probability} \\ \text{distribution}}]{} \alpha X + \beta Y \tag{1.66}$$

It is finally possible to prove that the convergence in the p-th order mean involves the convergence in probability and the convergence in distribution. This last convergence law is the weakest mode of convergence.[2]

1.6.1. *Law of large numbers*

The limit theorems known as the laws of large numbers are essential in probability theory: they allow us to link the abstract concept of probability to the concept of frequency and arithmetic average. For historical reasons, it is normal

---

[2] There is another mode of convergence said to be *almost surely* that we have not approached, but which the reader can find in books on probability theory. It represents the case for which the events where there is no convergence have probability 0.

practice to allow the existence of two laws of large numbers stated at different stages in the evolution of the probability theory. In this section, we will limit ourselves to referring to the first law, the *weak law of large numbers*, which states the convergence in quadratic mean and in probability of a series of random variables.

Let (X_k) be a series of random variables in uncorrelated pairs, all having the same mean m and the same variance σ^2. For any $n \geq 1$, we set out:

$$S_n = \sum_{k=1}^{n} X_k$$

$$\bar{X}_n = \frac{S_n}{n}$$

[1.67]

thus the weak law of large numbers states that the sequence $\bar{X}_n$ converges in a quadratic mean towards m.

EXAMPLE 1.25.– Demonstrating the weak law of large numbers is quite simple. According to example 1.15, we obtain:

$$\mathbb{V}(S_n) = \sum_{k=1}^{n} \mathbb{V}(X_k) = n\,\sigma^2; \; \mathbb{V}(\bar{X}_n) = \mathbb{V}\left(\frac{S_n}{n}\right) = \frac{\mathbb{V}(S_n)}{n^2} = \frac{\sigma^2}{n}$$

In addition, as $\mathbb{E}(\bar{X}_n) = \frac{1}{n}\sum_{k=1}^{n} \mathbb{E}(X_n) = \frac{n\,m}{n} = m$, the variance of $\bar{X}_n$ is written

$\mathbb{V}(\bar{X}_n) = \mathbb{E}\left[\left(\bar{X}_n - m\right)^2\right] = \frac{\sigma^2}{n}$, which leads to $\lim\limits_{n\to\infty} \mathbb{E}\left[\left(\bar{X}_n - m\right)^2\right] = \lim\limits_{n\to\infty} \frac{\sigma^2}{n} = 0$: this is the expected result. □

The law of large numbers confirms that for all $\varepsilon > 0$ then $\lim\limits_{n\to\infty} P\left(\left|\bar{X}_n - m\right| > \varepsilon\right) = 0$ but this tells us nothing about the speed with which this convergence takes place. In practice, it is traditional to impose a probability which must not be exceeded the difference between $\bar{X}_n$ and m. It is therefore very important to determine the value n_0 from which this objective is reached, hence the use in bounding $P\left(\left|\bar{X}_n - m\right| > \varepsilon\right)$. To this effect, two inequalities are put forward, one with a general range but not very precise (Bienaymé-Chebychev inequality), the other more effective but more limited (Bernstein inequality). We will only mention the first inequality here, which is the most general and most well-known. The *Bienaymé-Chebychev inequality* states that, if the variables are not correlated, then[3]:

$$P\left(\left|\bar{X}_n - m\right| \geq \varepsilon\right) \leq \frac{\sigma^2}{n\,\varepsilon^2} \qquad\qquad [1.68]$$

EXAMPLE 1.26.– (adapted from [MAT 95]) Some investigations on a suspension cable made of parallel wires leads to distinguish two zones on a given cable cross-section. The first, A, is on the cable's periphery and is made of 78 wires. It concentrates the largest number of broken wires observations. The second, B, is located inside the cable and is made of 7,618 wires. The whole of Zone A is inspected (78 wires) and fifteen broken wires were observed. In zone B, only 1,057 wires were inspected and only three ruptures were recorded. The probability of broken wires can be obtained by using these inspections, $P_{fA} = 15/78 \approx 0.19$ and $P_{fB} = 3/1057 \approx 0.003$. The number of broken wires for each zone N_{fA}, N_{fB} can be modeled by a binomial distribution with parameters $\left(P_f, n_f\right)$ each worth $\left(0.19; 78\right)$ and $\left(0.003; 7618\right)$ respectively. The sum of the two variables formed in this way is a new variable N_f which represents the total number of broken wires in the cable. Its mean and variance are then given by (using equations 1.40):

$$\mathbb{E}(N_f) = \mathbb{E}(N_{fA}) + \mathbb{E}(N_{fB}) = n_{fA}\, P_{fA} + n_{fB}\, P_{fB} \approx 37$$

$$\mathbb{V}(N_f) = \mathbb{V}(N_{fA}) + \mathbb{V}(N_{fB}) = n_{fA}\, P_{fA}\left(1 - P_{fA}\right) + n_{fB}\, P_{fB}\left(1 - P_{fB}\right) \approx 34$$

With these probabilities being estimated on one cross-section only, it would be useful to establish a conservative value of the number of broken wires to be taken into account in the calculations. For this purpose, the Bienaymé-Chebychev inequality can be used. It is applied to the variable N_f ($n = 1$); if we fix an error of 1% on the number of broken wires (in other words, the probability of having a number of broken wires higher than the value used in the calculations is 1%), then the inequality gives:

---

3 From this equation, it is possible to give the trivial result for $n = 1$, $P\left(\left|X_1 - m\right| \geq \varepsilon\right) \leq \dfrac{\sigma^2}{\varepsilon^2}$,

a valid expression for the X_k variables. This is the most well-known form of the Bienaymé-Chebychev inequality.

$$P\left(\left|N_f - \mathbb{E}(N_f)\right| \geq \varepsilon\right) \leq \frac{\sigma^2}{\varepsilon^2} = 0.01$$

meaning $\varepsilon \approx 58.31$. As a consequence, the value of broken wires with a 1% probability of being exceeded, which must be taken into account when assessing the cable strength, is:

$$\hat{N}_f = \mathbb{E}(N_f) + \varepsilon \approx 37 + 58.31 \approx 96 \qquad \Box$$

1.6.2. Limit theorems

The law of large numbers specifies the asymptotic behavior of certain random variables. Another fundamental aspect is studying the asymptotic behavior of the corresponding distributions. In fact, the variables are often only described by their distribution and the real problems lead, most naturally, onto probability calculations based on these distributions. This highlights the interest of being able to approach a variable's unknown or complicated distribution by a simpler or more adapted distribution.

There are many limit theorems but the *central limit theorem* is far from being the most important. There are many versions of this theorem: this idea is owed to Laplace, but the first rigorous demonstrations of this are due to Liapounoff. This theorem states that, for a sequence (X_k) of arbitrary independent random variables, the sum S_n converges towards a normal distribution. A particular case is a series of variables with the same distribution, of mean m and variance σ^2; it is therefore possible to demonstrate that the distribution of variable $U_n = \dfrac{S_n - nm}{\sigma\sqrt{n}} = \dfrac{\sqrt{n}}{\sigma}(\bar{X}_n - m)$ converges towards the standard normal distribution.

EXAMPLE 1.27.– (adapted from [ANG 84]) Let us consider the example of a bridge slab made of precast elements of the same theoretical length L. From experience, the error e on a precast element's size is uniformly distributed between $\pm a$. For a given element, the density function of the difference is:

$$f_e(x) = \begin{cases} 1/2a & -a \leq x \leq a \\ 0 & \text{elsewhere} \end{cases}$$

Let us consider two precast elements linked to one another. The density function of the cumulated error is thus (assuming them to be independent between the two slabs):

$$f_{e_1+e_2}(x) = \begin{cases} \displaystyle\int_{-a}^{x+a} f_{e_1}(u)\, f_{e_2}(x-u)\, du & -2a \le x \le 0 \\[2mm] \displaystyle\int_{-a}^{-x+a} f_{e_1}(u)\, f_{e_2}(x-u)\, du & 0 \le x \le 2a \\[2mm] 0 & \text{elsewhere} \end{cases}$$

$$= \begin{cases} \dfrac{1}{4a^2}(2a+x) & -2a \le x \le 0 \\[2mm] \dfrac{1}{4a^2}(2a-x) & 0 \le x \le 2a \\[2mm] 0 & \text{elsewhere} \end{cases}$$

For three elements, the error on length $3L$ is modeled by the density function:

$$f_{e_1+e_2+e_3}(x) = \begin{cases} \dfrac{1}{16a^4}(3a+x)^2 & -3a \le x \le -a \\[2mm] \dfrac{3}{4} - \dfrac{1}{8a^2}x^2 & -a \le x \le a \\[2mm] \dfrac{1}{16a^4}(3a-x)^2 & a \le x \le 3a \\[2mm] 0 & \text{elsewhere} \end{cases}$$

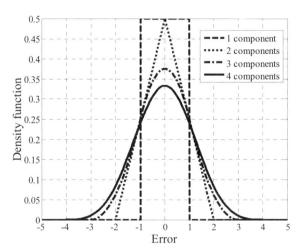

The previous figure highlights the asymptotic convergence in law of the sequence $\left(S_n = \displaystyle\sum_{k=1}^{n} e_k \right)$ towards a normal law $(a = 1)$. □

In an identical way, if we set out $\Sigma_n^2 = \dfrac{1}{n-1}\displaystyle\sum_{i=1}^{n}\left(X_i - \bar{X}\right)^2$, we can show that Σ_n^2 asymptotically tends in quadratic mean towards σ^2: the variable $\dfrac{\Sigma_n^2 - \sigma^2}{\sigma^2\sqrt{\dfrac{(\gamma_2 - 1)}{n}}}$

converges in law towards the normal centered reduced law. In the latter case, if the variables are Gaussian, we can also demonstrate that $\dfrac{(n-1)\Sigma_n^2}{\sigma^2}$ tends asymptotically in law towards the law χ_{n-1}^2.

EXAMPLE 1.28.– Let us consider a structure modeled as a linear oscillator with one degree of freedom. The response of this structure to an earthquake can be represented as the superposition of independent impulsions of random amplitude γ_i, from the same distribution and regularly spaced. In this case, the structure's response is reduced to a sum of the following type:

$$X_n = X(n\,\Delta t) = \sum_{i=1}^{n} \gamma_i\, h\big((n-i)\,\Delta t\big)$$

where $h(t)$ is the impulsive response to unitary acceleration. Δt represents the time interval (seconds, for example) between two pulses. If the number of pulses is large enough, the central limit theorem leads us to model the response by a normal distribution with a zero mean and a standard deviation σ_n. In the same way, the speed is written as:

$$\dot{X}_n = \sum_{i=1}^{n} \gamma_i\, g\big((n-i)\,\Delta t\big)$$

where g is the derivative of function h. $\dot{X}_n$ also behaves as a normal distribution with a zero mean and standard deviation $\omega_0\,\sigma_n$. ω_0 is the structure's natural circular frequency. When the excitation is stationary (which is presumed in this case), an interesting and surprising result can be obtained: the speed and displacement at the same moment are independent from each other (this is after the effect of the initial conditions has disappeared). Moreover, $\sigma_n = \sigma$ is a constant.

The amplitude of the oscillator's vibration is given by the expression:

$$A_n = \sqrt{\frac{1}{\omega_0^2}\dot{X}_n^2 + X_n^2} = \sigma\sqrt{\left(\frac{1}{\omega_0\,\sigma}\dot{X}_n\right)^2 + \left(\frac{1}{\sigma}X_n\right)^2}$$

The variables in brackets and squared are standard normal variables. The sum of the square of the two normal variables being a variable with the chi-square distribution, its squared root, i.e. A_n/σ is also χ_2^2 distributed. Its density function is then:

$$f_{A_n}(a) = \frac{a\,\exp\left(-\frac{1}{2}\left(\frac{a}{\sigma}\right)^2\right)}{\sigma^2} \quad a \geq 0$$

This distribution is better known as the Rayleigh distribution. It is a distribution of extreme values that will be described in section 1.8.   □

1.7. Modeling random variables

In the presence of a sample, the question is, therefore, knowing which information can be taken from the sample concerning the phenomena being observed. This accessible information is known as *statistics*. In fact, it deals with applications whose values are called *statistical information*. After having proceeded to build a sample, it is desirable, if not essential, to summarize its informative content by means of the numerical values taken by a certain number of statistics. The most interesting qualitative information generally involves parameters which are the characteristics of the random variable associated with the studied phenomenon. Most often, it involves a histogram of outcomes, or parameters such as the *mean* or the *standard deviation*. Statistics are also random variables (this is why we must speak of the sample's mean and standard deviation). As a consequence, we can only obtain estimated values of the *true* characteristic parameters of the random variable from a sample.

There are two main estimation methods: *point* and *interval estimation*. Point estimation may characterize a descriptor L of a random variable by a unique number λ determined by a statistic $\hat{\lambda}$. $\hat{\lambda}$ is then the estimator of L. The function of the sample's size n is written λ_n. Interval estimations seek to define an interval $]a,b[$ where the value L is likely to be found with a probability α which is fixed in advance.

In the two previous estimation methods, it is logical to correctly estimate a parameter L, to look for an estimator λ_n such as:

$$\forall n,\ \mathbb{E}(\lambda_n) = L \qquad\qquad [1.69]$$

i.e. *unbiased*. The unbiased nature of an estimator thus expresses that the mean value of the estimator is equal to parameter L, independently of the sample's size. However, this property does not specify whether the sample's individual values are close to the parameter or not.

An estimator may also be *consistent*; therefore, as more and more data becomes available, the estimator must converge towards the parameter's value.

$$\lim_{\infty} \lambda_n = L \qquad\qquad [1.70]$$

An estimator must finally be *convergent in probability* towards L, in the sense of the Bienaymé-Chebychev definition:

$$\lim_{\infty} \mathbb{V}(\lambda_n) = 0 \qquad\qquad [1.71]$$

This property means that the error induced by the estimator decreases with the sample's size.

A priori, many estimators of parameter L must exist: which ones to choose, according to which criteria? Amongst these estimators, it seems natural to keep back the one with the minimal variance for a sample with a given size. Such an estimator is then called the *efficiency estimator*. The efficiency of an estimator refers to the variance of the latter; if everything is equal, then, an estimator λ_n will be more useful than another, if it accepts a smaller variance. Finally, an estimator must be able to make use of all the information contained within the sample; it is, then, *sufficient*.

1.7.1. *Point estimation*

In practice, it is impossible to get estimators which verify the full set of the previous properties. Many methods for building satisfactory point estimators have been proposed. The most classical method is the *method of moments*[4]. Unquestionably, the most intuitive when there are, for example, *r* parameters to be determined, the method consists of identifying the theoretical expressions of *r* non-zero moments of the considered distribution with the empirical moments calculated from the sample. We obtain a system with *r* unknown estimators, whose solution leads to *r* estimators in principle. The method of moments makes it possible to give well-known approximations of the mean, standard deviations and the correlation coefficient when a sample of size *n*, $(x_i)_{1\leq i\leq n}$ is available:

$$\mathbb{E}(X) \approx \bar{X}_n = \frac{1}{n}\sum_{i=1}^{n} x_i \qquad\qquad [1.72]$$

$$\mathbb{V}(X) \approx \Sigma_n^2 = \frac{1}{n-1}\sum_{i=1}^{n}\left(x_i - \mathbb{E}(X)\right)^2 \qquad\qquad [1.73]$$

$$\rho(X,Y) \approx \Gamma_n(X,Y) = \frac{\displaystyle\sum_{i=1}^{n}\left[x_i - \mathbb{E}(X)\right]\left[y_i - \mathbb{E}(Y)\right]}{\sqrt{\displaystyle\sum_{i=1}^{n}\left[x_i - \mathbb{E}(X)\right]^2}\sqrt{\displaystyle\sum_{i=1}^{n}\left[y_i - \mathbb{E}(Y)\right]^2}} \qquad\qquad [1.74]$$

These expressions are based on limit theorems. The choice of $n-1$ for calculating the variance (and not n which would be more intuitive) is based on the fact that this estimator is an unbiased estimator of the variance, which is not the case when the ratio n is used. In fact, according to the limit theorem on variance, we obtain that Σ_n^2 converges in distribution to the normal distribution with a mean of σ^2 and a standard deviation of $\sigma^2\sqrt{\dfrac{(\gamma_2 -1)}{n}}$. When n is "large enough", the normal law is weakly dispersed ($\sigma^2\sqrt{\dfrac{(\gamma_2 -1)}{n}} \xrightarrow[n\to\infty]{} 0$) and therefore Σ_n^2 converges in a quadratic mean towards σ^2 (here we find again the weak law of large numbers). The estimator is then unbiased and converges in probability.

---

4 There is another method called the *method of maximum likelihood.* This is not described in this chapter.

EXAMPLE 1.29.– Let us use example 1.1 and presume that after the first crack appears, the beam test is monitored until failure of the beam. Let $\left(F_{rup,i}\right)_{1\leq i\leq15}$ be the 15 failure loads thus determined:

47	41	43	46	42
48	45	44	42	46
42	43	43	47	46

The means of the loads after the first crack appears, and the rupture loads, are given by (to the nearest kN):

$$\bar{F}_{15} = \frac{1}{15}\sum_{i=1}^{15} F_i = 33 \text{ kN}$$

$$\bar{F}_{rup,15} = \frac{1}{15}\sum_{i=1}^{15} F_{rup,i} = 44 \text{ kN}$$

A measure of the load dispersion can also be given by the standard deviations:

$$\Sigma_{F,15} = \sqrt{\frac{1}{14}\sum_{i=1}^{15}\left(F_i - \mathbb{E}(F)\right)^2} \approx 8 \text{ kN}$$

$$\Sigma_{F_{rup},15} = \sqrt{\frac{1}{14}\sum_{i=1}^{15}\left(F_{rup,i} - \mathbb{E}(F)\right)^2} \approx 2 \text{ kN}$$

The coefficients of variation are deduced directly from the previous values: $\mathbb{CDV}(F) \approx 24\%$ and $\mathbb{CDV}(F_{rup}) \approx 4.5\%$. These coefficients demonstrate that the load inducing the first crack is notably more dispersed than the load at the rupture. The study can be taken further by studying the skewness and Kurtosis coefficients:

$$\mathbb{G}_1(F) \approx \frac{\frac{1}{15}\sum_{i=1}^{15}\left(F_i - \mathbb{E}(F)\right)^3}{\sigma(F)^3} \approx 0.67$$

$$\mathbb{G}_1(F) \approx \frac{\dfrac{1}{15}\displaystyle\sum_{i=1}^{15}\left(F_{rup,i} - \mathbb{E}(F)\right)^3}{\sigma(F)^3} \approx 0.11$$

The value of these coefficients is positive, which means that the left tail distributions are longer than the right (which is not surprising because these values are still positive). Obviously, since dispersion is weaker on the rupture load, the tail length is shorter, which is described by the value of the skewness coefficient.

The Kurtosis coefficient shows a flattening of the distribution. For the two previous cases, we obtain:

$$\mathbb{G}_2[F] \approx \frac{\dfrac{1}{15}\displaystyle\sum_{i=1}^{15}\left(F_i - \mathbb{E}[F]\right)^4}{\sigma[F]^4} \approx 1.75$$

$$\mathbb{G}_2\left[F_{rup}\right] \approx \frac{\dfrac{1}{15}\displaystyle\sum_{i=1}^{15}\left(F_{rup,i} - \mathbb{E}[F]\right)^4}{\sigma[F]^4} \approx 1.43$$

The bigger the value is, the more noticeable the flattening will be. Generally, it is traditional to compare this value to the value 3, which is characteristic of the normal distribution. If the Kurtosis coefficient is higher than 3, the variable is then flatter than the normal distribution. In the two examples, they are less flattened.   □

EXAMPLE 1.30.– Let us now study the data in examples 1.1 and 1.29 together. Let us outline the cloud of sample points (see the following figure). This cloud very clearly shows an absence of linear correlation between the two sets of data. Calculating the correlation coefficient confirms this, since it can be estimated with equation [1.74] at +0.03. The low value of this coefficient highlights the low correlation between the loads inducing a first crack, and the rupture loads.

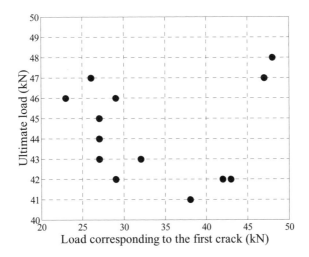

Load corresponding to the first crack (kN)

□

In certain situations, only limited information such as maximum and minimum values is known. If extra information may help to postulate on the type of probability distribution, it is possible to express the mean and variance according to these maximum and minimum values.

1.7.2. *Interval estimation*

When we have opted for an estimator λ of a parameter of L for a studied phenomenon, and when we have obtained an estimation λ of L, it is, then, often rather simple to define an interval $]a(\lambda), b(\lambda)[$ which covers the exact value of parameter L with a confidence probability p. More precisely, with the sample values and the value of λ calculated, we must rationally determine $]a(\lambda), b(\lambda)[$ such as

$$\mathcal{P}(a(\lambda) < L < b(\lambda)) = p.$$

1.7.2.1. *Estimation of the mean*

We have seen that whatever the variable X, then $\dfrac{S_n - n\,m}{\sigma\sqrt{n}} = \dfrac{\sqrt{n}}{\sigma}\left(\overline{X}_n - m\right)$

converges in distribution towards the standard normal distribution. Consequently, it is logical to determine the confidence interval with the standard normal variable. Through the symmetry of this distribution, we only need to obtain the interval $]a, b[$ with a confidence level p to look for the value $u_{\frac{1-p}{2}}$ such as:

$$P\left(-u_{\frac{1-p}{2}} < \frac{\overline{X}_n - m}{\sigma/\sqrt{n}} < u_{\frac{1-p}{2}}\right) = p = -\Phi\left(-u_{\frac{1-p}{2}}\right) + \Phi\left(u_{\frac{1-p}{2}}\right) = 1 - 2\Phi\left(-u_{\frac{1-p}{2}}\right) \qquad [1.75]$$

(where $\Phi(.)$ is the distribution function of the standard normal distribution) which is equivalent to:

$$\Phi\left(u_{\frac{1-p}{2}}\right) = \frac{1+p}{2} \qquad [1.76]$$

Let us note $\overline{x}_n$ an observation or outcome (also called sample mean) of the variable $\overline{X}_n$. Estimating the mean m with a confidence probability of p is therefore carried out by a confidence interval:

$$m \in \left]\overline{x}_n - \frac{u_{\frac{1-p}{2}}\,\sigma}{\sqrt{n}}, \overline{x}_n + \frac{u_{\frac{1-p}{2}}\,\sigma}{\sqrt{n}}\right[\qquad [1.77]$$

This equation presumes that the standard deviation is known, which is rarely the case in practice. For large samples, this standard deviation may be replaced by an observation of $\sqrt{\Sigma_n^2}$.

Confidence interval [1.77] is only valid for large samples. If the variables are normal, determining a confidence interval for samples of size n requires us to use a Student's variable with $n - 1$ degrees of freedom: $T_{n-1} = \dfrac{\overline{X}_n - m}{\Sigma_n/\sqrt{n}}$. Student's distribution is symmetrical:

$$P\left(-t_{n-1,\frac{1-p}{2}} < T_n < t_{n-1,\frac{1-p}{2}}\right) = p \qquad [1.78]$$

Estimating the mean m with a confidence level p is therefore carried out by using a confidence interval:

$$m \in \left]\overline{x}_n - \frac{t_{n-1,\frac{1-p}{2}}\,s_n}{\sqrt{n}}, \overline{x}_n + \frac{t_{n-1,\frac{1-p}{2}}\,s_n}{\sqrt{n}}\right[\qquad [1.79]$$

where s_n^2 is an observation of Σ_n^2.

EXAMPLE 1.31.– Let us consider a series of tests on 20 bars. The estimated average (mean) value of the strength is 1,800 kN, the standard deviation being evaluated at 23 kN. The 95% confidence interval is therefore:

$$1,800 \pm \frac{23}{\sqrt{20}} t_{19,0.025} = 1,800 \pm 5.14 \times 2.09 \approx 1,800 \pm 10.74 \text{ kN} \qquad \square$$

EXAMPLE 1.32.– A series of tests on concrete cores is carried out to assess the compressive concrete strength. This should be distributed according to a normal distribution. We are looking to determine the number of tests which need to be carried out so that the 90% confidence interval gives an error of only 10% on the mean value. It has been noticed that the coefficient of variation for the concrete's strength remains approximately constant and equal to 15%, whatever the mean.

According to equation [1.79], we obtain:

$$\bar{x}_n \pm \frac{t_{n-1,\,0.05}}{\sqrt{n}} \frac{s_n}{\sqrt{n}} \leq \bar{x}_n \pm 0.1\,\bar{x}_n \text{ meaning } \frac{t_{n-1,\,0.05}}{\sqrt{n}} \frac{s_n}{\sqrt{n}} = \frac{t_{n-1,\,0.05}}{\sqrt{n}} 0.15\,\bar{x}_n \leq 0.10\,\bar{x}_n$$

i.e. $\dfrac{t_{n-1,\,0.05}}{\sqrt{n}} \leq \dfrac{0.0}{0.15} \approx 0.67.$

For $n = 6$, we obtain: $\dfrac{t_{n-1,\,0.05}}{\sqrt{n}} \approx \dfrac{1.48}{\sqrt{6}} \approx 0.60 < 0.67.$ It is suitable to perform 6 tests in order to obtain a maximum error of 10% with a confidence of 90%. $\square$

1.7.2.2. Estimation of the variance

For variance, the problem becomes more complex. However, an interesting case can be presented, where the random variables X_k are normal independent variables. In the sections on limit theorems, we saw that in this case, $\dfrac{(n-1)\,\Sigma_n^2}{\sigma^2}$

asymptotically tends in distribution to the chi-square distribution χ_{n-1}^2. Due to the dissymmetric nature of this distribution, we are first of all led to define an interval $]\hat{\chi}_{n-1}^2{}', \hat{\chi}_{n-1}^2{}''[$ (the symbol $\wedge$ denotes that we are dealing with a value and not a random variable) such as:

$$P\left(\chi^2_{n-1} < \hat{\chi}^2_{n-1}{}'\right) = \frac{1-p}{2}$$

$$P\left(\hat{\chi}^2_{n-1}{}'' > \chi^2_{n-1}\right) = \frac{1-p}{2}$$

[1.80]

The confidence interval on the standard deviation with confidence level p is, then, given by:

$$P\left(\hat{\chi}^2_{n-1}{}' < \chi^2_{n-1} = \frac{(n-1)\Sigma^2_n}{\sigma^2} < \hat{\chi}^2_{n-1}{}''\right) = p$$

[1.81]

i.e. $\sigma^2 \in \left]\dfrac{(n-1)\,s^2_n}{\hat{\chi}^2_{n-1}{}''}, \dfrac{(n-1)\,s^2_n}{\hat{\chi}^2_{n-1}{}'}\right[$ where s^2_n is an observation of Σ^2_n.

1.7.3. *Estimation of fractiles*

With the difference between the means and variances, the fractiles correspond to values with low probabilities of occurrence. But, generally, the samples are often reduced in size, which means that this estimation should be approached with caution. There are, however, different statistical approaches to improve the estimation of the fractiles of a variable:

$$P\left(X \leq \mathbb{F}_p(X)\right) = p$$

[1.82]

1.7.3.1. *Order method*

The most general method does not need any assumptions about the distribution of a variable X. Let us consider an ordered sample $x_1 \leq x_2 \leq \cdots \leq x_{n-1} \leq x_n$; an estimation of the p-th order fractiles is given by:

$$f_p = x_{k+1}; k \leq n\,p < k+1$$

[1.83]

This method requires a large amount of observations to be reliable.

1.7.3.2. *Covering method*

This is the most commonly used technique. It consists of determining an interval for the fractile with a given confidence level α:

$$P\left(f_p < \mathbb{F}_p(X)\right) = \alpha$$

[1.84]

An estimator of the p-th order fractile is then given by:

$$f_p = x_n - k_{n,p} \, s_n \tag{1.85}$$

where $\bar{x}_n$, s_n are respectively the sample mean and standard deviation of variable X using a sample with dimension n. The coefficients $k_{n,p}$ depend on the distribution for variable X, the sample's size n, the probability p and the confidence level α. Table 1.1 gives these coefficients for a probability of $p = 5\%$, a confidence level $\alpha = 95\%$, for a lognormal distribution with three coefficients of skewness γ_1.

γ_1	\multicolumn{9}{c}{Sample size n}
	3	4	5	6	8	10	20	30	∞
−1.00	10.9	7.00	5.83	5.03	4.32	3.73	3.05	2.79	1.85
0.00	7.66	5.14	4.20	3.71	3.19	2.91	2.40	2.22	1.64
+1.00	5.88	3.91	3.18	2.82	2.44	2.25	1.88	1.77	1.34

Table 1.1. Coefficient $k_{n,p}$ ($p = 5\%$, $\alpha = 95\%$)

EXAMPLE 1.33.– Example 1.29 enabled us to calculate the mean, the variance, and the skewness coefficient for the load inducing a first crack during load tests on 15 beams: 33 kN, 8 kN and 0.67. The order method cannot be used here, because the sample's size is too small ($np = 0.75$). The skewness coefficient is not zero, therefore we cannot retain the hypothesis for the normal distribution to describe the variable F. However, as a comparison for the following, with this hypothesis, the 5% fractile with a confidence level of 95% is given by the estimator:

$$f_p = \bar{x}_n - k_{n,p} \, s_n \approx 33 - \frac{2.40 + 2.91}{2} \times 8 \approx 33 - 2.655 \times 8 \approx 11.8 \text{ kN}$$

If we now assume the lognormal distribution for F, this fractile becomes:

$$f_p \approx 33 - \left[\left(\frac{1.88 + 2.25}{2} - \frac{2.40 + 2.91}{2} \right) \times \frac{(0.67 - 0)}{1 - 0} + \left(\frac{2.40 + 2.91}{2} \right) \right] \times 8$$
$$\approx 33 - 2.26 \times 8 \approx 14.92 \text{ kN}$$

This value is 26% higher than the value obtained using the normal assumption. This example shows the contribution of the distribution skewness. In the present case, the positivity of the skewness coefficient makes the fractile calculation more favorable. For a negative coefficient, the result would have been different.   □

1.7.4. *Estimation of the distribution*

A random variable is fully determined by its distribution function. Determining this distribution function can be achieved empirically by using histograms based on one or many samples (see Figure 1.1). If we wish to apply a predefined distribution model (normal, lognormal, etc.) to this distribution function, we must, then, perform a statistical *goodness-of-fit test* (for example, chi-quare test, Kolmogorov-Smirnov test) to validate or invalidate the hypothesis. An example is given in section 1.9.2.2.

An alternative consists of applying optimization methods which use the principle of maximum entropy [KAP 92]. This principle, due to Shannon within the framework of its information theory, consists of determining probability distributions from a random variable by only using the information available, e.g. its mean, its moments, its variation interval, etc. Let X be a random variable with density function $f_X(.)$. The *entropy* of variable X is a measure of the variable's uncertainty, and is defined by:

$$\mathbb{S}_X = -\int_{S_X} f_X(x) \log\left(f_X(x)\right) dx = -\mathbb{E}\left[\log\left(f_X(x)\right)\right] \qquad [1.86]$$

with $\mathbb{S}_X$ as the variation support for variable X. As an example, we will presume that X follows a normal distribution with a zero mean and a standard deviation σ_X:

$$\begin{aligned}
\mathbb{S}_X &= -\int_{S_X} f_X(x) \left[\log\left(\frac{1}{\sqrt{2\pi}\,\sigma_X}\right) - \log\left(\exp\left(-\frac{x^2}{2\,\sigma_X^2}\right)\right) \right] dx \\
&= -\log\left(\frac{1}{\sqrt{2\pi}\,\sigma_X}\right) - \frac{1}{2\,\sigma_X^2}\int_{\mathbb{R}} f_X(x)\, x^2\, dx \qquad\qquad [1.87] \\
&= -\log\left(\frac{1}{\sqrt{2\pi}\,\sigma_X}\right) - \frac{\sigma_X^2}{2\,\sigma_X^2} = \log\left(\sqrt{2\pi}\,\sigma_X\right) - \frac{1}{2}
\end{aligned}$$

with $f_X(x) = \dfrac{1}{\sqrt{2\pi}\,\sigma_X}\exp\left(-\dfrac{x^2}{2\,\sigma_X^2}\right)$.   Entropy $\mathbb{S}_X$ progressively increases according to the standard deviation ($\mathbb{S}_X \xrightarrow[\sigma_X \to +\infty]{} +\infty$); it tends towards $-\infty$ when the standard deviation tends towards 0 (i.e. when the uncertainty increases).

The principle for maximum entropy aims, then, to determine the density function which maximizes entropy. We are dealing with a problem of optimization, under the constraint of available information. Let us calculate that the following information is given by:

$$f_i = \mathbb{E}[g_i(X)] = \int_{S_X} g_i(x) f_X(x)\, dx, \quad i = 1, \cdots, m \qquad [1.88]$$

Determining density $f_X(.)$ amounts to maximizing the entropy [1.86] by respecting the constraints [1.88]. To solve this optimization problem, $(m+1)$ Lagrange multipliers $(-1 + \lambda_0, \lambda_1, \cdots \lambda_m)$ are introduced, and the Lagrangian function to be maximized under the constraints [1.88] is defined by:

$$\mathcal{L}(f_x) = \mathbb{S}_X - (\lambda_0 - 1)\left(\int_{S_X} f_X(x)\, dx - 1 \right) - \sum_{k=1}^{m} \lambda_k \left(\int_{S_X} g_i(x) f_X(x)\, dx - f_i \right) \qquad [1.89]$$

It is possible to demonstrate that this minimization gives [KAP 92]:

$$f_X(x) = \mathbf{1}_{S_X} \exp\left(-{}^t\Lambda\, G(x) \right) \qquad [1.90]$$

with:

$$\Lambda = \begin{pmatrix} \lambda_0 \\ \lambda_1 \\ \vdots \\ \lambda_m \end{pmatrix}; \quad G(x) = \begin{pmatrix} 1 \\ g_1(x) \\ \vdots \\ g_m(x) \end{pmatrix} \qquad [1.91]$$

By setting out:

$$\int_{S_X} f_X(x)\, dx = \int_{S_X} \exp\left(-{}^t\Lambda\, G(x) \right) dx = 1$$

$$\int_{S_X} g_i(x) f_X(x)\, dx = \int_{S_X} g_i(x) \exp\left(-{}^t\Lambda\, G(x) \right) dx = f_i, \quad i = 1, \cdots, m \qquad [1.92]$$

we obtain that:

$$\exp(\lambda_0) = \int_{S_X} \exp\left(-\sum_{i=1}^{m} \lambda_i\, g_i(x)\right) dx$$

$$f_i \exp(\lambda_0) = \int_{S_X} g_i(x) \exp\left(-\sum_{j=1}^{m} \lambda_j\, g_j(x)\right) dx$$

[1.93]

meaning that:

$$f_i = \frac{\displaystyle\int_{S_X} g_i(x) \exp\left(-\sum_{j=1}^{m} \lambda_j\, g_j(x)\right) dx}{\displaystyle\int_{S_X} \exp\left(-\sum_{i=1}^{m} \lambda_i\, g_i(x)\right) dx}$$

[1.94]

As an example, let us consider that the distribution of the variable to be represented has an infinite support and that the mean value and standard deviation are known:

$$F = \begin{pmatrix} 1 \\ f_1 \\ f_2 \end{pmatrix} = \begin{pmatrix} 1 \\ m_X \\ \sigma_X^2 + m_X^2 \end{pmatrix}; \quad G(x) = \begin{pmatrix} 1 \\ x \\ x^2 \end{pmatrix}$$

[1.95]

We obtain that:

$$f_X(x) = \mathbf{1}_\mathbb{R} \exp\left(-\lambda_0 - \lambda_0\, x - \lambda_1\, x^2\right)$$

[1.96]

which gives the following solution:

$$\Lambda = \begin{pmatrix} -\log\left(\dfrac{1}{\sqrt{2\pi}\,\sigma_X}\right) - \dfrac{m_X^2}{2\,\sigma_X^2} \\ 0 \\ \dfrac{1}{2\,\sigma_X^2} \end{pmatrix}$$

[1.97]

then finally:

$$f_X(x) = 1_{\mathbb{R}} \exp\left(-\log\left(\frac{1}{\sqrt{2\pi}\,\sigma_X}\right) - \frac{m_X^2}{2\,\sigma_X^2} + \frac{m_X}{\sigma_X^2}x - \frac{1}{2\,\sigma_X^2}x^2\right)$$

$$= 1_{\mathbb{R}} \frac{1}{\sqrt{2\pi}\,\sigma_X} \exp\left(-\frac{1}{2\,\sigma_X^2}(x - m_X)^2\right)$$

[1.98]

Here we are dealing with a normal distribution: it is the best distribution for maximizing entropy when we know that its support is infinite, and when we know its average and variance. In the same way, when only the support is known and bounded, the best distribution will be the uniform distribution upon this support.

Using maximum entropy principles to characterize the distribution of a variable may be impractical in current cases where often we only have the mean, the variance and information on the support, thus leading us to unrealistic truncated exponentials.

1.8. Distribution of extremes

Modeling variables in a reliability analysis often requires the introduction of distributions of extreme values. A typical situation is, for example, the largest aerodynamic load that a structure can sustain over a given period. Generally, only instantaneous information is available, and we need to predict the distribution of the largest or smallest values.

For example, let us consider the variable X as the maximum of n independent random variables Z_i, with the same distribution function $F_Z(.)$; the distribution function of X is written as:

$$F_X(x) = \mathcal{P}(Z_1 \leq x, \cdots, Z_n \leq x)$$

$$= F_{Z_1}(x) \cdots F_{Z_n}(x)$$

$$= (F_Z(x))^n$$

[1.99]

Let us assume that when the number of variables n tends towards infinity, then the distribution function $F_X(.)$ exists. This distribution function verifies the following stability principle:

$$\exists a_n \quad \exists b_n \quad F_X\left(\frac{x - a_n}{b_n}\right) = F_X(x)^n$$

[1.100]

This principle is very important, since it allows us to establish that only six asymptotic forms for the distributions of extreme value verify it, three for the maxima and three for the minima. The type I, type II, and type III asymptotic forms are called the Gumbel, Fréchet and Weibull distributions respectively.

The type I distribution function (Gumbel) has for cumulative probability function:

– minimum values:

$$F_X(x) = 1 - \exp\left[-e^{\alpha(x-u)}\right] \quad -\infty \leq x \leq +\infty \qquad [1.101]$$

$$\mathbb{E}(X) = u - \frac{\gamma(=0.577)}{\alpha} \qquad [1.102]$$

$$\mathbb{V}(X) = \frac{\pi^2}{6\,\alpha^2} \qquad [1.103]$$

– maximum values:

$$F_X(x) = \exp\left[-e^{-\alpha(x-u)}\right] \quad -\infty \leq x \leq +\infty \qquad [1.104]$$

$$\mathbb{E}(X) = u + \frac{\gamma(=0.577)}{\alpha} \qquad [1.105]$$

$$\mathbb{V}(X) = \frac{\pi^2}{6\,\alpha^2} \qquad [1.106]$$

Figure 1.5 gives the standardized form of the probability density function for the maximum values.

The type I distribution of extreme value is widely used for describing the strength of brittle materials. The argument justifying this model lies in the fact that failure occurs when "elementary" microscopic volumes break. This model type is also used to describe the maximum daily flows over one year. This choice may be debatable, because, on the one hand it relies on the independence of the daily flows, and on the other hand, it resorts to using a distribution which may take negative values! Nevertheless, this distribution has proven to be representative of the observed phenomena.

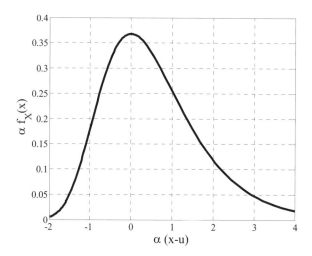

Figure 1.5. *Standardized density function for a Gumbel distribution (maximum values)*

EXAMPLE 1.34.– Annual maximum wind speeds were measured at a weather station in Nantes, France. The mean m_V and standard deviation σ_V are 30 m/s and 9 m/s, respectively. Determining parameters α, u is then directly deduced:

$$\alpha = \frac{\pi}{\sqrt{6}\,\sigma_V} \approx \frac{1.282}{9} \approx 0.42 \text{ i.e. } \frac{1}{\alpha} \approx 7.04 \text{ m/s}$$

$$u = m_V - \frac{0.577}{\alpha} = 30 - \frac{0.577}{0.142} \approx 25.94 \text{ m/s}$$

With these values, it is then possible to calculate the probability of exceeding a given value. If we take the value of 60 m/s, we obtain:

$$P(v \geq 60) = 1 - \exp\left(-e^{-0.142\,(60-25.94)}\right) = 0.008 \qquad \square$$

As for the type I (Gumbel), there are two expressions according to whether we are dealing with minimum or maximum values for the type II distribution. Actually, the distribution of minimum extreme values is of little use in practice and is, therefore, not mentioned here. Only the general expression of the type II distribution for maximum values (Fréchet distribution) is given:

$$F_X(x) = e^{-\left(\frac{u}{x}\right)^k} \quad 0 \le x \le +\infty \qquad [1.107]$$

$$\mathbb{E}(X) = u\,\Gamma\left(1 - \frac{1}{k}\right) \quad k > 1 \qquad [1.108]$$

$$\mathbb{V}(X) = u^2\left[\Gamma\left(1 - \frac{2}{k}\right) - \Gamma^2\left(1 - \frac{1}{k}\right)\right] \quad k > 2 \qquad [1.109]$$

Figure 1.6 gives the standardized form of the probability density function for the type II distribution of maximum extreme values.

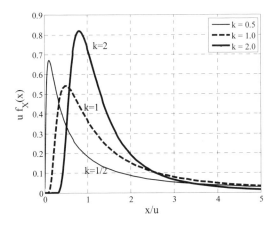

Figure 1.6. *Standardized density function of a Fréchet distribution (maximum values)*

The type II distributions have been particularly used in climatology for describing maximum annual wind speeds. Let us note that if X follows a type II distribution, then $\ln(X)$ follows a type I distribution with parameters $u_0 = \ln(u)$ and $\alpha = k$.

EXAMPLE 1.35.– It is sometimes difficult to choose between a type I or type II distribution to describe the distribution for maximum annual wind speeds. If we take the data for the previous example while determining the parameters for a Fréchet distribution, the coefficient of variation for the maximum wind speed can be calculated as follows:

$$\mathbb{CDV}(V) = \frac{9}{30} \approx 0.30$$

According to equations [1.108] and [1.109], the coefficient of variation for a Fréchet distribution is only a function of parameter k, making it possible to determine this parameter. For this, we will use the following graph.

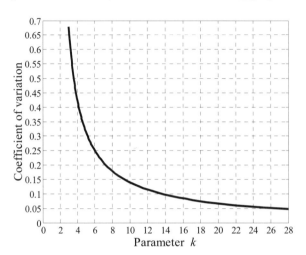

The value $k = 5.2$ can thus be determined, which (using equations [1.108] and [1.109]) leads us to calculate:

$$u = \frac{30}{\Gamma\left(1 - \dfrac{1}{5.2}\right)} \approx 25.96 \text{ m/s}$$

Let us calculate the maximum annual speed for a 2% probability of exceedence:

$$P(v \geq V) = 1 - \exp\left(-\left(\frac{25.96}{V}\right)^{5.2}\right) = 0.02$$

meaning 54.98 m/s. By using the Gumbel distribution from the previous example, we obtain:

$$P(v \geq V) = 1 - \exp\left(-e^{-0.142\,(V - 25.94)}\right) = 0.02$$

i.e. 53.42 m/s, which represents a difference of less than 3% between the two models.   □

The type III distribution of extreme values, or Weibull distribution, have for a cumulative probability function:

– minimum values:

$$F_X(x) = 1 - \exp\left(-\left(\frac{x-\varepsilon}{u-\varepsilon}\right)^k\right) \quad x \geq \varepsilon \qquad [1.110]$$

$$\mathbb{E}(X) = \varepsilon + (u-\varepsilon)\,\Gamma\left(1+\frac{1}{k}\right) \qquad [1.111]$$

$$\mathbb{V}(X) = (u-\varepsilon)^2\left[\Gamma\left(1+\frac{2}{k}\right) - \Gamma^2\left(1+\frac{1}{k}\right)\right] \qquad [1.112]$$

– maximum values:

$$F_X(x) = \exp\left(-\left(\frac{\varepsilon-x}{\varepsilon-u}\right)^k\right) \quad x \leq \varepsilon \qquad [1.113]$$

$$\mathbb{E}(X) = \varepsilon + (u-\varepsilon)\,\Gamma\left(1+\frac{1}{k}\right) \qquad [1.114]$$

$$\mathbb{V}(X) = (u-\varepsilon)^2\left[\Gamma\left(1+\frac{2}{k}\right) - \Gamma^2\left(1+\frac{1}{k}\right)\right] \qquad [1.115]$$

Figure 1.7 gives the general shape of this distribution's probability density function. The type III distributions are the distribution of extreme values including a lower boundary which cannot be exceeded. This is represented by parameter ε. This type of model was widely used by Weibull to describe fatigue resistance or the ultimate strength of a material. One of the difficulties is determining this threshold. For $\varepsilon = 0$, $k = 2$, we find the *Rayleigh distribution* again, introduced in example 1.28.

Let us note, finally, that if X follows a type III distribution (minimum values) then $\ln(X-\varepsilon)$ follows a type I distribution (minimum values) with $u_0 = \ln(u-\varepsilon)$ and $\alpha = k$ as parameters. In the same way, if X follows a type III distribution (maximum values), then $\ln(\varepsilon - X)$ follows a type I distribution (maximum values) with $u_0 = \ln(\varepsilon - u)$ and $\alpha = -k$ as parameters.

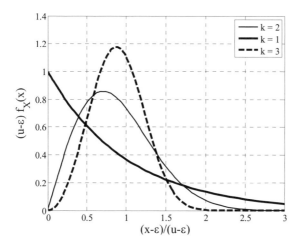

Figure 1.7. *Density function of a Weibull law*

EXAMPLE 1.36.– Fatigue tests on a welded joint with a given stress variation give the following table of fatigue cycles:

400,000	500,000
730,000	600,000
1,060,000	1,300,000
800,000	900,000

The cumulative distribution function of the number of fatigue cycles for the test's stress variation can thus be determined by classifying the previous table's values in increasing order, i being the order in this classification, and N the number of values (here, there are 8):

$$F(x_i) = \frac{i}{N+1}$$

Expressing the distribution function for a type III distribution of minimum extreme values can also be written as:

$$\ln\left(-\ln\left(1 - F_X(x)\right)\right) = \ln\left[\left(\frac{x - \varepsilon}{u - \varepsilon}\right)^k\right] = k\ln\left(x - \varepsilon\right) - k\ln\left(u - \varepsilon\right)$$

This implies that there is a linear relationship between $\ln\left(-\ln\left(1 - F_X(x)\right)\right)$ and $\ln\left(x - \varepsilon\right)$, with k being the gradient and $\ln\left(u - \varepsilon\right)$ the value at the origin.

We presume that $\varepsilon = 100,000$ cycles. It is, then, possible to outline the relationship between $\ln\left(-\ln\left(1 - F_X(x_i)\right)\right)$ and $\ln\left(x_i - \varepsilon\right)$:

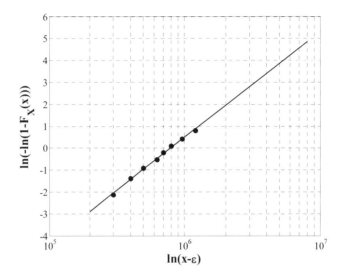

The correlation between $\ln\left(-\ln\left(1 - F_X(x_i)\right)\right)$ and $\ln\left(x_i - \varepsilon\right)$ is 99%, which highlights that the model for the type III distribution is *a posteriori* acceptable. Using this hypothesis, we may carry out an approximation of the gradient:

$$k = \frac{\ln\left(-\ln\left(1 - F_X(x_8)\right)\right) - \ln\left(-\ln\left(1 - F_X(x_1)\right)\right)}{\ln\left(x_8 - \varepsilon\right) - \ln\left(x_1 - \varepsilon\right)} \approx \frac{0.79 + 2.14}{14.00 - 12.61} = 2.11$$

The parameter u is obtained by noting that when $x \equiv u$, then $\ln\left(-\ln\left(1 - F_X(u)\right)\right) = 0$, which gives approximately $u - \varepsilon = 80,000$ cycles on the graph. □

1.9. Significance testing

There are many problems in current practice which are not expressed in terms of estimation, but in comparison. We are therefore often led to comparing two similar samples and to ask ourselves: from which threshold are these generally noticed differences significant, meaning, those which cannot be explained by the random sampling? In other terms, the question asked is to know whether the two samples represent the same phenomenon. With regard to manufacturing tests, the question of conforming to standards is frequently asked. From this threshold, we calculate whether or not the deviations between the characteristics measured on a sample, and their standard values, show a defect from the manufacturing process. We can also propose a comparison between a given distribution and a theoretical distribution i.e. verifying (by introducing an appropriate comparison threshold) whether the sample's distribution conforms to the theoretical distribution. In every case, the conventional procedure for drawing simple conclusions from observed statistical data is called *hypothesis testing*. This procedure generally favors one hypothesis over another, and is the beginning of the much vaster, *decision theory*. It means that we can rationalize choices which must be solved in the most uncertain situations of existence, and make decisions whilst limiting risks of errors on values which are judged to be acceptable.

The theory of hypothesis testing summarizes an uncertain situation as an alternative of two hypotheses, written as H and H'. H is, in fact, the only fundamental hypothesis, with H' being defined in relation to H, by accepting that:

– when we accept H, we reject H';

– when we reject H, we accept H'.

Most classical tests are related to Gaussian populations. On the contrary, if the samples are large enough, the central limit theorems allow us to still use the results from these Gaussian cases. Hypothesis tests are usually arranged into two large categories, each having two families:

– *conformity testing*:

    - tests for comparing with a standard: from these tests, we may determine if the parameters estimated from a sample differ or not from reference or expected values,

    - tests for comparing populations: from these tests, we may answer how "similar" to a control group (sample) is another sample;

– *validity testing* or *goodness-of-fit*:

    - adjustment tests: these can be used to check whether a sample has a distribution conforming to a theoretical distribution,

    - non-parametric tests: these are used for comparing samples extracted from unknown populations.

Amongst the tests for conformity, we will find all the confidence interval methods for the mean or variance. However, in the current practice of mathematical statistics, we often encounter random variables whose distributions are unknown.

With the aim of getting rid of this difficulty, non-parametric tests have been defined so as to verify that a model made to formalize certain hypotheses on these variables is indeed adequate. This concerns the form of unknown probability distributions (*goodness-of-fit*), the comparison of samples (*homogeneity tests*) or the link between many variables (*independence tests*).

In these tests, the favored hypothesis is that of the model, with the antagonistic hypothesis generally being badly defined. By construction, these *free tests* are valid, whatever the distribution of the random variables considered and are particularly *robust*. Some of these tests often require calculations which are simpler than their related parametric tests. For this reason, it is sometimes an advantage to substitute them for these parametric tests when obtained results are not very clear. They are sometimes criticized due to numerous errors of interpretation which can be made by neglecting their asymptotic behavior. The chi-square (or χ^2) test is the most powerful of all non-parametric tests. It simply uses various primary hypotheses and can be used as a test for homogeneity, independence, or goodness-of-fit. More suited to continuous cumulative distribution functions than the chi-square test, the *Kolmogorov* test is based on the concept of the distance between two cumulative probability functions.

1.9.1. *Type I and II errors*

By malapropism, hypothesis H is said to be true when we accept it, but false in the contrary case. This hypothesis is also called the *null hypothesis*. Actually, the conclusion is never completely certain. Accepting or rejecting gives behavior types which result from a reasoning which assesses the errors in terms of probabilities. In this way, two characteristic types of errors, called *type I and type II errors*, are introduced to quantify the error made in either accepting or rejecting a hypothesis.

The *type I error* is the error which rejects H although it is true:

$$\alpha = \mathcal{P}\left(\text{reject } H/H \text{ true}\right) \qquad [1.116]$$

The *type II error* is the error which accepts H when it is false:

$$\beta = \mathcal{P}\left(\text{accept } H/H \text{ false}\right) \qquad [1.117]$$

Table 1.2 summarizes the four outcomes for a hypothesis test, and the related probabilities.

	Acceptance of H	Rejection of H
H true	Right decision Probability $1 - \alpha$	Wrong decision Type I error Probability α
H' true	Wrong decision Type II error Probability β	Right decision Probability $1 - \beta$

Table 1.2. *Probabilities of acceptance or rejection*

Probability α is often called the test's *significance level* and $1 - \beta$ the *power* of the test. In practice, $\alpha = 0.1$ or 0.05. H' is called the *alternative hypothesis*.

1.9.2. *Usual tests*

Most tests relate to Gaussian populations. In the opposite case, if the considered samples are large enough, limit theorems can be applied and the principal results from the normal case can be still applied.

1.9.2.1. *Comparison tests*

It is often useful to test a mean, a variance or a proportion. Many examples can be envisaged, but for the sake of simplicity, in this chapter, only the most important examples will be presented.

Let us consider a normal variable with mean m and standard deviation σ.

$$\begin{aligned} H: &\quad m = m_0 \\ H': &\quad m \neq m_0 \end{aligned} \qquad [1.118]$$

Let there be a sample with size n. In section 1.7 we saw that the variable $\left(\bar{X}_n - m_0\right)/\left(\Sigma_n/\sqrt{n}\right)$ was distributed according to Student's distribution with $n-1$ degrees of freedom. Therefore we know the probability of $\bar{X}_n$ when hypothesis H is true. The test consists of seeking a confidence interval for $\bar{X}_n$ with a probability of $1-\alpha$ or more precisely for the variable $\left(\bar{X}_n - m_0\right)/\left(\Sigma_n/\sqrt{n}\right)$:

$$\begin{aligned} 1-\alpha &= \mathcal{P}\left(\text{accept } H/H \text{ true}\right) \\ &= \mathcal{P}\left(-t_{n-1,\frac{\alpha}{2}} \leq \frac{\bar{X}_n - m_0}{\Sigma_n}\sqrt{n} \leq t_{n-1,\frac{\alpha}{2}}\right) \end{aligned} \qquad [1.119]$$

After sampling, a mean and a standard deviation are calculated, $\bar{x}_n, s_n$. The test, then, consists of verifying:

$$\bar{x}_n \in \left] m_0 - t_{n-1,\frac{\alpha}{2}} \frac{s_n}{\sqrt{n}}, m_0 + t_{n-1,\frac{\alpha}{2}} \frac{s_n}{\sqrt{n}} \right[\quad m = m_0 \text{ is accepted}$$

$$\bar{x}_n \notin \left] m_0 - t_{n-1,\frac{\alpha}{2}} \frac{s_n}{\sqrt{n}}, m_0 + t_{n-1,\frac{\alpha}{2}} \frac{s_n}{\sqrt{n}} \right[\quad m = m_0 \text{ is rejected}$$

[1.120]

This result is close to that found in Equation 1.79 which consists of estimating the mean by a confidence interval.

EXAMPLE 1.37.– (adapted from [ANG 84]) A cement provider gives some bags of cement to a building site which only wishes to accept cements with no more than 0.3% moisture (symbolizing a bad cement) in 10% of cases. However, the provider thinks that he is delivering good quality cement when it has less than 0.1% moisture, and he does not want to throw away good quality cement in more than 5% of cases. If we presume that the previous moisture values are mean values, the type I and type II errors are $\alpha = 0.05$ and $\beta = 0.10$. If we presume that the moisture's standard deviation is estimated to be 0.1%, a good cement will be rejected for a sampling value L such as:

$$\frac{\sqrt{n}\,(L - 0.001)}{0.001} = -t_{n-1,5\%}; \quad \frac{\sqrt{n}\,(L - 0.003)}{0.001} \leq t_{n-1,10\%}$$

With the difference of expression [1.119], the probabilities are no longer connected to intervals, but to upper and lower limits for the type I and type II errors. Solving the previous equations gives $n = 4$ and $L = 0.22\%$. This means that, if a sample of four bags out of one batch gives average moisture values higher than 0.22%, then the batch must be rejected. In the opposite case, it will be accepted. □

As for the mean, the test on the variance is considered in a similar way. The test consists of either accepting or rejecting the value σ_0 for the standard deviation:

$$H: \quad \sigma = \sigma_0$$
$$H': \quad \sigma \neq \sigma_0$$

[1.121]

In this case, we saw in section 1.6.2 that the variable $(n-1)\Sigma_n^2/\sigma_0$ was distributed according to the chi-square distribution with $n - 1$ degrees of freedom.

The test, then, consists of seeking a confidence interval of Σ_n^2 with a probability of $1 - \alpha$ or more precisely for the variable $(n-1)\Sigma_n^2/\sigma_0^2$:

$$P\left(\hat{\chi}_{n-1}^{2\,'} < \chi_{n-1}^2 = \frac{(n-1)\Sigma_n^2}{\sigma_0^2} < \hat{\chi}_{n-1}^{2\,''}\right) = 1 - \alpha \qquad [1.122]$$

After sampling, a standard deviation s_n is calculated. Therefore, the test consists of verifying:

$$s_n \in \left]\frac{\hat{\chi}_{n-1}^{2\,'}\sigma_0^2}{(n-1)}, \frac{\hat{\chi}_{n-1}^{2\,''}\sigma_0^2}{(n-1)}\right[\qquad \sigma = \sigma_0 \text{ is accepted}$$

$$\qquad\qquad\qquad\qquad\qquad\qquad\qquad\qquad\qquad [1.123]$$

$$s_n \notin \left]\frac{\hat{\chi}_{n-1}^{2\,'}\sigma_0^2}{(n-1)}, \frac{\hat{\chi}_{n-1}^{2\,''}\sigma_0^2}{(n-1)}\right[\qquad \sigma = \sigma_0 \text{ is rejected}$$

This result is close to the confidence interval for variance estimation (equation [1.81]).

The type II error consists of calculating the probability of whether the null hypothesis $\sigma = \sigma_0$ is wrongly accepted. For this, it would be suitable to verify whether a value taken in the determined confidence interval leads to a too high type II error. This would lead us to accept the null hypothesis with a large error probability. Let σ_1 be this value, and let us set out $\lambda = \sigma_1/\sigma_0$. We then obtain:

$$\beta = P\left(\left[\Sigma_n^2 \le \frac{\sigma_0^2}{(n-1)}\hat{\chi}_{n-1}^{2\,'}\right] \cap \left[\Sigma_n^2 \ge \frac{\sigma_0^2}{(n-1)}\hat{\chi}_{n-1}^{2\,''}\right]\right)$$

$$= F_{\chi^2}\left(\frac{\sigma_0^2}{\sigma_1^2}\hat{\chi}_{n-1}^{2\,'}\right) - F_{\chi^2}\left(\frac{\sigma_0^2}{\sigma_1^2}\hat{\chi}_{n-1}^{2\,''}\right) \qquad [1.124]$$

1.9.2.2. Validity tests

The chi-square test is non-parametric and relatively simple to use. It can be used for homogeneity and independence tests as well as goodness-of-fit. It is well adapted to large samples. For small samples, it is preferable to use the Kolmogorov test.

Let us consider n observations of a random variable. The chi-square test consists of comparing the observed frequencies $n_1, n_2, \cdots, n_k$ of k intervals, with distribution's frequencies $e_1, e_2, \cdots, e_k$. The test's principle consists of evaluating the term:

$$D_i = \sum_{i=1}^{k} \frac{(n_i - e_i)^2}{e_i}$$ [1.125]

However, D_i is the realization of a distribution D which tends towards the chi-square distribution χ_{k-1}^2 when $n \to \infty$. However, if the parameters of this distribution are unknown and estimated from data, the previous convergence remains valid, subject to reducing the degree of freedom of χ_{k-1}^2 by 1 for each unknown parameter. On this basis, if k' represents the new degree of freedom, the test consists of verifying:

$$D_i = \sum_{i=1}^{k} \frac{(n_i - e_i)^2}{e_i} < \chi_{1-\alpha, k'-1}^2$$ [1.126]

in order to consider that the two distributions are identical for a significance level α.

The Kolmogorov-Smirnov (written K-S) test consists of comparing the cumulative distribution functions of the two distributions. If $F(x)$ and $F_n(x)$ are the reference distribution function and the function determined experimentally from a sample of size n respectively, then the Kolmogorov test studies the statistics of variable D_n:

$$D_n = \sup |F(x) - F_n(x)|$$ [1.127]

Clearly, if $D_n = 0$, then there is a perfect adequacy between the experimental distribution and the proposed distribution. Therefore, we understand the importance of analyzing the difference D_n: the validity will much better when the difference is reduced. The test's principle is, then, to verify whether, for a given significance level α, this difference remains acceptable:

$$\mathcal{P}\left(D_n \leq D_n^\alpha\right) = 1 - \alpha$$ [1.128]

The critical values D_n^α can be found in statistical tables according to n and α.

EXAMPLE 1.38.– Tests on cubic concrete samples give 143 values classified into eight intervals according to the following table:

46.5<	46.5-48.5	48.5-50.5	50.5-52.5	52.5-54.5	54.5-56.5	56.5-58.5	>58.5
9	17	22	31	28	20	9	7

Estimating the mean and the standard deviation from the 143 values gives 52.75 MPa and 3.95 MPa. To characterize a concrete's compressive strength, we retain either a normal (N) distribution or a lognormal (LN) distribution. The following table summarizes the chi-square values and the differences in absolute values between classes following these two models.

Class	n_i	e_i		$\dfrac{(n_i - e_i)^2}{e_i}$		Cumulative distribution function			$\left\|F(x) - F_n(x)\right\|$	
MPa	Exp.	N	LN	N	LN	Exp.	N	LN	N	LN
46.5	9	8.1	7.1	0.10	0.52	0.063	0.057	0.050	0.006	0.013
48.5	17	12.0	12.8	2.05	1.41	0.182	0.141	0.139	0.041	0.043
50.5	22	20.5	22.0	0.11	0.00	0.336	0.284	0.293	0.051	0.043
52.5	31	27.2	28.2	0.53	0.29	0.552	0.475	0.490	0.078	0.063
54.5	28	28.1	27.5	0.00	0.01	0.748	0.671	0.682	0.077	0.066
56.5	20	22.5	21.2	0.29	0.07	0.888	0.829	0.830	0.059	0.058
58.5	9	14.1	13.1	1.83	1.31	0.951	0.927	0.922	0.024	0.029
60.5	7	10.4	11.1	1.11	1.52	1.000	0.927	0.922	0.073	0.078
			Σ	6.01	5.12			Max	0.078	0.078

The chi-square test is carried out for a 5% type I error. As the parameters for these distributions are estimated, we must reduce the chi-square distribution by two degrees of freedom, which gives $k' = 8 - 2 = 6$, with 8 being the number of classes. The acceptance threshold value is then $\chi^2_{95\%,5} = 11.07$.

The previous table shows that the chi-square values for the two distributions may be retained for describing the concrete's compressive strength. As the test's value for the lognormal distribution is lower than the normal value, we may consider that the lognormal distribution is more appropriate. The Kolmogorov test shows identical values for both distributions; as the acceptance value $D_{143}^{5\%}$ is 0.11, the two distributions may be equally retained. □

1.10. Bayesian analysis

In practice, statistical parameters are rarely known and are estimated. A problem which is often encountered is updating these estimators by means of new samples. The Bayesian analysis offers some tools for *requalifying* them. Since the statistical parameters are estimated, the estimators must be considered as random variables (see section 1.7).

1.10.1. *A priori and a posteriori distributions*

Therefore, let Θ be a parameter of a random variable X, with probability density function $f'_{\Theta}(\theta)$. This function is based on the engineer's knowledge regarding the parameter Θ from existing information. The distribution of Θ, based on this information, is called the *a priori distribution*. Let us presume now that n independent observations of the variable X are available: $X = x_1, \cdots, X = x_n$. The objective is to use this sample X_n as efficiently as possible to fit a new distribution to the parameter Θ: $f''_{\Theta}(\theta)$ called the *a posteriori distribution*. This *a posteriori* distribution is written using the Bayes theorem for continuous variables:

$$f''_{\Theta}(\theta) = \mathcal{P}(\theta | X_n) = \frac{\mathcal{P}(X_n | \theta)}{\displaystyle\int_{D_{\Theta}} \mathcal{P}(X_n | \theta) \, f'_{\Theta}(\theta) \, d\theta} \, f'_{\Theta}(\theta) \qquad [1.129]$$

where D_{Θ} is the variation domain of parameter Θ. The term:

$$\lambda = \left(\int_{D_{\Theta}} \mathcal{P}(X_n | \theta) \, f'_{\Theta}(\theta) \, d\theta \right)^{-1} \qquad [1.130]$$

may be considered as a normalization constant so as to ensure that $f''_{\Theta}(\theta)$ defines a true probability density function, i.e. $\int_{D_{\Theta}} f''_{\Theta}(\theta) \, d\theta = 1$. Equation [1.129] is then written:

$$f''_{\Theta}(\theta) = \lambda \, \mathcal{P}(X_n | \theta) \, f'_{\Theta}(\theta) \qquad [1.131]$$

The term $\mathcal{P}(X_n | \theta)$ is called the *likelihood function*.

EXAMPLE 1.39.– Let us presume that the *a priori* distribution of the crack depths in a series of welding joints follows a triangular distribution:

$$f_X'(x) = \begin{cases} 66.7\,x & 0\text{ mm} < x \leq 0.1\text{ mm} \\ 10.005 - 33.35\,x & 0.1\text{ mm} < x \leq 0.3\text{ mm} \\ 0 & 0.3\text{ mm} < x \end{cases}$$

A non-destructive test is used to detect cracks. The probability of detecting a crack is given by the following probability distribution:

$$P_d(x) = 1 - \exp(-15\,x)$$

A weld is inspected by the magnetic particle method and no defect is detected. In this case, the *a posteriori* distribution for the presence of a crack is, after normalization by λ:

$$f_X''(x) = \lambda\left(1 - P_d(x)\right) f_X'(x)$$

$$= \frac{1}{0.2004} \begin{cases} 66.7\,x\left[\exp(-15\,x)\right] & 0\text{ mm} < x \leq 0.1\text{ mm} \\ [10.005 - 33.35\,x]\left[\exp(-15\,x)\right] & 0.1\text{ mm} < x \leq 0,3\text{ mm} \\ 0 & 0.3\text{ mm} < x \end{cases}$$

The following figure compares the three distributions. After inspection, if no defect is detected, the crack distribution shifts to smaller values.

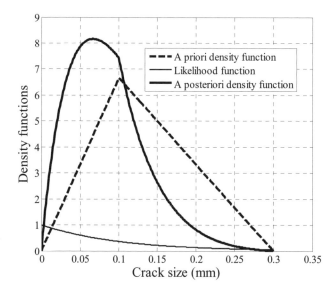

□

1.10.2. *Updating estimators*

When variable X is continuous and the observations are independent, the likelihood function is written as:

$$L(\theta|x_1, x_2, \cdots, x_n) = \mathcal{P}\left(X_n|\theta\right) = \prod_{i=1}^{n} f_X\left(x_i|\theta\right) \qquad [1.132]$$

Here, it is not possible to describe all the cases for random variables X; therefore, we will limit ourselves to the simplest case, which is a random normal variable with an unknown mean M and an unknown standard deviation Σ.

In this case, for parameter $\Theta = (M, \Sigma)$, the likelihood function can be clarified:

$$L(\theta|x_1, x_2, \cdots, x_n) = \prod_{i=1}^{n} f_X\left(x_i|\theta\right) = \prod_{i=1}^{n} \frac{1}{\sqrt{2\pi\sigma^2}} \exp\left[-\frac{(x_i - \mu)^2}{2\sigma^2}\right]$$

$$= \left(\frac{1}{\sqrt{2\pi\sigma^2}}\right)^n \exp\left[-\frac{1}{2\sigma^2}\sum_{i=1}^{n}(x_i - \mu)^2\right] \qquad [1.133]$$

with $\theta = (\mu, \sigma)$. If $\bar{x}_n, s_n^2$ are the observations of M and Σ based on the sample $(x_i)_{1 \le i \le n}$, then:

$$\bar{x}_n = \frac{1}{n}\sum_{i=1}^{n} x_i; \quad s_n^2 = \frac{1}{n-1}\sum_{i=1}^{n}(x_i - \bar{x}_n)^2 \qquad [1.134]$$

thus equation [1.133] changes into:

$$L(\theta|x_1, x_2, \cdots, x_n) = \left(\frac{1}{\sqrt{2\pi\sigma^2}}\right)\exp\left[-\frac{n(\mu - \bar{x}_n)^2}{2\sigma^2}\right]\left(\frac{1}{\sqrt{2\pi\sigma^2}}\right)^{n-1}\exp\left[-\frac{n-1}{2\sigma^2}s_n^2\right] \quad [1.135]$$

The likelihood function appears as the product of a normal distribution by a Gamma-normal distribution. It is, then, possible to show that this form can also be used for the *a priori* distribution $f'_\Theta(\theta)$ [BER 85].

$$f'_\Theta(\theta) = f' \left(\frac{1}{\sqrt{2\pi \dfrac{\sigma^2}{n'}}} \right) \exp\left[-\frac{n'}{2\sigma^2}(\mu - \bar{x}')^2 \right]$$

$$\times \frac{\left(\dfrac{n'-1}{2}\right)^{(n'-2)/2}}{\Gamma\left(\dfrac{n'-2}{2}\right)} \frac{2}{s'}\left(\frac{s'^2}{\sigma^2}\right)^{(n'-2)/2} \exp\left[-\frac{n'-1}{2\sigma^2}s'^2 \right]$$

[1.136]

where n', $\bar{x}'$ and s'^2 respectively are the size of the sample which allow us to build the *a priori* distribution, the *a priori* mean and variance. It is, then, easy to show that the *a posteriori* distribution has the same form as the *a priori* distribution with the following relationships:

$$n'' = n + n'$$

[1.137]

$$n''\bar{x}'' = n\,\bar{x}_n + n'\bar{x}'$$

[1.138]

$$(n''-1)s''^2 + n''\bar{x}''^2 = \left[(n-1)s_n^2 + n\,\bar{x}_n^2\right] + \left[(n'-1)s'^2 + n'\bar{x}'^2\right]$$

[1.139]

Determining marginal distributions for the mean and variance allows us to deduce the means and the variances of *a posteriori* estimators. Thus, for the mean M, the variable $(M - \bar{x}'')\sqrt{n''}/s''$ follows a Student distribution with $n''-2$ degrees of freedom. We then obtain:

$$\mathbb{E}(M'') = \bar{x}''; \quad \mathbb{V}(M'') = s''^2 \frac{n''-1}{n''(n''-2)}$$

[1.140]

In the same way, for the *a posteriori* estimator of the standard deviation, it is possible to show that the distribution is a chi-square distribution, which gives:

$$\mathbb{E}(\Sigma'') = s''\sqrt{\frac{n''-1}{2}} \frac{\Gamma\left(\dfrac{n''-3}{2}\right)}{\Gamma\left(\dfrac{n''-2}{2}\right)} \quad n'' > 3$$

$$\mathbb{V}(\Sigma'') = s''^2 \frac{n''-1}{n''-3} - \mathbb{E}(\Sigma'')^2 \quad n'' > 4$$

[1.141]

EXAMPLE 1.40.– Defects in a welded joint can be modeled by a Poisson distribution with an average occurrence rate of μ defects per meter. The applied test helps to detect 7 defects on a 15 m long welded joint. However, the quality for this type of joint leads us to estimate the average defect rate as 0.2 defects per linear meter with a 30% coefficient of variation.

We demonstrate that, if the distribution which describes the number of defects per meter of a welded joint is a Poisson distribution, then the parameter μ follows a Gamma *a priori* distribution [BER 85]. In this case, we obtain:

$$\mathbb{E}(M') = \frac{n'}{v'} = 0.2; \quad \mathbb{V}(M') = \frac{n'}{v'^2} = 0.3^2 \times \mathbb{E}(M')^2 = 0.3^2 \times \frac{n'^2}{v'^2}$$

which means that we can estimate the size of an *a priori* sample $n' = 11.11$ and $v' = 55.55$. The *a posteriori* distribution is also Gamma, which gives:

$$n'' = n' + n = 11.11 + 7 = 18.11; \quad v'' = v' + 15 = 70.55$$

The average rate for defects per linear meter is then updated to:

$$\mathbb{E}(M'') = \frac{n''}{v''} = \frac{18.11}{70.55} \approx 0.26; \quad \mathbb{V}(M'') = \frac{n''}{v''^2} = \frac{18.11}{70.55^2} \approx 0.0036$$

i.e. a coefficient of variation of 0.23.   □

1.10.3. *Bayesian networks*

Bayesian statistical techniques are often supported by Bayesian networks. A Bayesian networks (BN) is a graphical model that encodes probabilistic relationships among variables of interests [JEN 01]. This graphical model presents several advantages for data analysis. First, because the model encodes dependencies among all variables, it readily handles situations where some data entries are missing. Second, a Bayesian network can be used to learn causal relationships. Third, it is an ideal representation for combining prior knowledge and data. A Bayesian network for a set of random variables $X = (X_1, ..., X_n)'$ is a way of representing a joint probability density function among those variables. The network includes a graph structure $\mathcal{G}$ and a set of conditional probabilities $\mathcal{P}$. $\mathcal{G}$ is used to encode conditional dependencies among the variables. It is a n-nodes direct acyclic graph (DAG) in which the nodes are in one-to-one correspondence with the variables. For all i and j such as $1 \le i, j \le n$, an oriented arc linking X_i and X_j exists if and only if X_j is conditionally dependent X_i. The measure $\mathcal{P}$

describes the conditional dependencies that are represented by oriented arcs in the structure $\mathcal{G}$.

Since the BN structure is acyclic, the nodes of $\mathcal{G}$ can be topologically sorted in such way that if an arc exists from a node X to another node Y, then X must precede Y in the ordering. The acyclicity induces ancestral relations. For instance, if there is an arc from X to Y, then it is classically said that Y is a child of X and X is a parent of Y. For each i, $\pi(i)$ denotes the set of all parents of X_i and $\varepsilon(i)$ denotes the set of children of X_i. The joint probability distribution related to the set of variables $X = (X_1,...,X_n)$ can then be represented as follows:

$$P(X) = \prod_{i=1}^{n} P(X_i \mid X_{\pi(i)})$$
[1.142]

where $P(X)$ is the probability of X and $P(X_i \mid X_{\pi(i)})$ represents the conditional probability of X_i given the different parent variables $X_{\pi(i)} = (X_{i_1},...,X_{i_k})$.

From a practical point of view, the use of Bayesian networks is particularly interesting due to their property of inference. Let us consider a variable X_i with its related set of parents $X_{\pi(i)}$. The marginal distribution of the variable X_i can be calculated as follows:

$$P(X_i) = P(X_i \mid X_{\pi(i)}) \prod_{m=i_1}^{i_k} P(X_m)$$
[1.143]

The *inference property* consists of computing the updated probability distributions of the parents of X_i, conditioned by a given set of observations of the variable X_i. These observations can come from experimental measurements or experts' judgments. For all h such as $1 \le h \le k$, the Bayes theorem can be used to carry out these calculations as follows:

$$P(X_{i_h} \mid X_i) = \frac{P(X_i \mid X_{i_h})P(X_{i_h})}{P(X_i)}$$
[1.144]

Since the updated probability distributions related to all the parents of the variable X_i are determined, the updated probability distribution of X_i can be calculated.

Various topologies $\mathcal{G}$ can be used. This choice depends on the nature of the knowledge which is represented. The simplest is the V-structure where $\{X_i\}_{i\in\{1,...,n\}\setminus\{l\}}$ are the causal or parent variables and the variable X_l is the consequence or child variable. To completely define the Bayesian network, the knowledge of the marginal probability distribution of each variable $\{X_k\}_{k\in\{1,...,n\}}$ is required. For this purpose, they are represented by histograms. For causal random variables, their probability distribution functions are not conditioned by any other variable (parent nodes) and can be considered as given inputs. For the consequence random variable, the calculation of the probability distribution function is fully conditioned by the causal ones. This requires to know the conditional probability distribution $\mathcal{P}(X_l \mid X_k \; k \neq l)$. The conditional probability distribution functions are stored in a matrix called the *conditional probability table* (CPT).

Let us assume that the support of each probability distribution is divided into n_c classes. For causal random variables, the CPT is a $1 \times n_c$ matrix although for the consequence random variable, the corresponding CPT is $(n_c)^{n-1} \times n_c$. With the CPT and the inference property, the conditional probability distribution of any causal random variable can be updated with respect to a given observations set. In summary, the probability distributions of the causal (physical) variables are updated from the Bayesian inference statistical property based on the observations of the consequence variables.

EXAMPLE 1.41.– A set of concrete cores from different batches is exposed to low (E-L) and high (E-H) aggressive environments (salts); the concrete samples can be further spread in two groups depending on the diffusion coefficient values. D-L denotes the group with a low coefficient of diffusion (i.e. below a threshold value) and D-H denotes the group of cores with a high coefficient of diffusion. After a given duration, the depth of the chloride ingress inside concrete is measured. If this depth is smaller than a prescribed value, the concrete is said to be "non-contaminated (NC)"; if it is larger it is said to be "contaminated (C)". The conditional probability table is established based on the analysis of the set of concrete cores and is given.

	Low aggressive environment (E-H)		High aggressive environment (E-L)	
	D-H	D-L	D-H	D-L
Contaminated (C)	95%	85%	90%	2%
Non-contaminated (NC)	5%	15%	10%	98%

A new concrete sample for an existing structure is analyzed and the contamination depth is found to be smaller than the prescribed value. Some uncertainties are linked to this information and a 10% probability is specified to the contamination event. It is also assumed that the de-icing process on the bridge was not too high and a 20% probability is assigned to the event (E-H). Furthermore, a 20% *a priori* probability is assigned to the "high coefficient of diffusion" event. On this basis, we want to know if the concrete sample belongs to the D-L group (low coefficient of diffusion), i.e. to calculate $P(\text{D-L}/\text{NC})$.

The Bayesian Network (BN) has a V-structure with two parent nodes (coefficient of diffusion, environment) and one child node (contamination).

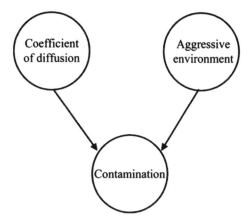

With the Bayes theorem and the BN structure, it becomes:

$$P(\text{D-L}/\text{NC}) = \frac{P(\text{NC}/\text{D-L})}{P(\text{NC})} P(\text{D-L})$$

Using the law of total probability gives (assuming that the events on "environment" and "coefficient of diffusion" are independent):

$$P(\text{NC}/\text{D-L}) = P(\text{NC}/\text{D-L,E-L})P(\text{E-L})$$
$$+ P(\text{NC}/\text{D-L,E-H})P(\text{E-H})$$
$$= 0.98 \times 0.8 + 0.15 \times 0.2 = 0.814$$

According to the Bayes theorem, $P(\text{D-L}/\text{NC}) \times P(\text{NC}) = 0.814 \times 0.1 = 0.0814$. Similarly, it becomes $P(\text{D-H}/\text{NC}) \times P(\text{NC}) = 0.12 \times 0.9 = 0.108$, since:

$$P(NC/D\text{-}H) = P(NC/D\text{-}H,E\text{-}L)P(E\text{-}L)$$
$$+P(NC/D\text{-}H,E\text{-}H)P(E\text{-}H)$$
$$= 0.1 \times 0.8 + 0.05 \times 0.8 = 0.12$$

Finally the desired calculation can be made:

$$P(D\text{-}L/NC) = \frac{P(NC/D\text{-}L)}{P(D\text{-}L/NC) \times P(NC) + P(D\text{-}H/NC) \times P(NC)} P(D\text{-}L)$$
$$= \frac{0.814}{0.0814 + 0.108} \times 0.2 \approx 0.86$$

In other words, there is an 86% probability that the concrete sample from the existing bridge has a low coefficient of diffusion.   □

1.11. Stochastic processes

1.11.1. *Basic principles*

A process $Y(t)$ is a collection of random variables indexed by a continuous or discrete parameter t. At $t = t_0$, $y(t_0)$ is an observation of $Y(t_0)$. For a fixed time t, there are many possible outcomes. By generalizing this for each instant, many time histories of observations $y_i(t)$ can be generated. Figure 1.8 illustrates the realization histories (index t represents time).

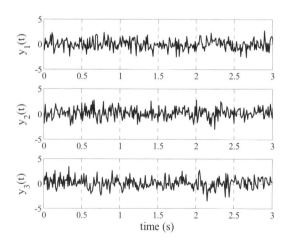

Figure 1.8. *Histories of a stochastic process*

In Chapter 5 which covers the specificities of resistance and loads, we will give a detailed account of the characteristics related to stochastic processes of continuous variables. In this chapter, we will only present the case of processes of discrete variables.

1.11.2. Markovian chains

In civil engineering, when studying certain parts of structures where modeling the degradation is difficult, it is often interesting to introduce special stochastic processes known as *Markov chains*.

EXAMPLE 1.42.– According to particular inspection procedures, the structural condition of a bridge may be classified into five categories: (1) very good; (2) good; (3) average; (4) bad; (5) dangerous. These five categories define the structure's possible states. After an exceptional event, the condition of the structure may remain the same or move towards another state. The transition from the initial state towards the next one over a period of time is characterized by a probability called the *transition probability*.   □

Let us consider a system able to take m possible states and let us assume that the transition from one state to another takes place at times $t_1, t_2, \cdots, t_n$. Let X_{n+1} be the state of the system at time t_{n+1}. In theory, this state may depend on the system's history, i.e. on the states it has known at $t_1, t_2, \cdots, t_n$:

$$P\left(X_{n+1} = x_{n+1} \middle| X_0 = x_0, X_1 = x_1, \cdots, X_n = x_n\right) \qquad [1.145]$$

where $x_0, x_1, \cdots, x_n$ represent the system's previous states. If the future state of the system is solely controlled by the present state, the conditional probability [1.145] is reduced to:

$$P\left(X_{n+1} = x_{n+1} \middle| X_0 = x_0, X_1 = x_1, \cdots, X_n = x_n\right) = P\left(X_{n+1} = x_{n+1} \middle| X_n = x_n\right) \quad [1.146]$$

When this condition is verified, we can say that process $\left(X_k\right)$ is a Markov chain. The *transition probability* from a state x_i at time t_m to a state x_j at time t_n is written:

$$p_{i,j}(n,m) = P\left(X_n = x_j \middle| X_m = x_i\right), n > m \qquad [1.147]$$

A Markov chain is said to be *homogeneous* if $p_{i,j}(n,m)$ depends only on the difference $t_n - t_m$. In this case, the transition probability of going from state x_i to state x_j over k periods Δt, is given by:

$$p_{i,j}(k) = P\left(X_k = x_j \middle| X_0 = x_i\right) = P\left(X_{k+s} = x_j \middle| X_s = x_i\right), \quad s \geq 0 \qquad [1.148]$$

This probability is deduced from the probabilities of transition $p_{i,j}$ between states over Δt. These transition probabilities can be used to generate a *transition probability matrix.*

$$[P] = \begin{bmatrix} p_{1,1} & p_{1,2} & p_{1,m} \\ p_{2,1} & p_{2,2} & p_{2,m} \\ & & \\ p_{m,1} & p_{m,2} & p_{m,m} \end{bmatrix} \qquad [1.149]$$

As the states are exclusive and exhaustive, the sum of probabilities for a transition matrix line $[P]$ is equal to 1.0. For a homogenous Markov chain, the initial state probabilities provide the only data necessary to predict the system's future behavior.

1.11.3. *State probability*

State probability is the probability that the system has one of the different state categories at a certain instant; at time t_n they are written:

$$\{q(t_n)\} = \{q_1(t_n), \cdots, q_m(t_n)\} \qquad [1.150]$$

On the one hand it is possible to show that:

$$\{q(t_n)\} = \{q(t_0)\}[P]^n \qquad [1.151]$$

and on the other hand that:

$$p_{i,j}(n) = \sum_k p_{k,j}(n-r)\, p_{i,k}(r), \quad 0 < r < n \qquad [1.152]$$

Equation [1.152] is called the *Chapman-Kolmogorov* equation.

EXAMPLE 1.43.– According the state categories defined in example 1.42, we consider that the transition matrix after an exceptional event is given by the following matrix:

$$[P] = \begin{bmatrix} 0.8 & 0.15 & 0.045 & 0.005 & 0 \\ 0 & 0.6 & 0.3 & 0.07 & 0.03 \\ 0 & 0 & 0.5 & 0.4 & 0.1 \\ 0 & 0 & 0 & 0.3 & 0.7 \\ 0 & 0 & 0 & 0 & 1.0 \end{bmatrix}$$

For a structure which is in a very good state, the state probability after two events is:

$$\{q(2)\} = \{1 \ \ 0 \ \ 0 \ \ 0 \ \ 0\} \begin{bmatrix} 0.8 & 0.15 & 0.045 & 0.005 & 0 \\ 0 & 0.6 & 0.3 & 0.07 & 0.03 \\ 0 & 0 & 0.5 & 0.4 & 0.1 \\ 0 & 0 & 0 & 0.3 & 0.7 \\ 0 & 0 & 0 & 0 & 1.0 \end{bmatrix}^2$$

$$= \{0.6400 \ \ 0.2100 \ \ 0.1035 \ \ 0.0340 \ \ 0.0125\}$$

□

A state is said to be *absorbing* if the probability $p_{j,j} = 1.0$. An interesting quantity is the average time T_j before absorption if the system leaves the initial state x_j (non-absorbing):

$$T_j = 1 + \sum_{i=1}^{m-1} T_i \, p_{j,i}$$

[1.153]

EXAMPLE 1.44.– By using the hypotheses from examples 1.42 and 1.43, we will now focus on the mean number of exceptional events to reach the "average" state for a structure which is in very good or good initial state.

For this, we can combine states 3, 4 and 5 into one new state 3b; the transition matrix then becomes:

$$[P_b] = \begin{bmatrix} 0.8 & 0.15 & 0.05 \\ 0 & 0.6 & 0.4 \\ 0 & 0 & 1.0 \end{bmatrix}$$

It is then sufficient to apply expressions [1.152]:

$$\begin{cases} T_1 = 1 + 0.8\,T_1 + 0.15\,T_2 \\ T_2 = 1 + 0.6\,T_2 \end{cases} \Rightarrow \begin{cases} T_1 \approx 7 \\ T_2 \approx 2.5 \end{cases}$$

These results show that on average, seven events are needed to make a structure in very good state change to an average or worse, state. On the other hand, less than three events are needed to make a structure change from being in a good state to a average or worse state. □

1.11.4. *Time between stages*

The period of time needed to go from one state to another in a Markov chain is a random variable. The number of transitions over a given period is not, therefore, deterministic.

The probability $p_i(t)$ of a system being found in a state i at time t will depend on the number of transitions or stages n and the state probabilities after n stages:

$$p_i(t) = \sum_{n=0}^{\infty} p_i(n)\, p_N(n,t) \qquad [1.154]$$

$p_N(n,t)$ is the probability of the number of stages N at time t. The probability of the number of transitions N exceeding a value n at time t is identical to the probability of the total duration period being less than or equal to t over $n+1$ stages:

$$F_N(n) = 1 - P(N > n) = 1 - P\left(T_1 + T_2 + \cdots + T_n + T_{n+1} \le t\right) \qquad [1.155]$$

Let us write $S_{n+1} = T_1 + T_2 + \cdots + T_n + T_{n+1}$ as the sum of the different transition times; we obtain that:

$$F_N(n) = 1 - F_{S_{n+1}}(t) \qquad [1.156]$$

We then deduce:

$$p_N(n,t) = F_N(n) - F_N(n-1) = F_{S_n}(t) - F_{S_{n+1}}(t) \qquad [1.157]$$

If the transition times are assumed to be independent and normally distributed according to $\mathcal{N}(\mu,\sigma)$, then S_n is also $\mathcal{N}(n\mu,\sqrt{n}\,\sigma)$, which gives:

$$p_N(n,t) = \Phi\left(\frac{t-n\mu}{\sqrt{n}\,\sigma}\right) - \Phi\left(\frac{t-(n+1)\mu}{\sqrt{n+1}\,\sigma}\right) \tag{1.158}$$

If $p_N(n,t)$ follows a Poisson distribution (the transition times then follow exponential distributions), we simply obtain:

$$p_N(n,t) = \frac{e^{-vt}(vt)^n}{n!} \tag{1.159}$$

where v is the average transition rate.

EXAMPLE 1.45.– By using the hypotheses from examples 1.42, 1.43 and 1.44, we presume that the number of exceptional events follows a Poisson distribution with an average rate of transition which is equal to 2 every 10 years. The probability of a structure suffering at least moderate damage over a period of 30 years depends on the number of exceptional events over this period.

By using equations [1.154] and [1.159], we find that:

$$P(D\geq 3) = \sum_{n=0}^{\infty} P(D\geq 3|n)\frac{\exp(-30\times 2/10)\times(30\times 2/10)^n}{n!}$$

$P(D\geq 3|n)$ is calculated from the reduced transition matrix from example 1.43. The following table summarizes the different calculations:

n	$p_N(n,30\text{ years}) = \dfrac{e^{-6}(6)^n}{n!}$	$P(D\geq 3|n)$	$P(D\geq 3|n)p_N(n,30\text{ years})$
0	0.0025	0.0000	0.0000
1	0.0149	0.0500	0.0007
2	0.0446	0.1500	0.0067
3	0.0892	0.2660	0.0237
4	0.1339	0.3804	0.0509
5	0.1606	0.4849	0.0779
6	0.1606	0.5762	0.0926
7	0.1377	0.6540	0.0900
8	0.1033	0.7190	0.0742
9	0.0688	0.7727	0.0532

n	$p_N(n,30 \text{ years}) = \dfrac{e^{-6}(6)^n}{n!}$	$\mathcal{P}(D \geq 3 \mid n)$	$\mathcal{P}(D \geq 3 \mid n)\, p_N(n,30 \text{ years})$
10	0.0413	0.8166	0.0337
11	0.0225	0.8524	0.0192
12	0.0113	0.8814	0.0099
13	0.0052	0.9048	0.0047
14	0.0022	0.9236	0.0021
15	0.0009	0.9388	0.0008
16	0.0025	0.0000	0.0000
			0.5405

Therefore, there is a 50% chance that the structure will suffer at least moderate damage over a period of 30 years.   □

EXAMPLE 1.46.– By using example 1.45, we now presume that the structure is inspected after each exceptional event and that its initial performance is completely restored before the next exceptional event. The probability of being subjected to at least moderate damage is then brought back to 0.05%. In such a strategy of repair and maintenance, the probability of undergoing such damage over a period of 30 years is given by:

$$\mathcal{P}(D \geq 3) = 1 - \mathcal{P}(D < 3) = 1 - \sum_{n=0}^{\infty} \mathcal{P}(D < 3 \mid n) \frac{\exp(-6)6^n}{n!}$$

$$= 1 - \sum_{n=0}^{\infty} (1 - 0.05)^n \frac{\exp(-6)6^n}{n!} = 1 - \exp(-6)\exp(6 \times 0.95) = 0.26$$

which is lower than the probability in example 1.42. In conclusion, the program for inspection/maintenance/repair enables us to procure an improved reliability for the structure over time.   □

Chapter 2

Structural Safety, Performance and Risk

2.1. Introduction

Society hopes that those using a structure – and to some extent, everything around it – may do so in complete safety. Society implicitly trusts the designers and managers, and thinks that in the absence of absolute and total safety, these people should be making sure that those risks are controlled, meaning, that risks are limited to an acceptable value. Generally, for civil engineering structures, failures and ruptures are rare, except for cases of major natural disasters[1].

This vision of safety is, however, truncated; the exceptional nature of a global or local failure in a structure must not hide other questions concerned with user safety. Thus, the appearance of corrosion in reinforced concrete elements may cause the concrete to delaminate, which would be a danger for users around the structure, and this is even before its load carrying capacity is concerned. With regard to this structure, we are dealing more with a problem concerning its serviceability or durability, rather than its structural safety. These concepts of durability, serviceability and structural safety, which will be defined and studied with more detail in section 2.6.2, are grouped under the concept of *performance*.

Engineers are not generally familiar with the concepts of risk, because such notions are not explicit in their practice of designing and assessing bridges. However, for the last ten years, there has been a growing interest in rationalizing the decisions made for managing existing structures (see [CRE 02]). This explains in particular the large amount of studies and research in the domain of structural

---

1 By failure, we do not necessarily mean the structure is ruined. Failure can also be local.

reliability and safety. It also increasingly recognized that the majority of phenomena dealt with in civil engineering are not perfectly known. The objective of standards for the design of new structures (and in some extent, for the assessment of existing structures) is no longer to reach a level of absolute safety, but to ensure an allowable risk level, which is coherent with user safety and a realistic economic cost to maintain this level of safety.

2.2. Safety and risk

Safety and risk are two closely linked concepts. However, they are different: risks are quantifiable, whereas safety is not.

2.2.1. *Concepts of safety*

In current usage, the word "safe" has different meanings and is applied to material goods as well as to people: "the bridge is safe to cross", "flying is safer than driving". In civil engineering, the term is used in a more restrained way [DES 95]: "a structure is considered "safe" if its performance (load carrying, deformation, excessive vibrations, etc.) is ensured in normal operational conditions, while protecting human lives, avoiding injuries or unacceptable economic losses, and if it is sufficiently robust so as to continue to fulfill part of its functions under exceptional operational conditions whilst avoiding any ruin with catastrophic consequences".

This definition is misleading: it lets us believe that it is possible to know the circumstances which may lead to failure or collapse at any time in the operating period of a structure. This is obviously wrong: exceptional circumstances may be very different, often to larger extents than foreseen (a high amplitude earthquake) or completely unforeseen (a terrorist attack, for example). Estimating the safety of a structure at a given time can only rely on confidence based on a certain number of elements: regulations and standards used for its design, the materials used, the implemented quality assurance, and the experience of the engineers judging this safety. The fact still remains that all future aspects of a structure are unknown, and real safety is hidden. Such a definition of safety is only meaningful when the possible and accepted situations, where the structure can be found, and the dangerous situations (as we will see in section 2.2.2, expressed through their consequences), can be defined. Safety also, then, aims to "measure" the margins which separate these two families of situations. The main point of the previous definition is that safety must be perceived as an objective to be reached through principles guiding the design, the construction, the operation and the maintenance of the structure.

2.2.2. *Concept of risk related to a danger or threat*

Risk is a concept which can be defined as a joint measure of the occurrence of a hazard and the consequences induced by its realization. Thus, the collapse of a structure could be described by an occurrence probability $\mathcal{P}$ and the resultant consequence D, which is often a mean value expressed in monetary units or human lives. The previous definition remains qualitative and must be clarified in detail to be quantitatively easy to handle: a simple but classic expression of risk R is:

$$R = \mathcal{P} \, \mathbb{E}(D) \tag{2.1}$$

Such a definition, however, creates a dilemma: what to do when the probability is very small, and yet the consequences are very large $(\mathcal{P} \to 0, \mathbb{E}(D) \to \infty)$? Generally, such conditions often lead us to reject the activity which conditions the risk, because the maximum consequences are judged as unacceptable, even if the occurrence probability is very low. D is related to an undesired or feared event that is traditionally called a *danger* or a *threat*[2]. Risk is, therefore, a measurement of danger relating to its occurrence. Unfortunately, usage of the word means that we – wrongly – confuse risk and occurrence probability, leaving out the induced consequences. This semantic slip may be found in the usual approaches for assessing risks.

EXAMPLE 2.1. – (adapted from [ANG 84]) Repairing a pile bridge in the middle of a river requires a cofferdam to be installed. Let us presume that a flood occurrence follows a Poisson distribution with 1.5 as its average annual rate. The rising of the water level during a flood follows an exponential distribution, allowing an average of 2 m above the river's average level. We presume that weather alerts will allow people to evacuate staff, which decreases the consequences of a cofferdam flood, with the only costs being delays and pumping, valued at €25,000. If h is the cofferdam height, the risk of the dam flooding is therefore given by expression [2.1] with:

$$R = \mathbb{E}(\text{cofferdam flooding}) \, \mathcal{P}(\text{flood})$$

$$= 25,000 \int_{h}^{\infty} \frac{1}{2} \exp\left(-\frac{x}{2}\right) dx = 25,000 \exp\left(-\frac{h}{2}\right)$$

This risk is related to a single flood. If we are interested in the flood risks over two years, we must take into account the occurrence probability of the floods:

---

2 Danger is traditionally defined as a potentially harmful effect, endangering people, goods, and the environment.

$$R = \sum_{k=0}^{\infty} \mathbb{E}(\text{cofferdam flooding}|k \text{ floods}) \, \mathcal{P}\left(k \text{ floodings in 2 years}\right)$$

$$= \sum_{k=0}^{\infty} k \, 25,000 \exp\left(-\frac{h}{2}\right) \frac{(1.5 \times 2)^k}{k!} \exp(-1.5 \times 2)$$

$$= 25,000 \exp\left(-\frac{h}{2}\right) \sum_{k=0}^{\infty} \frac{3^k}{(k-1)!} \exp(-3)$$

$$= 25,000 \exp\left(-\frac{h}{2}\right) 3 \exp(-3) \underbrace{\left(\sum_{k=1}^{\infty} \frac{3^{k-1}}{(k-1)!}\right)}_{\exp(3)}$$

$$= 75,000 \exp\left(-\frac{h}{2}\right)$$

□

2.2.3. *Risk assessment*

According to how the risk is assessed, it may be classified into three categories: an observed, calculated or perceived risk. Each category relates to a different way of treating uncertainties. The *observed* risk deals with extrapolating a certain number of occurrences of observed events. Car accidents are an example of this. Generally, the observed risk can only be a rough estimate of the failure probability. In fact, firstly, in cases where failure or collapse are very rare (bridges, dams, etc.), it is necessary to group together many, very different, structure types (prestressed concrete, metal, etc.) in one single total category to obtain a number of sufficient data to be analyzed. Secondly, there is no differentiation with regard the cause of failure. These two problems show that assessing an observed risk is sensitive to the context and the categories used to determine it. An observed risk may be more misleading than it seems at first sight.

The *calculated* risk particularly relies on calculating failure probability, based on an analysis of this structure or that element. It may naturally be brought back to an average value over a certain number of structures, but only by calculating the failure probabilities for a certain number of different scenarios. With the difference of the observed risk including all the sources of failure, the calculated risk depends on the choice of failure modes and their causes. Some are certainly removed, such as human error, which can have a significant contribution sometimes. The assessment can be high or low: some causes are not considered in the calculations, whereas in others, safety hypotheses may be generated. This is typically the case for the nuclear industry. It is therefore not surprising that observed and calculated risks greatly differ. Brown

[BRO 79] uses the case of large bridges where the observed failure frequencies are estimated as being between 10^{-2} and 10^{-3}, where calculated failure probabilities (i.e. subjacent to design regulations) are in the order of 10^{-6}!

The *perceived* risk is a different concept. It is simply a perception of the risk, or a part of it, by society. Today, this is a concept which is greatly taken into account by environmental activists (site choice for a chemical factory, waste treatment center, etc.). The *Royal Society* [ROY 92] provides a well-documented analysis of the causes influencing risk perception by individuals. They are all real, but not always reasonable, at least in terms of rationality for an engineer.

2.2.4. *Hazard*

Hazard is often confused with risk, but in truth they are two different concepts entirely. A hazard is a threat, the possibility that something harmful might occur. Risk, on the other hand, is linked to the possibility of the threat itself. For example, an avalanche may threaten a road. The avalanche is a potential hazard. But there is no risk if the authorities are aware of this hazard and close the road. This confusion between risk and hazard can be seen in the terminology for natural risks. It would sometimes be preferable to talk about natural hazards. A hazard, however, is an essential component of a risk. Thus, managing risks relies on managing hazards on the one hand, and the vulnerability faced with these hazards on the other hand.

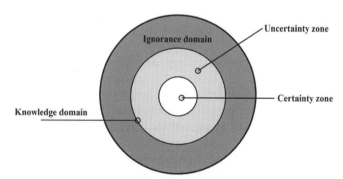

Figure 2.1. *Domains of knowledge and ignorance, and certainty and uncertainty zones (adapted from [DES 95])*

In section 2.2.2 we saw how structural safety was apprehended by the notion of risk. This notion introduces the consequences with regard to the structure's failure and the probabilities when such an event may occur. Probabilistic and statistical methods are natural tools for modern structural safety. Desroches [DES 95] defines two domains (Figure 2.1):

– *Knowledge domain* in which it is possible to describe all the functioning (and dysfunctioning) states precisely, along with their consequences on the environment. This is also the case for environmental effects on the structure;

– An *ignorance domain*, on the functioning states.

If in the knowledge domain we can hope to evaluate the probability of a failure mode with more or less precision (see section 2.2.1), we do not need *a priori* to estimate the probability of an event which is not, or is badly, qualitatively defined. In other words, as stated by Desroches [DES 95], "it is absurd to try to probabilize the…unknown".

The knowledge domain can be structured into two complimentary zones:

– the uncertainty zone;

– the certainty zone.

The uncertainty zone is related to qualitative knowledge about the functioning states of the structure associated with random knowledge of each state for a given situation. For example, this is the failure probability of a reinforced concrete section under bending at a given time when it is subjected to corrosion.

The certainty zone relates to deterministic knowledge of all the functioning states and their consequences. Generally, this knowledge is theoretical (as for physical laws which are infinitely reproducible and proven by experience, such as Newton's law) or statistical (values deduced from observations).

Making a decision in the uncertainty domain is achieved by applying the *principle of practical certainty*, relating to the consideration of relatively impossible events. This is stated as follows [DES 95]: "if the probability of a non-specific event in a given experiment is small enough, we can almost be certain that, when the experiment is carried out *just once,* the considered event will not take place".

This principle cannot be demonstrated but is confirmed by every day experiences. This quasi-certainty or quasi-uncertainty of an event occurring based on experience is misleading with regard to structural safety. In fact, the real period of observation may vary, and may be smaller than the length necessary for observing the feared, considered event. From this, it follows that this event cannot be excluded from the risk analysis just because it is judged impossible by the only justification that it has never actually been observed[3].

---

3 The principle of practical certainty in fact follows from the law of large numbers.

These notions of known or unknown hazards, certain or uncertain, can be found in the classification proposed by Schneider [SCH 94]. Schneider then distinguishes hazards, or *accepted* or *tolerated dangers*, from *residual dangers*.

Tolerated or accepted dangers are represented by the risks that the engineers is aware of, but which he has considered as acceptable on the basis of a risk analysis (these concepts will be developed further in sections 2.3 and 2.4).

Residual dangers relate to unknown dangers (fatigue for old metal structures, etc.), unnoticed dangers (dynamic coupling, etc.), neglected dangers (bad design or monitoring, negligence with regard to natural or operation risks) or dangers where inappropriate or defective corrective measures have been retained (see Figure 2.2).

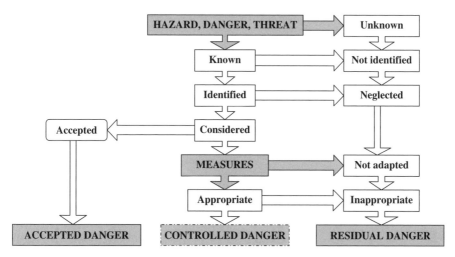

Figure 2.2. *Danger classification (adapted from [SCH 94])*

2.3. Risk evaluation and acceptable risk

2.3.1. *Risk assessment*

It is often traditional to reduce the risk to that of *failure probability*. This is misused, in the sense that it does not take a lot of factors into account, including the most important: the consequences induced by the hazard or danger. If the concept of failure probability (or one of its reformulations, like the semi-probabilistic format in standards) is greatly used in structures' calculations, then other approaches are, however, possible.

The *fatal accident rate* (FAR) is an assessment of the risk of death that an individual will be faced with when performing a particular activity, or when in a given situation. It is defined by the relation:

$$\text{FAR = probability of death for an individual per hour of exposure} \qquad [2.2]$$

The FAR index is useful for comparing different activities. It may also be relevant for assessing implications in terms of the safety of different designs, or for managing safety options. Although it does not seem to have been used in the context of structural calculations except by Menzies [MEN 97], a minimum FAR index for any person using a structure could be used as an indicator of its performance. To fix such ideas, Table 2.1 provides a few FAR index values for various activities (these values are not representative of a particular year, but are an amalgamation of various pieces of information).

The FAR indicator must, however, be considered carefully. In fact, it gives the probability of a person exposed to death, per year. But, another way of defining it is to reduce it to the number of deaths per exposure hour. Thus, air transport is safer than car transport in terms of death/person/year, but not if this mode of transport is analyzed in terms of death/exposure hour.

Activity	Death/person/year $(\times 10^{-6})$ (death/hour of exposure)	Activity	Death/person/year $(\times 10^{-6})$ (death/hour of exposure)
Sport		Natural hazards	
Caving	45	Hurricanes	0.4
Mountaineering	1,500 (*30*)	Tornados	0.4
Parachuting	1900	Lightning	0.5
Drowning	170 (*3.5*)	Earthquake	2.0
Domestic accidents		Transport	
Poisoning	20	Rail	15 (*0.08*)
Burns	40	Air	24 (*1.2*)
Falls	90	Car	200 (*0.7*)
Smoking	1 000		
Buildings		Industrial activities	
Structural failure	0.1 (*0.00002*)	Chemistry	85
Fires	16 (*0.001*)	Farming	110
		Construction	150 (*0.07*)
		Mines (coal)	210
		Mines (non coal)	750
		Oil	1,650
		Fishing (high sea)	2,800

Table 2.1. *FAR indexes for activities*

2.3.2. Acceptable risk

When a risk has been quantified, we must decide whether it is acceptable or not. The definition and qualification of this value – or, in other words, of this acceptable risk – is a difficult problem. Two levels must be distinguished: *personally acceptable risk* and *socially acceptable risk*. The first constitutes the smallest acceptable risk level whereas the second is an amalgamation of the personal considerations.

Psychological research shows that society is far from thinking in terms of failure probabilities or with any abstract concepts in risk evaluation. Research reveals that this evaluation is classified according to a ranking of increasing importance, according to whether it is a matter of voluntary exposure, the number of people exposed to the hazard, the possibility of reducing the consequences by appropriate measures when danger arises, or fatal direct or indirect consequences (collateral damage).

Within this framework, *voluntary risks* must be distinguished from *involuntary risks* and the exposure time (see Table 1.1). Figure 2.3 gives a classification of the risk levels by order of probability and acceptability. Factor η has been called the political factor by Vrijling [VRI 92]. It varies from 10 for the case of complete freedom of choice, to 0.1 for an imposed risk. It is related to the accepted probability for an accident associated with a given activity a:

$$P_{f,a} = \frac{\eta\, 10^{-4}}{P_{df,a}} \qquad\qquad [2.3]$$

where $P_{df,a}$ is the probability of death in activity a.

A possible approach for fixing an acceptability threshold is to use failure statistics. However, these are often biased through subjectivity. Thus, many failures are often accounted for as failure or collapse due to exceeding a limit state, whereas it is often a question of violating operational limit states. As was highlighted in the definition of observed risks, these statistics are often tainted by the grouping together of structural failures both in construction and when in operation (two groups which should be distinguished) and by the absence of information on the smallest failures. Finally, acceptable risk is a projection of collective, social or economic perception of danger (i.e. the feared event): therefore it cannot be uniquely defined, and depends on the times and countries where it is considered.

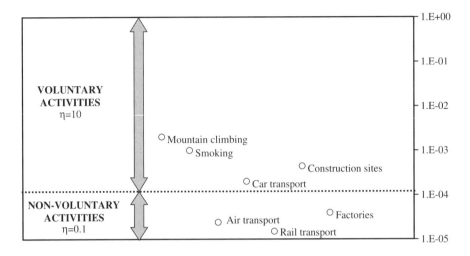

Figure 2.3. *Personal risks and acceptability thresholds*

Society's aversion to confronting risks is related to the number of people concerned and the context of injuries: the loss of a large number of lives in one accident (like an airliner crash) will have more repercussions than the same number of deaths distributed over many accidents (like traffic accidents). Thus, the acceptable risk level for structural failure causing multiple fatalities and injuries will be clearly lower than that for a construction failure with only injuries. This suggests the acceptable risks related to structural failure should really depend on the construction's characteristics, the operation type and the causes for failure. Therefore, Menzies [MEN 97] proposes that we retain three approaches to define an acceptable risk for bridges: comparison between FAR indexes, societal attitude and tolerable risks. Other approaches, based on the decision theory, may also be introduced but they are still currently under research.

2.3.2.1. *Tolerable risk*

Tolerable risk is based on the concept that public authorities must do everything in their power to reduce risks, provided that the costs are not disproportionate to the expected gains. The analysis of risks run by society, like those given in Table 2.1, has led to some industries being greatly concerned by risks (nuclear power plants, chemical installations, etc.) to define acceptable risk levels on the sole basis of consequences. This is the principle for *F-N curves* (*Fatalities/Numbers of accidents*) which expresses accident frequency in relation to its amplitude, expressed in amount of damage or numbers of deaths (Figure 2.4).

These diagrams define three areas. There is an upper limit to the risk, beyond which the risk is no longer tolerated, and a lower limit, below which the risk is no longer important and no precaution is necessary. Between these two limits, there is a zone (ALARP – *as low as reasonably practical*) where the risks can be brought back to a level as low as it is reasonable to be in practice, i.e. without inducing extra prohibiting costs. Although such an analysis is intuitive, a delicate implementation, as much theoretical as philosophical, still remains to be performed.

For railway lines, this approach has led to an annual acceptable probability P_f^0 of 10^{-6} for the users of this mode of transport. By extension, this value can be taken for the acceptable risk in relation to the failure of a rail bridge. In principle, this level of risk does not concern the individual, who is aware of the threat but does not feel concerned [BAI 02].

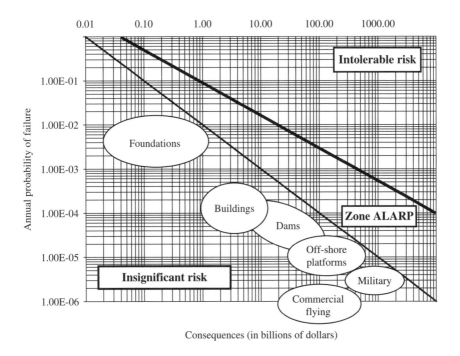

Figure 2.4. *Example of a F-N diagram*

Otway [OTW 70] also proposes a risk classification according to a societal approach described in Table 2.2.

Individual probability [death/pers/year]	Societal attitude
10^{-3}	*Intolerable risk*: corrective measures must be taken immediately to reduce it. Otherwise the activity must be abandoned.
10^{-4}	*Risk reduction* by public authorities (laws, regulations, etc.).
10^{-5}	*Warnings* against the risk (drownings, domestic accidents, etc.).
10^{-6}	*Resignation or not worrying* when faced with risk (natural risks, etc.).

Table 2.2. *Risk tolerance classification according to Otway [OTW 70]*

EXAMPLE 2.2.– The analysis of societal attitudes by Otway has led to defining an annual failure probability of 10^{-6}, which is lower for bridges, whereas a P_f^0 value of 10^{-5} seems acceptable.   □

A few expressions are then proposed for calculating the target failure probability:

$$P_f^0 = 10^{-4} \mu\, t_L\, n^{-1} \qquad\qquad [2.4]$$

or:

$$P_f^0 = 10^{-5} A\, W^{-1}\, t_L\, n^{-1/2} \qquad\qquad [2.5]$$

where t_L and n respectively are the structure's service life and the average number of individuals in, or around, the structure during the service life, and μ, A, W are social activity or warning factors (Table 2.3). Introducing the power of ½ is explained in the Allen's formula by introducing concepts from the utility theory.

μ		A		W	
Public buildings, dams	0.005	Post-disaster activity	0.3	Progressive failure with warning	0.1
Domestic, office, marketing, industry	0.05	Normal activities (bridges)	3.0	Invisible progressive failure	0.3
Bridges	0.5	Normal activities (buldings)	1.0	Brittle rupture without warning	1.0
Towers, offshore, pylons	5.0	Greatly exposed structures (off-shore, etc.)	10.0		

Table 2.3. *Social, activity and warning factors*

The previous expression must, however, be handled with care. This is why the traditional procedure consists of calculating failure probabilities by using existing regulations or standards and using these probabilities (possibly modified to take into account extra risk factors) as reference values. Thus, current regulations provide P_f^0 values often between 10^{-3} and 10^{-6}.

2.3.2.2. *Cost optimization*

Looking for the acceptable risk value, in fact, passes by a compromise between what the main contractor is willing to pay if he were to take into account the occurrence of danger and the resultant safety measure, and what he would have to pay out afterwards if he were to ignore them:

– costs for repairing human and material damage;

– costs of unavailability;

– media impact perhaps leading to the structure's closure.

Following this reasoning, if we consider the total cost of safety, namely:

– the cost of studies and safety devices, called "*a priori* costs";

– cost of accidents, called "*a posteriori* costs".

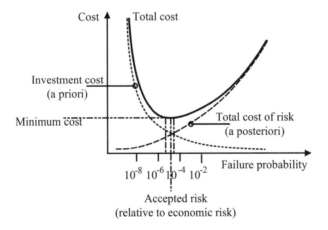

Figure 2.5. *Maintenance costs and optimal probability*

It is then possible to look for the best total cost according to the failure probability (Figure 2.5). The (*a priori*) investment $C(P_f)$ is naturally an increasing function of the failure probability P_f (Figure 2.5). Let us presume then that the structural failure leads to an *a posteriori* cost equal to that of the investment (reconstruction cost), raised by direct costs, D (losses of human lives, etc.), $C(P_f) + D$. As the failure probabilities relate to annual failure probabilities, the interest or discounted rates can be omitted (for analyzes over long periods, this is obviously no longer possible and the calculations must be changed).

The average total cost is therefore the sum of the investment and the mean cost of failure:

$$C(P_f) + \left[C(P_f) + D \right] P_f \tag{2.6}$$

At this stage, it would be judicious to notice that the mean cost includes the term $\left[C(P_f) + D \right] P_f$ which is none other than a risk, since it conjugates failure consequences with an occurrence probability of the danger. This is why economic approaches are often called *risk-based approaches*. The acceptable failure probability will therefore be the value which minimizes the average total cost:

$$\frac{dC}{dP_f}(P_f)\left[1 + P_f \right] + C(P_f) + D = 0 \tag{2.7}$$

For civil engineering structures, value D will be large enough for P_f to be low, which leads us to ignore P_f before 1:

$$\frac{dC}{dP_f}(P_f) + C(P_f) \simeq -D \tag{2.8}$$

Ditlevsen and Madsen [DIT 97] suggest that capital investment variation may be approached in most cases by a linear function of the reliability index variation:

$$C(P_f^2) - C(P_f^1) = a\left(\beta_2 - \beta_1 \right) \tag{2.9}$$

which turns equation [2.8] into:

$$-D \simeq \frac{dC}{d\beta}\frac{d\beta}{dP_f} + C = -a\sqrt{2\pi}\ e^{\beta^2/2} + C \tag{2.10}$$

since $P_f = \Phi(-\beta)$ with $\Phi'(u) = \varphi(u) = \frac{1}{\sqrt{2\pi}}e^{-u^2/2}$. As $C(P_f) = a\beta + C_0$, the previous equation becomes:

$$\frac{D+C_0}{a} \simeq \sqrt{2\,\pi}\; e^{\beta^2/2} - \beta \simeq \sqrt{2\,\pi}\; e^{\beta^2/2} \qquad\qquad [2.11]$$

The reliability index solution for this equation is therefore the target reliability index:

$$\beta_0 \simeq \sqrt{2\ln\left(\frac{D+C_0}{a\sqrt{2\,\pi}}\right)} \qquad\qquad [2.12]$$

For a ratio $(D+C_0)/a$ providing an optimal reliability index of 4.5, a variation in this ratio by a factor of 10 only modifies the optimal reliability index by 10%.

The procedure for optimizing the average total cost (or risk) can also be used to look for optimal parameters other than acceptable failure probability. The following example illustrates this:

EXAMPLE 2.3.– Taking example 2.1 by presuming that the construction cost of a cofferdam is a function of its height:

$$C_c = C_0 + 3,000\, h$$

where C_0 is the irreducible cost for building a cofferdam to a height which is equal to the average level of the river. Let us remember that h is the height of the cofferdam above this average level. In this case, the total average cost related to the flood risk over two years (building period) is:

$$C_T = C_c + R = C_0 + 3,000\, h + 75,000\, \exp\left(-\frac{h}{2}\right)$$

Since the water level follows an exponential law, the probability of flooding may be directed related to the water height by the ratio:

$$p = \int_h^\infty \frac{1}{2} e^{-x/2}\, dx, \text{ i.e. } h = -2\ln(p)$$

The total cost may then be expressed directly according the probability of flooding. By differentiating this cost in relation to p, the optimum technical and economic failure probability (i.e. cofferdam flood) is given by:

$$\frac{dC_T}{dp} = -6,000\frac{1}{p} + 75,000 \equiv 0 \ \text{ or } \ p_{opt} = 0.08$$

The optimum height of the cofferdam relating to this optimal failure probability is therefore $h_{opt} = -2\ln(0.08) \approx 5.05$ m.

$\square$

2.4. Risk-based management

2.4.1. *Strategies*

There are many possible strategies, direct or indirect, to reach a given risk level:

– *Regulations.* Design and assessment regulations or standards are the main approaches through which society is assured of structural safety. Mostly, they decide what must be done, which loads should be used, which safety coefficients should be introduced.

– Quality assurance. This is a matter of eliminating or minimizing human mistakes. It deals with uncertainty, and to a certain extent, ignorance. However, it does not entirely remove the possibility of making mistakes. Its use is necessary, but not enough to ensure safety: it ensures that information and documentation are complete and easily located, but it has little to do with the intrinsic quality of the information.

– Fighting against ignorance, uncertainty and complexity. For ignorance, the general fighting strategy is formation, for uncertainty it is reducing limits, and for complexity, it a question of simplification.

In parallel with these approaches, the process known as *risk analysis* is particularly useful: it represents a technique for identifying, characterizing, quantifying and assessing these hazards and their consequences, in order to try to answer the following questions:

– Which event may lead to the system's failure?

– How does it occur?

– What is its level of possibility?

– What will the consequences be?

2.4.2. *Risk analysis*

This analysis can be qualitative or quantitative, but in any case, it must include the following stages:

– identification of accident scenarios (or dysfunctioning) leading to equipment failure;

– identification of degradation mechanisms and potential failure modes;

– assessing the occurrence of each degradation mechanism and each mode;

– assessing the consequences (in cost or utility);

– determining the risks;

– classifying the risks;

– defining acceptable risks.

This process has been used in many industrial sectors (nuclear, chemical, etc.) and can be changed according many levels of detail. Thus, for complex chemical installations, the risk analyses are very sophisticated and consider a very wide range of accident scenarios resulting from different initiating events. For simpler systems, such as apparatus under pressure, the variety of hazards and consequences is more limited and easier to identify.

In practice, the previous stages are available in four large chapters which are grouped together under the title *dependability* [NAH 01, COX 09]:

– system analysis;

– qualitative risk analysis;

– quantitative risk analysis;

– detailed criticality analysis.

System analysis allows a summarized description of a system's functioning modes and the knowledge of the functions to be ensured. We, then, identify the functions and their characteristics, the performances with expected value criterion and external stresses. There are many techniques according to the concerned domain (mechanical, electronic, organizational, software, etc.). They consider a system from the point of view of its objective, and take into account all the factors concerning this system and its environment. One of the main principles is, therefore, separating the needs which have to be fulfilled from the solution chosen to fulfill it. This procedure may widen the range of techniques likely to be used and reach a better optimization for the proposed solutions. This is also characterized by listing and classifying information in technical and lawful domains.

Qualitative risk analysis is a deductive analysis whose results are:

– a basic function analysis of the equipment or the structure;

– the identification of errors and non-conformities (bad design or execution, damage, etc.);

– the risk qualification presented by each component;

– the identification of degradation mechanisms, potential failure modes and possible consequences;

– the description of possible degraded functioning modes;

– a first assessment of additional detailed analyses and investigations which should be envisaged in the framework in a detailed criticality analysis.

The results for this qualitative risk analysis must allow:

– an overview of the structure or system's shortcomings;

– an overview of the consequences induced by shortcomings in terms of dysfunctioning;

– an initial idea of the functioning in degraded mode after dysfunction or after a safety measure put into place due to danger;

– an initial justification of the detailed analyses and investigations.

Quantitative risk analysis will complete the qualitative risk analysis by a quantitative approach, and will definitively rule over the opportunity to carry out (or not) detailed investigations by auscultation or calculation, i.e. to proceed to a *detailed criticality analysis.* This analysis aims to determine means of detecting shortcomings and to evaluate the criticality in a detailed way.

Severity category	Definition	Occurrence	Definition
Catastrophic	Multiple deaths Structure destroyed or greatly damaged	Probable	Low return period (example: < 10 years)
Critical	Single death or multiple injuries Certain structural elements damaged with no consequences for the structure itself	Potential	Average return period (example: between 10 and 100 years)
Marginal	Minor injuries Certain structural elements displaying minor damages	Possible	Large return period (example: between 100 and 500 years)
Negligible	Marginal injuries No loss of structural safety for components and structure	Improbable	Very large return period (example: over 1,000 years)

Risk rating matrix	Catastrophic	Critical	Marginal	Negligible
Likely	4	4	4	2
Potential	4	4	3	2
Possible	4	3	2	1
Unlikely	3	2	1	1

1 – The risk is tolerable: no corrective action necessary.

2 – The risk is tolerable: corrective actions may be taken if their cost is moderate and are not to the detriment of other measures.

3 – The risk is at the tolerable limit: corrective actions must be sought.

4 – The risk is intolerable: many corrective solutions must be identified.

Table 2.4. *Example of a risk rating matrix*

In a risk analysis, the risk can then be assessed qualitatively or quantitatively, for a given failure scenario. For a *qualitative risk analysis,* the results are often roughly expressed as a *risk rating matrix*, also known as the *consequence matrix*.

– *catastrophic risk:* this relates to the consequences such as significant harm to society (death, permanent handicap), the total destruction of the structure and/or its environment;

– *critical risk:* serious consequences but no permanent injuries, partial destruction;

– *marginal risk*: consequences such as light injuries, structure rendered unavailable;

– *negligible risk*: consequences which do not prevent the structure from use.

The columns of the risk rating matrix represent the occurrence of a hazard, of a threat, or a danger for a given consequence level. A value can be associated with each term in this matrix, a value which only has a meaning in this matrix. We are dealing with a *semi-quantitative analysis*. These values are used only for setting out priorities, based on the judgment of experts; therefore, they are always debatable. The risk rating matrix may also be represented as a *Farmer diagram* (Figure 2.6). This type of diagram is particularly interesting for visualizing the domains with acceptable and unacceptable risks. Following the risk rating matrix, the unacceptable or intolerable risks are – arbitrarily – defined by note 4.

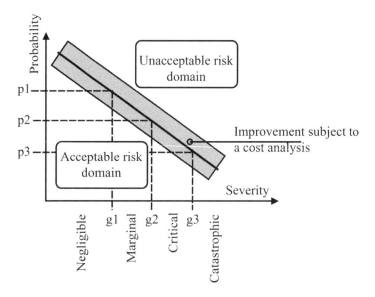

Figure 2.6. *Farmer's diagram*

EXAMPLE 2.4.– (adapted from [MAN 05]) A construction manager has implemented a policy for prioritizing maintenance on the basis of an approach founded on the criticality analysis. Four levels of occurrence are introduced:

Occurrence	Level	Target estimate
Certain	Certainty (event has already occurred and will reoccur)	1.00
High	Highly likely	0.85
Medium	Likely	0.45
Low	Possible but unlikely	0.10

The severity categories are divided into three classes, but distinguish safety and serviceability conditions:

Nature	Level	Description
Safety	High	High number of fatalities
	Medium	Low number of fatalities/high level of serious injuries
	Low	Low number of serious fatalities/injuries unlikely
Serviceability	High	Strategic road closure
	Medium	Secondary road closure/Strategic traffic restriction on the road/Traffic restriction on secondary road
	Low	Closure/Traffic restriction on local road

The risk rating matrix is the following:

Risk rating matrix	Consequences		
Occurrence	High	Medium	Low
Certain	C	C	AC
High	C	C	AC
Medium	C	AC	AC
Low	AC	NC	AC

NC: non-critical structure in relation to the considered performance (safety, serviceability).

MC: medium critical structure in relation to the considered performance (safety, serviceability).

C: critical structure in relation to the considered performance (safety, serviceability).

Classifying a structure according to its structural safety and its serviceability is finally used to prioritize intervention according to the following scale:

Priority	Risk	Priority rank
1	C - Structural safety	
2	MC - Structural safety	Obligatory intervention
3	C - Serviceability	
4	NC - Structural safety	Intervention desirable if not to the
5	MC - Serviceability	detriment of other measures
6	NC - Serviceability	Intervention not a priority

Let us consider an existing under-dimensioned structure for which a propping system and a tonnage limitation are necessary. This structure supports a national road (strategic) passing above a secondary road. The structure may fail according to four causes, with four respective probabilities: 45%, 45%, 85% and 85%:

– by vandalism on the prop: factors affecting this event are the absence of lighting on the structure, acts of vandalism around the structure;

– propping system failure: factors affecting this event are the system's bad dimensioning, etc.;

– overweight: an overweight vehicle in relation to the tonnage limits passes over the structure;

– overload: factors affecting these events are the under-estimation of the structure's residual load carrying capacity, and a degradation which was much quicker than predicted.

According to the probabilities associated with various events, the likelihood value is calculated as follows:

$$\mathcal{L}_f = 0.45 \times 0.45 \times 0.85 \times 0.85 \approx 0.15$$

which puts the likelihood rating in the *low* category.

Due to the nature of supported and crossed roads, the nature of the consequences regarding the safety and serviceability is *high*.

By bringing these two results into the risk rating matrix, we obtain that the structure is classed as MC with regard to the two performance requirements, which gives the manager the option to intervene.   □

For many components or elements of the structure, failure may be an event which initiates other incidents with consequences which are much more serious than for the element itself. In both cases, causes of failure or its effects must be assessed during the risk quantification. Many techniques are available, and here we will cite some of them [COX 09]:

– failure modes and effects analysis (FMEA) (and its derivatives, such as the failure mode, effect, and criticality analysis (FMECA), and the hazard and operability study (HAZOP);

– success diagram method (SDM);

– event tree method (ETM);

– fault tree method (FTM);

– cause-consequence analysis (CCA).

In truth, all these methods are either based on an inductive or deductive process. In the first case, we move away from the precise to the general. In other words, faced with a system and a failure, we will study in detail the effects or consequences of this failure on the system. The FMEA, FMECA, HAZOP, CQTM, etc., methods are inductive processes.

In the deductive approach, the reasoning is the opposite: by assuming the system failed, we will look for the causes. The main deductive method is the FTM. This method may be used qualitatively for diagnosing the different events leading to failure, and quantitatively for calculating failure probability. As previously mentioned, this method is deductive and consists of, from the top event, descending the tree structure to the basic (or elementary) events. Graphic representation relies on a certain number of symbols describing the combination of these events. We will return to this representation in Chapter 3 by illustrating the way in which a reliability block diagram (RBD) can be changed into a FTD. Figure 2.7 gives some of the traditional symbols used in graphic representations of the FTM.

Function	Symbol	Description
or		The output event occurs if at least one of the input events occurs.
and		The output event occurs if all input events occur.
k out of n	k/n	The output event occurs if k inputs out of n input events occur.
top		Top event and also event description
basic event		A basic initiating fault (or failure event).
undeveloped event		An event which is not further developed. It is a basic event that does not need further resolution.

Figure 2.7. *Examples of symbols for the FTM*

EXAMPLE 2.5.– If we use example 2.4, in trying to describe the causes which generate each event leading to the structure's failure, the fault tree diagram can be outlined:

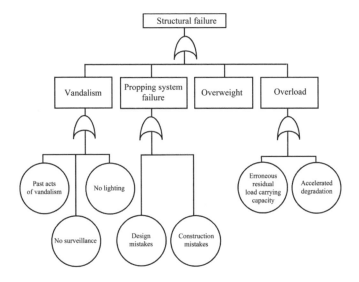

2.4.3. *Legal point of view for a risk-based approach*

A crucial question concerns considering the risks within a legal framework. Traditionally, the law is often based on deterministic safety measures and instructions, for that which concerns hazardous activities. As a consequence, in some cases, it requires safety measures which are not justified due to their low efficiency, or does not prescribe in situations where they could prove to be very useful. This explains the tendency for introducing legal risk-based frameworks to correct the limitations of actual frameworks. In Switzerland, between 1996 and 1999 [SEI 01] such work was launched on nine cases so as to assess its practical implementation (storage of explosives in military bases, storage of explosives on civil sites, transportation of dangerous materials, safety of nuclear reactors, fire protection, road accidents, construction safety, nuclear waste management, chemical waste storage sites). In this context, the law does not prescribe precise safety measures in the domain, but acceptable risk levels instead. The aim, then, is to respect this acceptable threshold by using appropriate means. Looking for an acceptable risk level is expressed in terms of consequences and their occurrence probability (i.e. following the expression [2.1]). In particular, this allows us to compare the effects of various safety measures. In truth, traditional safety procedures also take risks into consideration, but in a non-quantitative way. The legal questions raised by such a procedure concern the capacity to quantify risks and define acceptable risks.

We have seen that risk, and therefore consequences or prejudices, can be subjective. The legal definition of risk must therefore specify what must be understood by these consequences. It is a question of protecting individual and collective rights, which is not simple because these rights may prove to be competitive in some cases. Moreover, the law only aims to warn against prejudices to the other, not to oneself: each person is, in fact, responsible for their own actions towards themselves. As a consequence, only external damage is dealt with by the law. This separation between internal and external damage is not clear. This distinction, for a given accident, depends on the voluntary character of the activity, on the possibility of each person reducing the risks, or warning against the accident. Four categories can therefore be defined:

– *category 1*: voluntary exposure to risk to satisfy our own desires (example, dangerous sports);

– *category 2*: high degree of self-determination with direct individual benefits (example, driving);

– *category 3*: low degree of self-determination with individual benefits (example, work);

– *category 4*: involuntary exposure without direct benefits (example, location of a dangerous industrial site).

The criteria of risk acceptability are a function of these categories and differ according to whether it is a question of individual or collective risks. In the first case, this criteria are expressed in terms of acceptable probabilities whereas in the second case, the limits are given either by applying an ALARP (as low as reasonably practicable) approach or by an approach taken from the decision theory looking to optimize safety measures: the risk is acceptable if the cost of reducing it is larger than the cost run by the risk.

Such an approach does not substitute a political decision, but offers a methodological tool in which political options can be explicitly incorporated.

2.5. Examples of failure: bridges

The collapse of a bridge is generally a spectacular event, most of the time leading to losses of human lives or injuries. Table 2.5 gives an outline of the most remarkable – and tragic – bridge collapses from the last 50 years. Figure 2.8 gives a few notable examples.

Construction	Construction	Year	Collapse type
Elbow Grade Bridge	USA	1950	Collapse after erection
Duplessis Bridge	Canada	1951	Structural failure, collapse, 4 dead
Harrow Station	UK	1952	Train impact on a footbridge, 112 dead, 340 injured
St. Johns Station	UK	1957	Collision between two trains with impact on the construction's pile, 90 dead, 173 injured
Narrows Bridge	Canada	1958	Design defect, collapse in construction, 15 dead, 20 injured
Bristol Bridge	UK	1960	Collision between two boats on a bridge pile, 5 dead.
King Street Bridge	Australia	1962	Span collapse by brittle rupture in cold weather during a 47 t convoy
General Rafael Urdaneta Bridge	Venezuela	1964	Boat collision against many bridge piles, 7 dead, 2 injured
Antwerp	Belgium	1966	Landslide, 2 dead, 16 injured
Punta Piedras	Venezuela	1966	Overload, 20 dead
Silver Bridge	USA	1967	Broken cable due to fatigue and corrosion, 46 dead, 9 injured
Udine	Italy	1968	High waters, scouring
Montenegro	Montenegro	1968	Overload, 6 dead, 21 injured
Cleddau Bridge	UK	1970	Collapse due to pushing, 4 dead
Hamburg	Germany	1970	Pylon and bridge rupture, following strong ocscillations due to the wind
West Gate Bridge	Australia	1970	Collapse during construction, 4 dead

Table 2.5. *Some bridge failures between 1950 and 2007*

South Bridge Koblenz	Germany	1971	Box girder buckling, 13 dead
Katerini	Greece	1972	High waters, scouring, 1 dead
Milford Haven	UK	1972	Box girder buckling during construction
Yarra	Australia	1972	Box girder buckling during construction, 35 dead
Viaduc sur la Sorge	Switzerland	1973	Pushing in the direction of a steep slope (6.5%), slide
Charleroi	Belgium	1974	Train derailment and collision with bridge, 17 dead, 80 injured
Pinzgau	Austria	1974	Footbridge collapse when a group of school children were crossing, 8 dead, 16 injured
Kempton	Germany	1974	Failure of temporary pilework during concrete casting, 9 dead
St-Paul	USA	1975	Crossing crack at mid-span (bad welding and low temperature)
Tasman Bridge	Australia	1975	Span collapse after impact with a cargo on pylons, 12 dead
Reichsbruecke	Austria	1976	Pile collapse, 1 dead
Moscow	Russia	1977	Insufficient rehabilitation/repair after rupture in 1940, 20 dead, 100 injured
Granville Railway Bridge	Australia	1977	Derailment and collision of a passenger train with the bridge piles, 83 dead, 210 injured
San Sebastian	Spain	1978	Rupture during a gathering of people, 7 dead
Duisbourg	Germany	1979	The mechanical shovel of a bulldozer struck the bridge, 8 dead
Humber	UK	1980	Damage of three sections during construction
Sunshine Skyway Bridge	USA	1980	Collapse after ship collision, 35 dead, 1 injured
Almöbron (Tjörnbron)	Sweden	1980	Boat impact on a bridge pile with arch collapse, 1 injured
Chicago	USA	1982	Overloading on inadequate propping system, 13 dead, 18 injured
Brajmanbari	Bangladesh	1982	Collapse of a bridge when a bus was crossing, 45 dead
Ile Cebu	Philippines	1983	Overload, 20 dead
Mianus River Bridge	USA	1983	Collapse linked to corrosion, fatigue and delayed maintenance
Ynys-y-Gwas Bridge	UK	1985	Corrosion of the pre-stressed section leading to unforeseen collapse
Schoharie Creek Thruway Bridge	USA	1987	Foundation scouring and bad static scheme, 10 dead
Aschaffenburg Bridge	Germany	1988	Collapse of the nose recovery system when reaching pile level, 1 dead, 10 injured
Los Angeles	USA	1989	Collapse when taking the scaffolding down to lower a precast voussoir, 5 injured
San Francisco-Oakland Bay Bridge	USA	1989	Earthquake, 1 dead
Tennessee Hatchie River Bridge	USA	1989	Deterioration of wooden foundation piles, 8 dead
Cypress Street Viaduct	USA	1989	Collapse during earthquake, 42 dead
Los Mochis	Mexico	1989	Scouring when a train was crossing, 103 dead, 200 injured
Ness	UK	1989	Foundation scouring

Table 2.5. *(continued)*

Hiroshima	Japan	1991	Stability (sliding), 14 dead
Changson	Korea	1992	Pile collapse leading to full collapse during the opening, 1 dead, 7 injured
Kilosa	Tanzania	1992	High waters when a train was crossing, 100 dead
Pont Cicero	Italy	1993	Scouring, 4 dead, 1 injured
Claiborne Avenue Bridge	USA	1993	Barge collision on a bridge pile
CSXT Big Bayou Canot rail bridge	USA	1993	Barge impact, train derailment, 47 dead, 103 injured
Haeng Ju	Korea	1994	Failure of temporary sheet piling during construction through lack of concrete in foundation
Afrique du Sud	Korea	1994	Failure of lattice elements by fatigue, 32 dead, 17 injured
Palau	Micronesia	1996	Failure due to excessive reinforcement by pre-stressing, bad concrete, 2 deaths, 4 injured
Jakarta	Indonesia	1996	Failure due to premature removal of temporary sheet piling, 100 dead, 105 injured
Injaka	Mpumalanga	1998	Collapse during deck launching, as a result of insufficient traction resistance, 14 dead
Eschede	Germany	1998	Train derailment, removing a bridge pile, 100 dead, 105 injured
Seo-Hae	Korea	1999	Partial collapse during construction following the rupture of the nose recovery system
Passerelle Yarkon	Israel	2000	Badly built and overloaded, 2 dead, 64 injured
Concorde	USA	2000	Pre-stressed cable corrosion due to bad injection
Ponto de Ferro	Portugal	2001	Foundation scouring, sand extraction and boat impacts, 70 dead
Queen Isabella Causeway	USA	2001	Impact of 4 barges on 3 sections, 8 dead, 13 injured
Webbers Falls Bridge	USA	2002	Barge shock on a bridge pile, 14 dead
Interstate 95 Howard Avenue	USA	2003	Partial collapse due to a truck fire, 1 injured
Kinzua Bridge	USA	2003	Tornado
Loncomilla Bridge	Chile	2004	Bad geotechnical conditions, 8 injured
Route Caracas-La Guaira	Venezuela	2005	Landslide
Almuñécar	Spain	2005	Accident during construction, 6 dead
Pont de la Concorde	Canada	2006	Bad design and maintenance, 6 dead
Al-Sarafiya Bridge	Iraq	2007	Collapse due to explosion, 10 dead, 26 injured
Highway 325 Bridge (Xijiang River)	China	2007	Ship impact, 8 dead
Minneapolis I-35W Bridge	USA	2007	Investigation in progress, 13 dead, 111 injured
Tuo River Bridge	China	2007	Investigation in progress, 34 dead, 22 injured
Harp Road Bridge	USA	2007	Collapse under convoy
Northern Bypass	Pakistan	2007	Investigation in progress, 10 dead
Can Tho Bridge	Vietnam	2007	Investigation in progress, 60 dead

Table 2.5. *(continued)*

Tuo River Bridge

Minneapolis I-35W Bridge

Highway 325 Bridge

Sully-sur-Loire

Webbers Falls Bridge

Ponto de Ferro

Figure 2.8. *Examples of total or partial bridge collapses*

Bailey *et al.* [BAI 02] analyzed 138 cases of bridge failure which have been indexed according to the Schneider classification into *accepted dangers* and *residual dangers* [SCH 94]. According to their analyses, 40% of collapses occur during construction or over the first few operating years. This highlights the importance of testing during the project and construction phase, a conscious reception but also a close monitoring over the first working years.

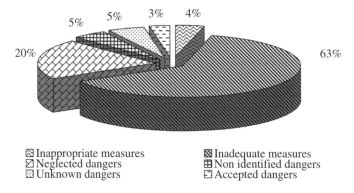

Figure 2.9a. *Accepted and residual dangers for operated bridges (from [BAI 96])*

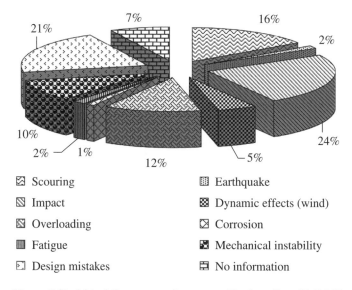

Figure 2.9b. *Main failure causes for operated bridges (from [BAI 96])*

For bridges under operation at the time of collapse, a large percentage of failure comes from inadequate measures (Figure 2.9a), which confirms the *important responsibility of the engineer* in the case of failure [BAI 02]. Figure 2.9b shows the distribution according to the technical causes of analyzed failures. The collapses due to corrosion or fatigue occur at late age in the structures' lifespan; this highlights the utility and importance of appropriate monitoring. Failures induced by natural causes generally affect old bridges, for which the dangers had gone unnoticed or only considered by inadequate measures. Bailey groups together these various causes of failure into three categories: human causes (72%), environmental causes (19.4%) and other (8.4%).

Shetty [SHE 97] also proposes studying road bridges according to the consequences of their failure, which he divides into four categories:

– *human*: loss of human lives or injuries;

– *environmental*: for example, dam failure;

– *delays*: inaccessibility to a bridge may lead to detours which generate extra costs and environmental damage (pollution);

– *economic*: direct material loss (construction cost, removing debris, toll closures, etc.).

2.6. From safety to performance

As was highlighted in the introduction, safety is only one aspect of the structural performance. This concept will only be significant in relation to the functions which justify the structure's construction.

2.6.1. *Functions of a structure*

Any civil engineering structure is intended to fulfill one or many functions. A bridge or a tunnel ensures the continuity of a route. For a bridge, this primary function of continuity is doubled by secondary functions such as the crossing of electric cables, telecommunications or pipe works. The structure's function may also be clarified by specifying the nature of the traffic to be carried, the number of lanes, the lifespan, etc. A dam's role is to regulate water or to produce energy. A canal bank must retain earth, prevent erosion and guide ships. Most functions of a civil engineering structure are permanent and active, but some of these functions are passive and sometimes activated on demand (like a lifting bridge, for example). A structure is a complex system which is generally made of many elements, which are possible to classify according to their structural and functional importance. On the system's upper level, one or many functions must be fulfilled, and on a lower level, the system's elements may lose their functions due to external loads, human errors or degradation processes. This distinction will not be without consequences with regard to maintenance activities, which must be implemented on the various elements. In fact, the maintenance strategies carried out on the elements must be defined when the element suffers loss of function, but also (and maybe especially) where there is risk of losing an infrastructural function [CRE 02].

The technical description of a structure is an incontestable aspect for a good understanding of its function (Figure 2.10). It is particularly indispensable in a system analysis (see Section 2.4.2). Therefore, the case of a bridge can be broken down into two aspects. The first is the structure which ensures the role of the load carrying sub-system. The second relates to the bearing system (piles, abutments, foundations). Between these two parts, bearing devices can help get back certain movements. Different equipment (waterproofing system, water draining, etc.) is also important for maintaining structural performance. For the safety of those using the structure and their comfort, additional elements may be incorporated: such as safety railings, safety barriers or lighting.

For each main function of the structure or infrastructure, we must break down the elements or put them into sub-systems so that maintenance can be carried out. If it is a simple structure, the causes and consequences of a component failure can easily be identified. In the opposite case, breaking down the elements must be subject to a more precise analysis: identifying failure modes and analyzing their effects will be performed.

In a construction, we can distinguish different types of elements:

– civil engineering elements,

– hydraulic, and mechanical elements which can move,

– electric or electronic elements.

Parts specific to civil engineering have often static behaviors (this is not exclusive: some structural elements, which are very flexible, are clearly sensitive to vibrations and some structures are mobile by construction, such as mobile bridges). They can be distinguished by their material (concrete, wood, metal, brick, stone). Their functions may be to support, to anchor, to protect, etc. Mechanical parts are, by definition, moving parts which, according to their functions, are divided into purely mechanical or hydraulic systems. The electric or electronic elements are divided according to their function: lighting, power supply, monitoring, information system, etc.

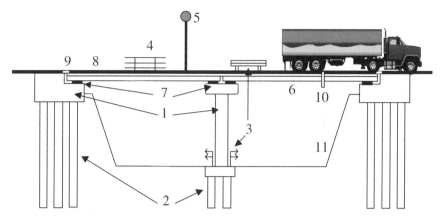

1.Piles and bearing devices; 2. foundation; 3. safety barriers; 4. safety railing; 5. lamp; 6. deck; 7. bearing device; 8. roadway; 9. roadway join; 10. rain water drain; 11. bank.

Figure 2.10. *Main parts of a bridge*

2.6.2. *Performance*

The functions of a structure generally describe the requirements to be fulfilled. The structure's ability to fill these requirements is then known as *performance.* Most official texts group these requirements (such as the European directive 89/106, relating to construction products, or the Eurocode EN 1990) into *safety* or *integrity*, *serviceability* and *durability.*

2.6.2.1. *Design working life*

The EN 1990[4] defines the *design working life* as assumed period for which a structure or part of it is to be used for its intended purpose with anticipated maintenance but without major repair being necessary. Defining a design working life for a structure may be misleading, because even in practice not all elements will necessarily have the same working lifetime. In the case for bridges, waterproofing systems, expansion joints, and bearing devices are generally replaced on a regular basis. Imposing a working life of 100 years on a bridge means that the usage lifetime is not applied to these elements, but to the structural system (deck, piles, foundation). Appendix A2 of the EN 1990 specifies three categories of expected indicative service lifetime for replaceable structural elements and equipment when the working lifetime is fixed at 100 years. Thus, for elements which

---

4 The *model codes for service life design* published by the FIB [FIB 06] call this *service design life.*

can be easily replaced or repaired, this working lifetime is 10 years; for elements which are difficult to replace or repair, it is 25 years. Table 2.6 represents the indicated working lifetimes proposed by the EN 1990 for various structure categories.

Category	Design working life	Examples
1	10	Temporary structures (scaffolding, etc.)
2	10-25	Replacable structural parts (bearing, gantry girders, etc.)
3	15-30	Agricultural structures, etc.
4	50	Buildings, hospitals, schools, etc.
5	100	Bridges, monumental buildings, etc.

Table 2.6. *Indicated design working lives according to the EN 1990*

The *real lifespan* is the period at the end of which the structure is recognized as being structurally or functionally obsolete. This concept is important, above all for a structure which has already had heavy repairs or been reinforced. In other words, if a structure which is structurally deficient continues to be in operation, it is in a state between serviceability performance and structural safety. It should be noted here that a structure's lifespan can be extended if it is maintained under reduced operating conditions. However, such a situation is not always desirable with the structure being difficult to manage.

Finally, these various lifespans must not be confused with the *reference period* which is the time interval generally chosen as a basis for assessing different actions in terms of statistics.

2.6.2.2. *Structural safety*

A structure must be in a position to resist all loads imposed during its construction and operation periods under normal situations. This implies the capacity of the structure's materials to ensure the static equilibrium to normal loading, and the absence of irreversible or accumulative damage. Following the definition given for safety in section 2.2.1, we must also expect adequate behavior towards abnormal conditions (fire, explosions, impact, etc.). In this case, the structure must be capable of maintaining its general shape and stability. In exceptional situations, large deformations and irreversible damage are acceptable, insofar as the users' safety is not changed. The structure's re-opening will, however, depend on the conclusions of a detailed inspection and on the repairs carried out. In the majority of cases, not respecting structural safety requirements will lead to rupture, or collapse of the structure itself or one of its parts.

Load/defect	Examples
Different or unforeseen loading situations	– changes to operational loads – accidental actions (vehicle crashes, terrorist acts, vandalism, etc.)
Insufficient or erroneous structural system	– absence of longitudinal blockage on the bearing devices – structural system change in relation to original design
Insufficient or erroneous calculation models	– failure modes or mechanisms ignored – erroneous hypotheses on the static system
Insufficient or erroneous calculations	– design errors – erroneous use of calculation programs
Material errors	– production errors (bad concrete formulation, etc.)
Project errors	– drawing errors
Execution errors	– use of the wrong materials – inappropriate construction procedures
Geometric imperfections	– bad dimensions – bad assemblies
Unforeseen compaction	– foundation settlements
Unforeseen degradations	– corrosion for metal structures – delamination, corrosion for concrete structures

Table 2.7. *Examples of unintentional loads or defects*

These exceptional, unforeseen or accidental situations also condition what we call a construction's *robustness*. Building structures with progressively longer spans, more and more flexible, etc., may lead to failures with huge consequences in terms of human and economic loss, which may be disproportionate in relation to the initial events. The robustness analysis of a structure, defined by Eurocodes EN1990 and EN1991-1-7 [EN 02, EN 06], is based on an assessment of the capacity of a structure "that it will not be damaged by events such as explosion, impact, and human errors, to an extent disproportionate to the original cause". Following this principle, the redundancy and the non-linear properties of the structure must be taken into account.

The purpose of these requirements is, then, to reduce the structure's sensitivity to unintentional loads and defects which are not covered by regulations, standards or recommendations. This is the approach adopted in the Danish regulation [DS 03] which stipulates that "a construction will be robust if the parts of the structure which are essential for structural safety are not sensitive to accidental loads or defects, or if the structure's global failure cannot be initiated by local failure". Table 2.7 summarizes a few examples of possible exposures against which the structure's robustness will be assessed.

However, to correctly apprehend the concept of robustness, it is necessary to distinguish robustness and *vulnerability*. A structure will be sufficiently robust if it can sustain damage (linked to the failure of one its elements, for example). On the other hand, a structure will be vulnerable if it is sensitive (i.e. subject to damage) to one or many hazards. If a structure is judged to be vulnerable on an unacceptable level, then it will not be very robust. Studying vulnerability may enlighten the subject when there is a lack of robustness.

The interest in an approach to vulnerability is that it only seeks to identify the events which lead to a failure scenario, regardless of its occurrence. This analysis is, of course, limited. However, determining failure scenarios is a vitally important stage in the application of system reliability analysis techniques (see Chapter 3). On the contrary, for cases where it is hard or even impossible to quantify event occurrences (rock avalanches, terrorist attacks, etc.), studying vulnerability is the only approach in order to identify the potential weaknesses of a structure and to make appropriate decisions. It is, then, an essential addition for reliability techniques in a strategy for managing structural risks. Procedures and methods for dependability analysis provide the risk analysis tools available to the engineer (see Appendices B and C in EN 1990). Identifying failure or fault trees may allow for an assessment of the structure's weakness and therefore its vulnerability when faced with unforeseen situations.

The robustness of a structure may be reached by means of appropriate materials, structure's static scheme, etc. and by the specific design of essential elements. Table 2.8 gives a few examples of these criteria, which, when analyzed, may increase the robustness. An essential element is defined as a limited part of the structure for which failure leads to global structural failure or a significant extent of damage. We will come back to this aspect in Chapter 3.

Criteria	Examples
Loads	− specifying acceptable loads − taking account of all the accidental loading scenarios imaginable
System	− favoring parallel systems − favoring systems which are not sensitive to settlements
Statically indeterminate systems	− force redistribution
Ductility	− favoring ductile materials and joints
Solidity	− wide geometries, reduced slender ratio, over design
Investigation and inspection	− critical analysis of essential elements during the design phase so as to identify important aspects for the system's reliability and robustness − ensuring accessibility of essential elements during construction and operation (inspections)

Table 2.8. *Examples of criteria able to increase robustness*

On the one hand, robustness involves verifying the system's "stability" to internal or external forces, and on the other hand, identifying the functioning modes in degraded conditions.

Robustness may be linked to another concept; that of a *damage tolerant structure*. This characteristic is applied to structures which are not very sensitive to small modifications to their initial condition with regard to respecting structural safety requirements. However, a *damage intolerant structure* will require efficient control of the material quality, and appropriate maintenance and inspections.

2.6.2.3. *Serviceability*

This deals with the requirements necessary for maintaining that the structure remains functioning. Not respecting these requirements rarely leads to concerns for users' safety[5], but may generate direct or indirect costs linked to the structure's operation. In most cases, it concerns requirements on the structure's deformability in relation to permanent loads (creep, etc.), variable loads (sags, etc.), dynamic effects (resonance and comfort, etc.), but also esthetic aspects, etc.

Not respecting these requirements can lead to very different situations, according to whether the structure's behavior is reversible or not after eliminating the conditions which led to not respecting the requirements (see section 2.6.2.5). This leads us to approach the serviceability requirements in three different ways:

– no infringement of the requirement is accepted;

– specific duration and frequency of infringement are accepted (this is only for reversible situations);

– specific long-term infringement is accepted for reversible situations.

2.6.2.4. *Fatigue*

For structures loaded by fluctuating actions, fatigue failure may occur over loading levels which are significantly lower than levels when static failure would normally be expected. This is, in particular, the main characteristic of accumulative damage caused by fatigue. Fatigue is generally dealt with differently from structural safety for many reasons:

---

5 Not respecting certain serviceability requirements may, however, put users' safety in danger: this is the case for a too large deformation on a bridge, making it dangerous during high speed traffic.

– fatigue loading is different from other loading since it is dependent on the amplitude and variation of loads, under service conditions, and according to time (through the number of cycles);

– effects related to fatigue constitute local degradation in the material which may be benign if the cracks relax the residual stresses (and therefore stop progressing) or destructive if the cracks lead to more severe stress conditions, accelerating crack growth;

– if the material presents sufficient toughness, crack propagation can be detected by regular inspections before the element fails.

2.6.2.5. *Durability*

This is often difficult to define and in practice has had many definitions. A satisfying definition consists of linking durability with the two other concepts of performance, namely structural safety and serviceability: durability represents the aptitude of an entity (structure or element) to maintain its performance of structural safety and serviceability in given service and maintenance conditions over the defined design working life. This is the meaning given by the EN 1990, which, in fact, states that a structure must be projected in such a way so that its eventual deterioration, during the design working life, does not lower its performances below those expected, taking into account the environment and the level of maintenance. It is worth recalling here that durability does not guarantee the structure an infinite lifespan, but is simply an objective for quality, directing the design of the structure as well as the material [AFG 04].

When a structure is being designed, it is accepted that the structure will be built, put into place and operated by conforming to the adopted hypotheses. However, the structure may acquire certain characteristics over time, either because of changes in service loads, or because of degradation phenomena which cause a lower resistance (see section 2.6.4). These changes introduce many uncertainties. This is why it is preferable to predict the structure's lifespan using a probabilistic approach. For a degraded structure, models of material degradation must be treated in a probabilistic way, and introduced into the design model or calculated separately. This may then describe the evolution over time of the structure's degradation profile, and the main stages in the evolution of performance (see Chapter 6).

Figure 2.11 shows the evolution of performance over time; the performance is likely to evolve over time due to many phenomena (see section 2.6.3). This performance can be measured in many ways: mechanically, financially, etc. It can even increase over the first years of the structure's life, as in the case for concrete structures for which resistance improves over time. It may also stay constant for some years, which is the case for metal structures protected against corrosion. As in all examples, however, performance will have a tendency to degrease after some or many years, following various degradation profiles. Inspections will reveal that certain serviceability requirements are no longer respected (excessive crack opening, fatigue cracks, etc.), often highlighting irreversible problems. At a certain point, "normal" maintenance will neither stop nor slow down the loss of performance, and more thorough repairs will be needed to prevent the minimal performance threshold from being reached within the expected working life [SIL 05].

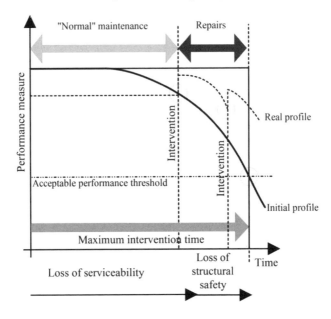

Figure 2.11. *Evolution profiles of performance over time*

At this stage, it is appropriate to raise the question of ambiguity on the term durability. There are two distinct concepts with regard to durability: either we are referring to the *material's* durability, or the *structure's* durability. Material durability is verified by its capacity to retain its characteristics and integrity over the design working life of the structure. The structure's durability clearly depends on the material's durability, but does not completely rely on the quality of material used. This is why the structure's durability (global or local) consists of fulfilling the structural safety and serviceability performances in the planned service and maintenance conditions.

2.6.2.6. *Limit states*

Performance requirements cover structural behaviors, classed as being either acceptable or to be avoided; the separation between these two domains of behaviors is then described by *limit states*. Exceeding these limit states represents, then, not respecting the performance requirements. For structural safety requirements, limit states are called the *ultimate limit states*. For serviceability requirements, we are dealing with *serviceability limit states*.

As for risks, the concept of performance is only useful if it is quantifiable, therefore, characterized by one or many measures. There are many performance measures according to the nature of the structure and its functions. Maintenance will always try to ensure that the possible structural behaviors do not reach unwanted behaviors. The *limit state criteria* aim to protect the structure from dysfunctioning via *safety margins*. There are many possible approaches: safety coefficients, partial safety factors, acceptable failure probabilities which will be approached in Chapter 3.

2.6.3. *Evolution of structural functionality*

Structural functionality is not necessarily constant over time. The initial functions, meaning those which motivated the structure in the first place, may be modified either voluntarily due to a desired change of the functions, an extended lifespan, or involuntarily due to external loads or degradations.

2.6.3.1. *Loss of initial performance*

There are three main causes which can lead to a loss of initial performance[6]:

– ageing;

– human causes due to mistakes or inappropriate actions;

– external causes.

It is clear that even an in depth inspection can never guarantee against performance loss. Inspections are, then, useful for detecting an unexpected function loss, or for limiting consequences over time.

2.6.3.2. *Ageing*

Ageing in the elements of a civil engineering structure may lead to sudden or gradual loss of resistance, and therefore to partial or complete loss of performance in the element. According to the structure's capacity to respond to change, one or many of the structure's functions could be affected.

---

6 Experience shows us that these three causes are equally important in the loss of structural performance.

2.6.3.3. *Environment (external load)*

There are many examples of external causes which can be given: pollution, vehicle or ship impacts, vandalism, terrorism, earthquakes, storms, etc. According to the frequency of occurrence and the size of the structure, some of these external loads must be taken into account during design. Exceptional loads can be included in extreme loads. Let us also point out that, for a bridge for example, an increase in traffic also acts as an external cause affecting the structural performance due to an increase in the loads carried.

2.6.3.4. *Human errors and actions*

Human errors may intervene at different points over the structure's life:

– in the design (for example, error in the reinforcement area);

– in the construction (for example, bad quality concrete);

– in the operation (for example, overloading);

– in the inspection (for example, undetected defects or damage);

– in the maintenance (for example, excessive use of de-icing salts).

2.6.3.5. *Change of functions*

It is also possible for the initial functions to be modified, not because of deficiencies in the structure, but through the will of the main contractor or manager. Thus, re-using certain railway bridges for high speed tracks leads us to ask questions on the structure's capacity to support dynamic effects. There is, then, a change of function (change from standard rail traffic to high speed), because there has been a change in the external loads. This change in function therefore requires the structure to be requalified with regard to its new functions. This problem arises for many civil engineering structures, where regulations and operations are evolving (seismic regulations, for example).

2.6.3.6. *Extending service lifetime*

Civil engineering structures are generally designed to be used and be usable over a given working or reference period (100 years, for example, for a standard road bridge, 50 years for buildings, or 30 years for nuclear plants). This service lifetime is often implicit in regulations, for assessing variable loads. This may be questioned, particularly in cases when the structure reaches the end of this period in a satisfactory condition. It is then necessary to requalify the structure, i.e. requalify the structure's capacity to fulfill its functions over an extended service lifetime.

2.6.4. *Consequences of performance losses*

Essentially, a structure guarantees one or many functions. A change in its performance will therefore require maintenance actions in order to remedy it. These procedures play an essential role in the maintenance of certain civil engineering structures because it involves *production loss* (electric plants, oil platforms, water networks). In a similar way, for others, the loss of performance leads to *accessibility loss*. This loss of accessibility results in indirect financial losses (waiting times, detours, etc.) or direct losses (toll closures).

Losing a structure also entails two different consequences: *material loss* and *immaterial loss*. Material loss affects the structure's value. Immaterial loss is, by definition, difficult to quantify: it concerns human lives whose economic value, *a priori* morally priceless, is also estimated by its contribution to the gross national product (GNP), as with material losses. Introducing these costs is necessary in any cost/benefit economic approach, or in life cycle cost analysis. In fact, many researchers over the last 10 years have preached that reducing risks should include a financial assessment of the cost in human lives [RAC 02]. We have already mentioned the irrationality of the risks taken by individuals, and the difference between individual risk and collective risk. This situation induces measures for controlling risks which are often non-uniform, non-systematic and, really, managed by personal or group interests. The professional ethics in engineering should make these risks quantifiable and assessed according to *rational, transparent, communicable, and economically realistic principles* [RAC 05]. It is within this framework that, very recently, the concept of *life quality index* was introduced in order to deduce a criterion for the acceptability of risks.

Deteriorations on the structure affect its resistance and may lead to a loss of function. Structures cannot escape this rule, but must continually fulfill, and in complete safety, the services for which they were built. It is, then, necessary to systematically and attentively monitor their condition and their service conditions, and to execute, in good time, preservation, maintenance or repair procedures which keep these structures in a good working condition.

2.6.5. *Generalization of the concept of risk*

The concept of risk is often associated with a catastrophic failure caused by a threat or serious danger (flood, earthquake, avalanche, etc.), i.e. the occurrence of a loss of structural safety. Conceptually, nothing prevents us from applying this notion to the general framework of structural performance, including durability and serviceability. In fact, an individual risk will, then, be defined as the product of the probability of performance loss by the cost generated by this loss. The point of this formalization means that the risk can be reconsidered (as insurers always do) as an average cost of loss. This approach helps us to understand the importance that a risk function constitutes mathematically; its minimization (generally under constraints) favors the determination of optimal intervention strategies.

2.7. Human errors

Uncertainties resulting from human intervention in the design, construction and use of structures and their components stem from *human errors and interventions.* These two sources of uncertainty are generally simultaneous. Human error can be separated into many categories: on the one hand, variability itself of any human action or decision, the usual errors which intervene in any design, construction and operation processes which are considered as being acceptable by the procedures, and on the other hand, errors of ignorance or omission.

Cause	%
Bad assessment of loads or structural behavior	43
Errors in drawings and calculation	7
Bad drafting of documents or instructions	4
Not respecting the constraints of documents or instructions	9
Bad execution of construction procedures	13
Bad use, degradation	7
Variations in loads, materials, quality of working staff	10
Other	7

Table 2.9. *Main failure causes*

Pugsley [PUG 73] considered the main factors as the source of potential accidents to be:

– using new or unusual materials;

– new or unusual construction methods;

– new or unusual structures;

– lack of experience of the teams responsible for design and construction;

– the level of research and development;

– financial climate;

– industrial climate;

– political climate.

Accompanying measures	Test measures
Education	Cross checks
Work	External test and inspection
Reduced complexity	Legal or contractual sanctions
Staff choice	

Table 2.10. *Intervention strategies for reducing human errors*

These factors influence human behavior and therefore the performance of human activity. The relationships connecting the different factors to behavior are difficult to establish. There are some empirical results, however, in domains such as nuclear, aeronautics and chemical industries. Many existing structures are maintained in service in spite of minor errors made during the design and construction phases. One reason for this is the conservative nature of the regulations which lead to over-design the structures in many cases.

There are many ways of reducing uncertainties linked to human activity. Table 2.10 presents a few of them. These approaches must not be considered in isolation. In any project, these are only additional techniques for reaching the desired objective which, in most cases, is a project presenting the desired serviceability and safety at a reasonable cost. These procedures for reducing human error associated with appropriate management structures and organization, with acceptable controls and safety measures are grouped together under the generic title of *quality assurance*. In the particular context of a project, quality assurance procedures must take into account and apprehend safety, serviceability and durability objectives. In a similar way to other projects in domains other than civil engineering, one possible approach is to define a safety plan based on a detailed analysis of hazard scenarios. For civil engineering structures, these plans must include, in particular [SCH 81] the following essential aspects:

– precise definition of the infrastructure's functions;

– definitions of actions, responsibilities and duties;

– appropriate information circuit;

– list of items to be validated;

– definition of acceptable risks;

– inspection and maintenance plans;

– technical use instructions.

In addition, it is necessary to return to the information at each stage of the project. An excessive formulation of quality assurance procedures may nonetheless generate a huge mass of documents. It is, then, useful to dimension these procedures for each project. It goes without saying that controlling human errors must ensure an *in fine* improvement in robustness.

If we follow the principles of quality assurance, it seems clear that those responsible for a project must not only focus on reducing risks but also on managing hazards and their consequences. This management is particularly important in situations such as "small probability – large consequences". Nuclear accidents and earthquakes are part of these situations. Even if these hazards cannot be controlled or reduced, the after-event must also be managed correctly. Melchers [MEL 99] in particular quotes the Kobe earthquake in 1994, where it was stated that controlling the immediate consequences of a rare earthquake might be impractical, because it could lead to costly rehabilitation procedures for existing structures. On the other hand, structures with appropriate safety measures must be put in place, just like rescue operations and health controls.

Chapter 3

Performance-based Assessment

3.1. Analysis methods and structural safety

The idea of structural safety, in its most primitive form, must stem from the very practice of construction. As far back as 2000 BC, Hammourabi was starting to discover this concept; the man whose famous code has been preserved to the present day, and kept in the Louvre Museum (Figure 3.1). However, this autocrat knew nothing about probability. His vision of safety made the builder himself responsible for the collapse of structures, and liable for less severe punishment but for the same reasons, in the case of partial ruin.

Over many centuries, the builder was left to his own devices, to his professional skills, to his experience and the experience of his predecessors, with limits often being determined by accidents or the observation of collapsed structures.

This empiricism did not allow, however, for the development new structures with new materials. The emergence of construction science, with structural mechanics and strength of materials, only occurred much later and only progressively. The disappearance of empiricism to the gain of engineering science was greatly helped by the development of metal construction. However, even at this stage, the concept of "structural safety" was still not mentioned in technical literature, and the use of reduction coefficients applied to resistance appeared as the true expression of safety. The adopted safety *principle*, known as *allowable stress*, consisted of verifying that the maximum calculated stress in any section of any part of the structure, and in the case of the least desired loads, remained lower than a stress said to be allowable. This stress was deduced from the ultimate resistance R_{rup}, by dividing the latter by a conventionally fixed *safety coefficient K* (see section 3.1.1). The design method based on the allowable stress principle was used at the beginning of this century, without really setting out a definition of these allowable stresses. Their values had been arbitrarily fixed on the basis of the mechanical properties of the materials used. The consideration of improving steel production, and structural design and construction led to an increase in the allowable stress calculation, by reducing the safety coefficients K. Attempts at improving design rules based on the allowable stress principle, in order to obtain a better definition of loads and resistances, revealed the dispersed nature of data and the results. It became clear that there was a need for tools to approach this variability.

Figure 3.1. *Hammourabi code (The Louvre, Paris)*

Moreover, a material's strength is not necessarily the most important variable. It is, however, the most important for a material with brittle behavior like cast iron, for example. But it is not, however, for ductile materials such as mild steel or aluminum, whose resistance limit is accompanied by large deformations which are unacceptable for a structure.

The yield stress is, then, a mechanical characteristic which is at least as important as the ultimate strength. The principle of allowable stress does not take into account mechanical adaptation phenomena (plasticity, creep, etc.) or the diversity of loads applied to the structures (combinations of loads, fatigue, etc.).

In fact, two problems were raised by the safety formulation following the principle of allowable stress [CAL 91]:

– replacing the allowable stress criteria by other criteria (limit states);

– rationalizing the methods of introducing safety.

This explains why many engineers have tried to approach the problem from a different angle, by defining safety by a probability threshold.

In 1928, in an international forum, Professor Streletsky stated that the concept of a safety coefficient was devoid of any real meaning. But this only managed to awaken a weak echo in the world of research and construction. A few scientists, leading studies on the assessment of material and structural resistance, developed in particular the basic concept of random events, thus pointing out a rupture with classical structural design rules [MAY 26, WEI 39]. Prot and Levi, in many correspondences between 1936 and 1953, made great efforts to try to define combination rules and, therefore, to determine failure probability [PRO 36, PRO 53]. Wierzbicki [WIE 39] believed in particular that failure probability must be comparable between civil engineering structures and human activity. At a competition organized by the Royal Swedish Academy in 1938, Kjellmann and Wästlund [KJE 40, WAS 40] especially, defended the introduction of a probabilistic approach for apprehending safety problems in construction cost assessments.

In 1945, Freudenthal created an institute for studying reliability and fatigue at the University of Columbia, and thus promoted the emergence of a probabilistic approach to safety. But, it was only at the third AIPC congress (Liege, 1948) that the real attack on deterministic thinking was started up by three French people: Marcel Prot, Robert Levi and Jean Dutheil. Driven by them, the concept of probabilistic safety was born. This was the turning point where the foundations of construction calculation methods found themselves being shaken up and replaced by new bases and new methods. However, it was only really in the 1960s that the need for scientifically defining safety margins came about. The development of mathematical tools, however, did not manage to change ways of thinking. The collapse of structures was rare, and when this was the case, it could be explained by human error. Moreover, the reliability theory was, then, mathematically and numerically complex. The small amount of statistical information available on the different variables did not allow for correct models. Also, most engineers considered it preferable to use an irrational procedure which worked, rather than a more rational and complicated approach, on the whole with a completely irrational implementation. It is in this way that a particular effort was made to level out the various difficulties. There are many names associated with these developments: Basler in 1961, Cornell in 1967 and Hasofer in 1974. Based on the work of Basler, Cornell [COR 67] introduced a reliability index from which Lind [LIN 73] showed the possibility of deducing safety coefficients. This stage therefore promoted the emergence of a semi-probabilistic procedure for structural safety. This is now present in most calculation standards for new structures. The probabilistic approach has, therefore, essentially helped in the development of semi-probabilistic formats. We owe, however, the most shining breakthrough of the direct application of probabilistic approaches [MAD 89, MAD 90] to Scandinavian oil engineering, in the construction of oil platforms. This explains why the names of many Northern European engineering researchers are still associated with developments in the reliability theory for calibrating standards, assessing and managing structures [THO 84].

3.1.1. *Allowable stress principle*

In the allowable stress approach, the structural safety principle therefore consists of verifying that the maximum stress S calculated at a given section and for a combination of unfavorable loads remains lower than an allowed stress R_{adm}. This value is deduced from the material's strength R_{rup}, by dividing this value by a conventionally fixed safety coefficient K. The justifications consist of ensuring inequality:

$$S \le R_{adm} = \frac{R_{rup}}{K} \tag{3.1}$$

Structural safety is therefore defined by the safety coefficient K. This coefficient takes values between 1.5 and 2.0 for structural elements, and 3.0 in soil mechanics. This higher value is explained by the largest dispersion in the mechanical behavior and resistance of soils.

This safety measure presents advantages such as easy implementation, but is still insufficient. Firstly, a structural element shows safety coefficient K when the stresses issued from maximum service loads S are such that, when making them increase in proportion from 1 to K, failure (or rupture) is reached. This assumes that all loads vary in the same ratio, whereas some of them (dead load, for instance) show a low variability or their maximum values are perfectly simultaneous. As a result, K is not the same real uniform safety measure for all elements.

The use of stresses (and no other load effects) shows an enormous advantage for measuring safety by comparison with normal mechanical characteristics (elasticity, ultimate, fatigue, etc.). This is only valid if the geometric and material linearization hypotheses are verified, and are satisfactory for materials under fatigue loading, but unacceptable under static loads (yielding) or in the presence of instability phenomena (large displacements).

3.1.2. *Limit states and partial factors*

This analysis method, known as the *semi-probabilistic approach*, relies on the same concept as that for allowable stresses, but in a more detailed way. It was developed in order to avoid the main errors of the allowable stress method, and to try to have a more realistic vision of safety. Set up by the *Comité euro-international du Béton* (CEB) in the 1950s, and introduced by the *Fédération Internationale de la Précontrainte* (FIP) in its models from the 1970s, it was also adopted by the *European Convention for Constructional Steelwork* (ECCS), then in various standards. The principles of current standards have largely taken on this procedure of calibrating safety coefficients, following three main ideas:

– Recalling that rupture criterion is not limited to stress assessment.

– Probabilizing with suitable precision. The evaluation method for a variable consists of statistically determining a characteristic value and incorporating uncertainties which have not been taken into account, into a fixed coefficient.

– Ignoring the dispersion of certain data by using deterministic values.

By doing so, this procedure must lead to better safety estimations and to a more economic design.

3.1.2.1. *Performance limit states*

The idea of limit states is to describe the functioning modes of an element (of a structure), i.e. the functions for which it is intended. It is in this way that the ultimate and serviceability limit states are distinguished.

Ultimate limit states characterize extreme service conditions (rupture, loss of equilibrium, instability, fatigue, excessive displacements, etc.), while the *serviceability limit states* represent conditions where the element (or the structure) is out of use but can be repaired (vibrations, excessive deformation, excessive cracking, degradations, etc.). Eurocode 1 [EN 03], however, distinguishes two ultimate limit states, *static equilibrium* and *resistance*. These two concepts are often synthesized into the concept of structural safety. We saw in Chapter 1 how durability is also a performance function. Until recently, it was rare for durability to be defined by a limit state function; most of the time it has been expressed in qualitative and subjective terms. In the framework of concrete, the AFGC guide [AFG 04] defines durability indicators (for concrete material) which constitute a certain step towards a performance based approach for durability.

A limit state allows us to define normal service situations from situations which we would prefer to avoid. The limit state criteria aims to specify conditions which lead to exceeding these limit states; they therefore seek to introduce a *safety margin* so that we do not reach unacceptable service conditions. Safety coefficients for the allowable stress principle constitute one of the first attempts at introducing a safety margin, by reducing the material's strength. The most recent forms are based on the concept of the characteristic value and partial safety factor, or the acceptable failure probability.

3.1.2.2. *Performance verification*

Semi-probabilistic approaches or formats consist of checking that the value for the load effects S_d remain lower than the resistance value for calculation R_d. Following the considered limit state type, this resistance value will be understood as the calculation value of the stabilizing load effects (in opposition to the destabilizing ones for an ultimate limit state of static equilibrium), as true strength (for a ultimate limit state of structural resistance) or as a characteristic calculation value (strain, crack opening, etc. for a serviceability limit state).

$$S_d \leq R_d \tag{3.2}$$

S_d and R_d respectively result from loads, or strengths:

$$\begin{cases} R_d = \mathcal{F}\left(\cdots, f_{id}, \cdots\right) \\ S_d = \mathcal{F}\left(\cdots, F_{jd}, \cdots\right) \end{cases} \tag{3.3}$$

where the values f_{id} and F_{jd} relate to the element's resistance properties and an applied load. These values are called *design values*.

This method is therefore called semi-probabilistic so as to express the coexistence, within the same limit state criterion, of variables whose values are assessed statistically, and others whose values are assessed deterministically. The values of these determined variables are said to be characteristic, independently of their method of obtaining these values (deterministic or probabilistic). The resistances f_i are generally defined by their characteristic values f_{ic} which are fractiles with low probabilities of being exceeded. Loads are defined by the characteristic values F_{jc} which are either fractiles (for permanent actions), or loads estimated as not being exceeded more than once out of 10^n for an occurrence period over a structure's lifetime (variable actions). The semi-probabilistic approach introduces *design values* (f_{id}, F_{jd}) determined by – multiplying for actions, dividing for resistances – the corresponding characteristic values by *partial factors*:

$$\begin{cases} f_{id} = \dfrac{f_{ic}}{\gamma_{f_i}} \\ F_{jd} = \gamma_{F_j} F_{jc} \end{cases} \tag{3.4}$$

The calculation values of effects S_d and structural resistance R_d are deduced by the expressions:

$$\left\{ \begin{array}{l} R_d = \dfrac{1}{\gamma_{R_d}} \mathcal{R}\left(\cdots, \dfrac{f_{ic}}{\gamma_{f_i}}, \cdots \right) \\[3mm] S_d = \gamma_{S_d} \, S\left(\cdots, \gamma_{F_j} \, F_{jc}, \cdots \right) \end{array} \right. \qquad [3.5]$$

γ_{R_d}, γ_{S_d} represent partial coefficients on the one hand intended to cover the uncertainties of resistance models and geometric properties, and on the other hand, uncertainties of the structural model and loads.

The semi-probabilistic format introduces other coefficients. For example, in Eurocode 1 [EN 03], *combination coefficients* ψ_{F_j} and *conversion coefficients* η_{f_i} are associated with actions in order to take into account the conditions of their combinations and the effects of loading periods, effects of volume, scale, moisture, temperatures, etc. on the resistances. In this case, equation [3.5] changes into:

$$\left\{ \begin{array}{l} R_d = \dfrac{1}{\gamma_{R_d}} \mathcal{R}\left(\cdots, \eta_{f_i} \dfrac{f_{ic}}{\gamma_{f_i}}, \cdots \right) \\[3mm] S_d = \gamma_{S_d} \, S\left(\cdots, \gamma_{F_j} \, \psi_{F_j} \, F_{jc}, \cdots \right) \end{array} \right. \qquad [3.6]$$

This approach allows for easy calculation implementation, but leads to the use of uncertainty coefficients, linked to the deterministic choice of certain parameters and to fractiles for others. They must also cover uncertainties linked to minor parameters or those ignored during study. They are also calibrated so as not to move away from the previous design methods. Their main aim is to cover the possibilities for an unfavorable change in the basic parameters and to take into account the errors related to the parameter characterization model (the probability distribution used for calculating the fractile), and to the load effect calculation model from the basic parameters.

The partial coefficients γ_R and γ_S have values higher than 1.0 with products $\gamma_R.\gamma_S$ between 1.5 and 2.0. These values are based on probabilistic analyses but also on engineering judgment. For example, a lower partial safety factor will be used for steel rather than concrete, because steel variability is controlled better. In the same way, the partial factors of permanent loads will be lower than for live loads, due to reduced variability.

3.1.3. *Probability-based approach*

The partial factor and characteristic value approach is described as being semi-probabilistic, through the application of statistics and probabilities in the evaluation of data input, the formulation of performance assessment criteria and the determination of load and resistance factors. However, from the engineer's point of view, applying this approach still remains deterministic. It does not provide the relationships or methods which would allow him to assess the risk, or the residual load-carrying capacities in structural elements when applying a semi-probabilistic format.

Partial factors are, in fact, intended to cover a large set of uncertainties and may also prove not very representative of the real need for the performance assessment of a particular structure. For exceptional or damaged structures, assessing the reliability may be over or under estimated. The introduction and consideration of uncertainties seem to contribute to answering the need for rationalizing performance assessments. This is justified by various reasons:

– loading changes over time are often ignored;

– material properties are also likely to change (corrosion, fatigue, etc.);

– load effect combinations are badly introduced;

– the behavior of real elements is often different from test samples, on which the performances are measured;

– errors of structural behavior modeling are generally omitted.

An approach which takes into account uncertainties related to variables appears as a realistic criterion for assessing structural performance. From this fact, the probabilistic approach is, today, an alternative to semi-probabilistic formats. As a procedure, it:

– identifies the set of parameters which influence the expression of the limit state criterion;

– statistically studies the variability of each of the considered variables and their correlation for deducing the joint probability distribution;

– calculates the probability of exceeding the limit state criterion;

– compares the obtained probability to an acceptable probability.

This process is known as the *reliability theory*. As attractive as it is, the reliability theory is limited by many factors:

– some data is difficult to measure;

– the statistical data necessary is often non-existent;

– probability calculations are becoming increasingly unachievable.

These considerations determine what we may hope to achieve from the limits of a probability-based approach. These limits mean that the probabilities we are often talking about are, unfortunately, frequency estimations (generally unobservable), based on a changing set of partial information. These limits also result in assumptions (e.g. distribution type choice) which make them conventional. Thus, the result of a probabilistic approach will depend on the hypotheses based on uncertainties. If these hypotheses do not rely on reliable and realistic results, the performance estimation will be incorrect. In fact, the probabilistic approach is sometimes incorrectly used when the variables are badly modeled, in terms of probabilistic characteristics. The quality of data and the validity of statistical hypotheses determine when we would use a probabilistic approach in order to make decisions regarding the apparent performance of a structure.

In a probabilistic process, effect S applied to a structural element and R, the resistance variable, are described randomly, since their values are not perfectly known. If the verification of the criteria related to the limit state results in inequality:

$$S \leq R \tag{3.7}$$

with the failure being related to the exceedence of this limit state, the probability P_f of event $(R \leq S)$ will characterize the level of reliability in relation to the considered limit state:

$$P_f = \mathcal{P}(R < S) \tag{3.8}$$

The semi-probabilistic format, in fact, schematically replaces this probabilistic calculation by checking a criterion which makes the characteristic values of R and S and the partial factors intervene. The probabilistic approach has, therefore, essentially helped with developing semi-probabilistic formats [CAL 96]. As was noted in section 3.1, oil engineering was the first domain to initiate applications of a probabilistic approach for designing and maintaining oil platforms. This also required studying the reliability of metal tubular joints and the implementation of maintenance strategies [MAD 89], [MAD 90]. Applying the reliability theory has allowed the domain to define an exemplary inspection and assessment planning. Some applications in civil engineering structures can already be found in literature (see reference [IAB 01] in particular). The probabilistic approach will be developed further in sections 3.4, 3.5 and 3.4.

3.2. Safety and performance principles

Unless we go back to Hammourabi times, the word "safety" must be perceived of as the expression of a reasonable guarantee against structural collapse. The degree of safety depends on the relationship between the loads likely to lead to the structural failure, and the characteristics of the structure's resistance. But, neither one could be determined with precision and certainty. This means that the basic elements of the problem are random. However, if we can only observe imperfect objects, and if experiments are not guaranteed by absolute objectivity, must we lose all hope of finding a general principle by observing particular cases and defining valid general laws? Certainly not, and we are attempting to witness the different paths being offered to the engineer in order to answer this question.

Structural safety (and more generally structural performance) can be defined as a set of margins between situations where we would find a structure, and unacceptable situations which would be dangerous for it, or its users, or which would even compromise its normal use. In this sense, it only has meaning to the extent where the following can be defined for the structure:

– the set of normal situations;

– the set of related unacceptable situations and the seriousness of these situations in terms of consequences;

– the level of confidence of not reaching these unacceptable situations, taking into account quality assurance conditions during the construction phase, and maintenance conditions during the operation phase, knowing that these confidence levels represent the margins between normal and unacceptable situations.

To illustrate this definition, let us take the simple case of a bar under axial traction S, and let R be the variable representing the bar's resistance. The set of normal service situations for this bar relate to cases where the applied force S stays lower than resistance R. Vice versa, the set of unacceptable situations are those where the applied force S is higher or equal to resistance R. The level of confidence for not reaching these unacceptable situations may be defined as the risk incurred by a situation when a structure's situation approaches an unacceptable situation. This confidence level (and the seriousness of the incurred risks) is a piece of data which lets us qualify, or even quantify, the structural reliability:

$$S < R \qquad\qquad [3.9]$$

Therefore, we can say that the bar is found to be in the *safety domain* (functioning normally). However, if the previous inequality is not checked, we will say that the bar is found in the *failure domain.*

$$R < S \qquad\qquad [3.10]$$

When $R = S$, we can say that the bar has reached the *limit state*.

In Chapter 4 we will return to these essential concepts in the study of structural performance. We must, however, now clearly distinguish the way in which performance is taken into account for new structures from how it is introduced for existing structures.

3.2.1. *New structures*

The traditional procedure for designing a structure can be summarized as three essential stages:

1. Specifying and assessing loads.

2. Fixing the shape, the dimensions, and the materials in an initial approximation.

3. Checking that the structure's predicted behavior under loading is "satisfactory" with regard to predefined criteria.

This behavior is defined by displacements, stresses, strains, forces, etc. The structure will be calculated correctly if the fixed criteria are verified. If this is not the case, we reiterate stages 2 and 3. In the logical order of decision making, it is first necessary to list the mechanical loads that the structure is subjected to (stage 1). Their definitions depend on the specifications for foreseeable use and natural loads. Observing natural loads as well as the choice of service functions constitute an upstream level. It is, then, important to represent the mechanical effects of these loads and functions.

3.2.2. *Existing structures*

Here we will remind our readers that the rules written in design regulations or standards are only valid for certain contexts. Also, a structure must conform to design rules according to a number of points:

– the types of structures;

– the methods used in the structural analysis;

– the quality of the materials and of the construction;

– the current live load condition;

– the structural condition;

– the most common errors recorded.

To assess the load-carrying capacity of an existing structure, some situations may arise which make the standards not applicable, due to particular structural conditions or non-conforming construction details.

The partial factors for new structures are, in fact, intended to cover a large set of uncertainties and, then, may be little representative of the real need for performance assessments of a particular structure. They are supposed to take into account the evolution of materials and load effects in a fixed way. As mentioned before, for exceptional or damaged structures, the reliability assessment may be either over or under estimated. The introduction and consideration of uncertainties may contribute to fulfilling the need for rationalizing performance assessments. Regulations also include a large amount of generalization in terms of safety, performance and loading. This may be useful for studying design because calculations are becoming easy, and because the induced costs are marginal in the budget of a new construction. In the case of repair or rehabilitation, the required level of performance cannot always be achieved by general standards or regulations. The result of this may lead to expensive repair projects.

Establishing principles and procedures specific to structural assessment is therefore essential, because certain aspects of the assessment are based on an approach which greatly differs from the design, and requires a level of knowledge which goes far beyond the field of applying regulations.

3.3. Invariant measures

As section 3.1.1 has shown, the safety coefficient is defined as the ratio between a specified strength R and an applied and calculated load S:

$$K = \frac{R}{S}$$ [3.11]

If $K > 1$, then the element is safe. However, if $K < 1$, this means that the element is in the failure domain, or the domain where function is not recommended.

Opposing safety measures by safety factors is not just based on the need to consider limit states with many variables. It very quickly turned out that the concept of a safety coefficient presented an important limit in its introduction into regulations: a limit of *invariance* related to the way in which the resistances and loads are calculated.

EXAMPLE 3.1.– A classical example [DIT 97] to illustrate the lack of invariance is the case of a reinforced concrete beam subjected to a normal force $\mathfrak{N}$ and a bending moment $\mathfrak{M}_S$, defined in relation to the beam's neutral fiber.

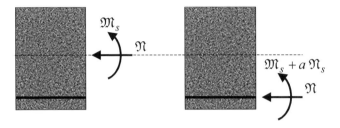

Let $\mathfrak{M}_R$ be the moment of resistance also defined in relation to the neutral fiber. An intuitive safety coefficient is then defined by the ratio:

$$K = \frac{\mathfrak{M}_R}{\mathfrak{M}_S}$$

Let us now consider these same forces relative to the reinforcement location at a distance a from the neutral fiber. The safety coefficient is then written as:

$$K = \frac{\mathfrak{M}_R + a\,\mathfrak{N}}{\mathfrak{M}_S + a\,\mathfrak{N}}$$

According to this equation, it is possible to demonstrate that if there is a value of a which gives $K > 1$, then $K > 1$ for all the values of a. On the other hand, the safety coefficient K varies between 1 (for $a \to \pm\infty$) and ∞ (for $a = -\mathfrak{M}_S/\mathfrak{N}$). Only the case of $K = 1$ does not vary in relation to a.   □

This example shows just how consideration of performance by safety coefficients is difficult to introduce into regulation. In fact, where the problem at hand is strictly identical, the approach leads us to define and calculate different safety coefficients. For this, then, we need to explicitly define the mathematical expressions of the calculation process. It is a pity that performance recommendations are not independent of the mathematical expressions relating to the calculation variables.

The semi-probabilistic format also shows no invariance. In fact, for the same limit state, the partial factors are not unique.

EXAMPLE 3.2.– Let us, then, use the same example and presume that the applied loads are distributed in loads related to permanent and live loads: $\mathfrak{M}_S = \mathfrak{M}_{S,P} + \mathfrak{M}_{S,E}$

and $\mathfrak{N}_S = \mathfrak{N}_{S,P} + \mathfrak{N}_{S,E}$. On the neutral axis, the semi-probabilistic format should be written as:

$$\frac{\mathfrak{M}_R}{\gamma_R} = \gamma_P \mathfrak{M}_{S,P} + \gamma_E \mathfrak{M}_{S,E}$$

On the reinforcement level, it is expressed by:

$$\frac{\mathfrak{M}_R + a \mathfrak{N}_{S,P} + a \mathfrak{N}_{S,E}}{\gamma_R} = \gamma_P \left(\mathfrak{M}_{S,P} + a \mathfrak{N}_{S,P} \right) + \gamma_E \left(\mathfrak{M}_{S,E} + a \mathfrak{N}_{S,E} \right)$$

By subtracting the first equation from the second, we obtain:

$$\left(\frac{1}{\gamma_R} - \gamma_P \right) \mathfrak{N}_{S,P} + \left(\frac{1}{\gamma_R} - \gamma_E \right) \mathfrak{N}_{S,E} = 0$$

Generally, $\mathfrak{N}_{S,P} > 0, \mathfrak{N}_{S,E} > 0$; the previous condition holds if and only if $\gamma_P \leq \dfrac{1}{\gamma_R} \leq \gamma_E$ or $\gamma_E \leq \dfrac{1}{\gamma_R} \leq \gamma_P$. With the exception of the case $\gamma_P = \gamma_E = \gamma_R = 1$, these two expressions are inconsistent with the convention $\gamma_P > 0, \gamma_E > 0, \gamma_R > 0$ in order to increase load effects and reduce strength. $\qquad\qquad\qquad\square$

This lack of invariance in relation to formulations of resistance and loads for a given problem are a forceful driving force in the development of alternative approaches for considering performance in regulations. This is especially the case for the probabilistic approach of the reliability theory which will be developed in the next section.

3.4. Reliability theory

3.4.1. *Basic problem*

In a probabilistic procedure, the effect S applied to a structural element and R, the resistance variable, are described as being random, since their values are not perfectly known. If the verification of the criterion related to the limit state results in inequality:

$$R \leq S \qquad\qquad\qquad\qquad\qquad\qquad\qquad\qquad\qquad\qquad [3.12]$$

with the component's failure being related to exceeding this limit state, then the probability P_f of the event $(R-S)$ will characterize the component's level of reliability in relation to the considered limit state:

$$P_f = P(R < S) \tag{3.13}$$

This probability is called the *failure probability*.

3.4.2. *Convolution integral*

Let us say that variables R and S are random variables, respectively with probability density functions $f_R(r)$ and $f_S(s)$. If these variables are independent, the joint probability density function $f_{RS}(r,s)$ verifies the property (see Chapter 1):

$$f_{RS}(r,s) = f_R(r) \ f_S(s) \tag{3.14}$$

which gives the failure probability:

$$\begin{aligned}
P_f &= P(R - S \le 0) \\
&= \int_D f_{RS}(r,s) \ dr \ ds \\
&= \int_{-\infty}^{+\infty} \int_{-\infty}^{s \ge r} f_R(r) \ f_S(s) \ dr \ ds
\end{aligned} \tag{3.15}$$

where D defines the *failure domain*, meaning the set of couples (r,s) such as $r \le s$ (Figure 3.2). If $F_R(r) = \int_{-\infty}^{r} f_R(y) \ dy$ is the cumulative distribution function of variable R, then equation [3.15] is written:

$$\begin{aligned}
P_f &= P(R - S \le 0) \\
&= \int_{-\infty}^{+\infty} F_R(s) \ f_S(s) \ ds
\end{aligned} \tag{3.16}$$

An alternative expression for the failure probability is:

$$P_f = \int_{-\infty}^{+\infty} [1 - F_S(r)] \ f_R(r) \ dr \tag{3.17}$$

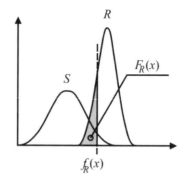

Figure 3.2. *Calculation of the failure probability*

Equation [3.17] is a convolution integral whose meaning is given in Figure 3.2. $F_R(s)$ is the probability of the resistance R being lower than s; the term $f_S(s)$ represents the probability of load effect S having a value between s and $s + \Delta s$ when Δs tends towards 0. By considering the value of s (i.e. by taking the integral on all the s values), the total failure probability is directly obtained.

3.4.3. *Normal variables*

Very few distributions of R and S lead to a direct integration of the convolution integral [3.17]. The simplest example introduces two normal variables with respective means and variances $m_R, V_R = \sigma_R^2$ and $m_S, V_S = \sigma_S^2$. Since the difference of two normal variables is a normal variable, the *safety margin* $M = R - S$ is a random variable with a mean $m_M = m_R - m_S$ and a variance $V_M = \sigma_M^2 = \sigma_R^2 + \sigma_S^2$. The failure probability is then given by:

$$P_f = \mathcal{P}(R - S \leq 0) = \mathcal{P}(M \leq 0)$$
$$= F_M(0)$$

[3.18]

when $F_M(.)$ is the cumulative probability function of variable M. Since we are dealing with a normal variable, we obtain:

$$P_f = F_M(0)$$
$$= \Phi\left(\frac{0 - m_M}{\sigma_M}\right) = \Phi\left(-\frac{m_R - m_S}{\sqrt{\sigma_R^2 + \sigma_S^2}}\right)$$

[3.19]

with $\Phi(.)$ as the cumulative probability function of a standard normal variable (zero mean and unitary standard deviation). The term:

$$\beta = \frac{m_M}{\sigma_M} = \frac{m_R - m_S}{\sqrt{\sigma_R^2 + \sigma_S^2}} \qquad\qquad [3.20]$$

is called the *reliability index*. This index is the inverse value of the coefficient of variation of variable M. It expresses the standard deviation between the mean value of M and 0. Figure 3.3 is called the *Warner diagram*. The larger this index is, the more safety is needed and therefore the failure probability is smaller. The relationship between failure probability and the reliability index is illustrated in Table 3.1. This table shows the importance of taking into account at least two significant figures to give a reliability index value. In fact, a difference of 0.5 points on the reliability index may lead to the failure probability being divided (or multiplied) by ten!

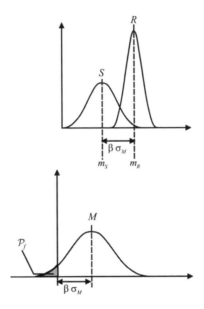

Figure 3.3. *Failure probability and reliability index*

β	0.0	1.28	2.33	3.09	3.72	4.26	4.75	5.20	5.61	6.00	6.36
P_f	$5\,10^{-1}$	10^{-1}	10^{-2}	10^{-3}	10^{-4}	10^{-5}	10^{-6}	10^{-7}	10^{-8}	10^{-9}	10^{-10}

Table 3.1. *Relationship between the reliability index and the failure probability*

EXAMPLE 3.3.– Let us consider a simply supported beam, subjected to a load of $Q = 3$ kN at mid-span with a 10% coefficient of variation. The beam is 5 m long and the resisting bending is $R = 10$ kN.m with a 15% coefficient of variation. The applied bending moment at mid-span is $S = Q\ L/4$. Therefore, we obtain:

$$\begin{cases} m_S = 3 \times \dfrac{5}{4} = 3.75 \ \text{kN.m} \\[2mm] \sigma_S = \left(\dfrac{5}{4}\right) \times 0.1 \times 3 = 0.375 \ \text{kN.m} \end{cases}$$

The reliability index is therefore given by:

$$\beta = \frac{m_R - m_S}{\sqrt{\sigma_R^2 + \sigma_S^2}} = \frac{10 - 3.75}{\sqrt{0.375^2 + 10^2 \times 0.15^2}} = \frac{6.25}{1.5461} = 4.04$$

Let us assume that the dead load from the design rules is deterministic and equal to $P = 1$ kN/m. The applied bending moment is therefore $S = PL^2/8 = 3.125$ kN.m. If this deterministic load is simultaneously applied to load Q, we obtain:

$$\beta = \frac{m_R - m_S}{\sqrt{\sigma_R^2 + \sigma_S^2}} = \frac{10 - 3.75 - 3.125}{\sqrt{0.375^2 + 10^2 \times 0.15^2}} = \frac{3.125}{1.5461} \approx 2.02$$

with S as the sum of the two applied load effects, in the present case.   □

3.4.4. *Geometric expression of the reliability index*

Let us transform each of the random variables R and S into standard normal variables R_0 and S_0:

$$\begin{cases} R = m_R + \sigma_R \ R_0 \\ S = m_S + \sigma_S \ S_0 \end{cases}$$ [3.21]

The safety margin M is expressed according to these new variables:

$$M = R - S = m_R - m_S + \sigma_R \ R_0 - \sigma_S \ S_0$$ [3.22]

The problem of calculating the failure probability is transformed from the *physical variable space* (Figure 3.4) described by variables (R, S) into the *standardized Gaussian space* (R_0, S_0) (Figure 3.5).

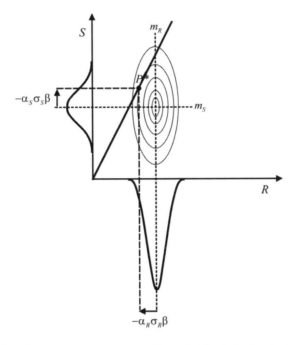

Figure 3.4. *Geometric representation of the reliability index in physical space*

The initial limit state $R - S = 0$ is then transformed into the limit state $m_R - m_S + \sigma_R\ R_0 - \sigma_S\ S_0 = 0$. Let us focus on this new limit state by calculating the point which is the closest to the origin in the standardized Gaussian space.

$$P_0^* \left| \begin{array}{l} -\dfrac{m_R - m_S}{\sigma_R^2 + \sigma_S^2}\sigma_R \\[2mm] \dfrac{m_R - m_S}{\sigma_R^2 + \sigma_S^2}\sigma_S \end{array} \right.$$

[3.23]

This point is at distance:

$$\dfrac{m_R - m_S}{\sqrt{\sigma_R^2 + \sigma_S^2}}$$

[3.24]

from the origin. This distance is none other than the reliability index β. The point P_0^* is called the *design point* in the standardized space (Figure 3.5).

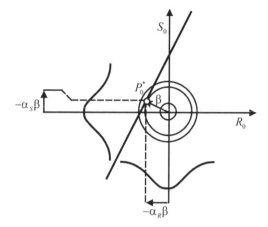

Figure 3.5. *Geometric representation of the reliability index in the standardized space*

At this point, the orthonormal vector on the limit state in the standardized Gaussian space $\{\alpha\}$, directed towards the safety domain, means that we can quantify the relative weight of the variables in the failure probability calculation:

$$\begin{cases} \alpha_R = \dfrac{\sigma_R}{\sqrt{\sigma_R^2 + \sigma_S^2}} \\[2ex] \alpha_S = -\dfrac{\sigma_S}{\sqrt{\sigma_R^2 + \sigma_S^2}} \end{cases}$$ [3.25]

We therefore obtain:

$$\overrightarrow{OP_0^*} = -\beta \begin{Bmatrix} \alpha_R \\ \alpha_S \end{Bmatrix} = -\beta \{\alpha\}$$ [3.26]

The point P_0^* is the point on the limit state qualified as the *most likely*, since the joint density value $\varphi_{R_0 S_0 (...)}$ is the highest. When brought back into the physical space, the design point is defined by:

$$P^* \begin{vmatrix} \dfrac{m_R \; \sigma_S^2 + m_S \; \sigma_R^2}{\sigma_R^2 + \sigma_S^2} \\[2ex] \dfrac{m_R \; \sigma_S^2 + m_S \; \sigma_R^2}{\sigma_R^2 + \sigma_S^2} \end{vmatrix}$$ [3.27]

3.4.5. *Joint distribution representation*

A different representation of the linear problem with two normal variables is given in Figure 3.6. In this representation, variables R and S are associated with the marginal distributions of the joint variable (R,S). This variable has the function $f_{R,S}(r,s)$ for density. This function takes the shape of a 2D bell, which is split into two sections, respectively corresponding to the failure domain and the safety domain. In this case, we are referring to the bivariate normal distribution (see Chapter 1). The failure probability is, then, given by equation [3.15].

Let us once again consider the basic case. With the variables being independent, the density of the joint distribution for R and S is simply written as the product of the densities of marginal distributions, i.e. from $f_R(r)$ and $f_S(s)$:

$$f_{R,S}(r,s) = f_R(r)f_S(s)$$
$$= \frac{1}{\sqrt{2\pi}\,\sigma_R}\exp\left(-\frac{(r-m_S)^2}{2\,\sigma_R^2}\right)\frac{1}{\sqrt{2\pi}\,\sigma_S}\exp\left(-\frac{(s-m_S)^2}{2\,\sigma_S^2}\right) \qquad [3.28]$$

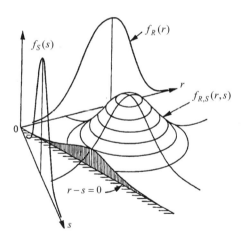

Figure 3.6. *Representation by a joint distribution*

3.4.6. *Limit state with more than two uncorrelated normal variables*

The formalism of the previous sections can be easily generalized for linear limit states which introduce more than two random normal uncorrelated variables. Let us take a linear limit state characterized by a failure hyperplane.

$$a_0 + \sum_{i=1}^{n} a_i\, X_i = 0 \qquad\qquad\qquad [3.29]$$

The safety margin $M = a_0 + \sum_{i=1}^{n} a_i\, X_i$ is still a normal variable with the mean and standard deviation:

$$\mathbb{E}(M) = a_0 + \sum_{i=1}^{n} a_i\, \mathbb{E}(X_i); \quad \mathbb{V}(M) = \sum_{i=1}^{n} a_i^2\, \mathbb{V}(X_i) \qquad [3.30]$$

because the variables are uncorrelated.

The reliability index is written:

$$\beta = \frac{\mathbb{E}(M)}{\sqrt{\mathbb{V}(M)}} = \frac{a_0 + \displaystyle\sum_{i=1}^{n} a_i\, \mathbb{E}(X_i)}{\sqrt{\displaystyle\sum_{i=1}^{n} a_i^2\, \mathbb{V}(X_i)}} \qquad\qquad [3.31]$$

By geometric interpretation, the direction cosines $\left(\alpha_i\right)_{1 \le i \le n}$ are given by the components of the orthonormal vector directed towards the safety domain of the limit state expressed in the standardized space:

$$\left(U_i = \frac{X_i - \mathbb{E}(X_i)}{\sigma_{X_i}} \right)_{1 \le i \le n} :$$

$$a_0 + \sum_{i=1}^{n} a_i\, X_i = a_0 + \sum_{i=1}^{n} a_i \left(\mathbb{E}(X_i) + \sigma_{X_i} U_i \right)$$

$$= \left(a_0 + \sum_{i=1}^{n} a_i\, \mathbb{E}(X_i) \right) + \sum_{i=1}^{n} a_i\, \sigma_{X_i} U_i = 0$$

[3.32]

We then obtain:

$$\alpha_i = \frac{a_i\,\sigma_{X_i}}{\sqrt{\sum_{i=1}^{n}\left(a_i\,\sigma_{X_i}\right)^2}} \qquad\qquad [3.33]$$

EXAMPLE 3.4.– A bridge is subjected to permanent loads, assessed at 3,223 kN.m. We presume that the resisting bending moment is normally distributed with a bias of 1.1, on the characteristic resistance of 11,395 kN.m with a 12% coefficient of variation.

We accept that the moment due to the permanent loads is normal distributed with a bias of 1.0 and a 10% coefficient of variation, and that the moment due to traffic loads is assessed at 2,600 kN.m (coefficient of variation 0.2 and normal distribution). Using these hypotheses and the assumption of independence between the different variables, the reliability index is worth:

$$\begin{aligned}\beta &= \frac{m_R - m_{S_P} - m_{S_T}}{\sqrt{\sigma_R^2 + \sigma_{S_P}^2 + \sigma_{S_E}^2}} \\ &= \frac{1.1\times 11,395 - 3,223 - 2,600}{\sqrt{\left(1.1\times 11,395\times 0.12\right)^2 + \left(3,223\times 0.1\right)^2 + \left(2,600\times 0.2\right)^2}} \\ &\approx \frac{6,711.5}{1,623.8} = 4.13\end{aligned}$$

The direction cosines are worth, respectively:

$$\alpha_R = \frac{\sigma_R}{\sqrt{\sigma_R^2 + \sigma_{S_P}^2 + \sigma_{S_T}^2}} \approx \frac{1.1\times 11,395\times 0.12}{1,623.8} = 0.926$$

$$\alpha_{S_P} = -\frac{\sigma_{S_P}}{\sqrt{\sigma_R^2 + \sigma_{S_P}^2 + \sigma_{S_T}^2}} \approx -\frac{3,223\times 0.1}{1,623.8} = -0.198$$

$$\alpha_{S_T} = -\frac{\sigma_{S_E}}{\sqrt{\sigma_R^2 + \sigma_{S_P}^2 + \sigma_{S_T}^2}} \approx -\frac{2,600\times 0.2}{1,623.8} = -0.320$$

The values of these direction cosines imply that the resisting moment and the live (traffic) load are the most important variables when calculating failure probability. □

3.4.7. *Limit state with correlated variables*

Calculating the reliability index for a linear safety margin with normal correlated variables may be carried out identically to the previous case, by writing:

$$\mathbb{V}(M) = \sum_{i=1}^{n}\sum_{j=1}^{n} a_i\, a_j\, \mathbb{COV}(X_i) \qquad\qquad [3.34]$$

Therefore the reliability index is given by:

$$\beta = \frac{\mathbb{E}(M)}{\sqrt{\mathbb{V}(M)}} = \frac{a_0 + \displaystyle\sum_{i=1}^{n} a_i\, \mathbb{E}(X_i)}{\sqrt{\displaystyle\sum_{i=1}^{n}\sum_{j=1}^{n} a_i\, a_j\, \mathbb{COV}(X_i)}} \qquad\qquad [3.35]$$

Obtaining the direction cosines is more difficult, because it requires us to write the limit state in a space with standard normal variables. For this, we must transform the correlated standardized variables $\{X_i'\} = \left\{\dfrac{X_i - \mathbb{E}(X_i)}{\sigma_{X_i}}\right\}$ into reduced uncorrelated standard normal variables $\{U\}$:

$$\{X'\} = [B]\{U\} \qquad\qquad [3.36]$$

The terms of the covariance matrix $\{X'\}$ are therefore related to the covariance matrix $\{U\}$ by the ratio:

$$\begin{aligned}
\left[\mathbb{COV}\left(\{X'\}\right)\right] &= \mathbb{E}\left[\{X'\}\ {}^t\{X'\}\right] \\
&= \mathbb{E}\left[[B]\{U\}\ {}^t\{U\}\ {}^t[B]\right] = [B]\,\mathbb{E}\left[\{U\}\ {}^t\{U\}\right]\ {}^t[B] \qquad [3.37] \\
&= [B]\ {}^t[B]
\end{aligned}$$

since the variables $\{U\}$ are uncorrelated and have unitary variance. The decomposition of covariance matrix $\{X'\}$ is similar to the Choleski decomposition. This is why it is mostly used for generating matrix $[B]$, it therefore allows the property of being a lower triangle.

We will finally note that the covariance matrix $\left[\text{COV}\left(\{X'\}\right)\right]$ is nothing other than the correlation matrix:

$$
\begin{aligned}
\text{COV}\left(X_i', X_j'\right) &= \mathbb{E}\left[\left(\frac{X_i - \mathbb{E}[X_i]}{\sigma_{X_i}}\right)\left(\frac{X_j - \mathbb{E}[X_j]}{\sigma_{X_j}}\right)\right] \\
&= \frac{\mathbb{E}\left[\left(X_i - \mathbb{E}[X_i]\right)\left(X_j - \mathbb{E}[X_j]\right)\right]}{\sigma_{X_i}\,\sigma_{X_j}} \\
&= \frac{\text{COV}\left(X_i, X_j\right)}{\sigma_{X_i}\,\sigma_{X_j}} = \rho_{X_i X_j}
\end{aligned}
\qquad [3.38]
$$

By introducing the correlation coefficients into the reliability index calculation, equation [3.35] becomes:

$$
\beta = \frac{a_0 + \displaystyle\sum_{i=1}^{n} a_i\,\mathbb{E}(X_i)}{\sqrt{\displaystyle\sum_{i=1}^{n}\sum_{j=1}^{n} a_i\,a_j\,\rho_{X_i X_j}\,\sigma_{X_i}\,\sigma_{X_j}}}
\qquad [3.39]
$$

EXAMPLE 3.5.– Let us take example 3.4. The resistance has been taken to be independent of the permanent loads. But, in truth, the resisting moment and permanent loads can be correlated positively (the structure's weight tends to increase with resistance). The variables are therefore taken as being correlated, as follows:

$$
\rho_{R,S_p} = 0.8; \quad \rho_{R,S_T} = 0.0; \quad \rho_{S_T,S_p} = 0.0
$$

The reliability index is therefore:

$$
\begin{aligned}
\beta &= \frac{m_R - m_{S_p} - m_{S_T}}{\sqrt{\sigma_R^2 + \sigma_{S_p}^2 + \sigma_{S_T}^2 + 2\rho_{R,S_p}\sigma_R\sigma_{S_p}}} \\
&\approx \frac{6,711.5}{1,874.3} = 3.58
\end{aligned}
$$

□

3.5. General formulation

3.5.1. *Failure component – failure mode*

Let us consider a structural element displaying perfect elastoplastic behavior, subjected to a compressive axial force F and to a torsion moment M (Figure 3.7). With the beam being compressed, it may buckle which implies failure at mid-span (maximum amplitude displacement). It may also fail due to shearing, attributed to an excessive torsion moment. This example allows us to define that which, in structural reliability, will be known as the *failure mode*. A *failure component* is defined by:

– a structural element which describes the geometry and mechanical properties, i.e. the place of the physical phenomenon;

– a set of loads or load effects;

– a failure mode and a model binding together the load effects and the resistance properties in a deterministic way;

– a characterization (deterministic or probabilistic) of all the variables and parameters of the previous model.

Figure 3.7. *Examples of failure modes*

This definition underlines that a structural element may have many failure modes.

3.5.2. *Safety margins – limit state functions*

A failure mode involves defining a failure criterion traditionally known as the *limit state function* or *safety margin*. This function, or margin, contains the basic variables characterizing the failure component's properties (resistance) and its environment (loads or load effects). If $\{Z\}$ is the vector of the n basic variables Z_i, then the limit state $g\left(\{Z\}\right)$ distinguishes three states:

- $g\left(\{Z\}\right) > 0$: safety;

- $g\left(\{Z\}\right) = 0$: limit state;

- $g\left(\{Z\}\right) < 0$: failure.

The basic or physical variables define a space with dimension n, and the limit state function describes a hypersurface with dimension $n-1$ and divides the space into a safety domain D_S and a failure domain D_R (Figure 3.8). The limit state does not only include the structural characteristics which guarantee the carrying-capacity for the various load effects (for a predefined period), but also the serviceability or durability criteria. In practice, these state functions call upon many parameters or variables, which are difficult to apprehend, vary over time and are not necessarily independent.

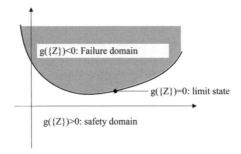

Figure 3.8. *Limit state – safety domain – failure domain*

3.5.3. *Calculation methods*

If $M = g\left(\{Z\}\right)$ is the safety margin for a failure component, the *failure probability* is a global measure determined by the density integral of the physical variables' joint probability density function over the failure domain D_R:

$$P_f = \mathcal{P}\left(g\left(\{Z\}\right) \leq 0\right) = \int_{D_R} f_Z\left(z_1,....,z_n\right) dz_1 \cdots dz_n \qquad [3.40]$$

If the physical or basic variables are independent, the joint probability density function is equal to the product of the marginal density functions for each variable:

$$f_Z\left(z_1,...,z_n\right) = \prod_{i=1}^{n} f_{Z_i}\left(z_i\right) = f_{Z_1}\left(z_1\right) \cdots f_{Z_n}\left(z_n\right) \qquad [3.41]$$

Assessing these integrals is often a difficult task, except for the particular cases of linear limit state functions. Various methods can be used. It is traditional practice to group these into *methods* called *level II and III*[1].

The *level III methods* are based on simulation techniques which are often very time consuming for calculations. The Monte-Carlo simulation methods are, for example, easy to use but require a powerful computer for the type of problems encountered in structure calculations. With the Monte-Carlo simulation methods, the probability density function and the statistical parameters associated with the safety margin are estimated. Random sampling (carried out using a pseudo-random generator available on most computers) is used to obtain a sample set of the random vector $\{Z\}$. The safety margin value is, then, assessed for this sample set in order to establish whether failure has occurred. This process is repeated many times and the failure probability is estimated from the fraction of the sample values, leading to the failure being divided by the total number of trials. Following this approach, the failure probability is evaluated directly without an algorithm. The process described here is the most direct but also the most approximate. It requires large samples in order to estimate, with a sufficient degree of confidence, the failure probability. Let us note here that the required sample (and the corresponding number of trials) increases when the failure probability decreases. There are simple rules to be found, such as $N > C/P_f$ where N is the dimension of the required sample, and C is a constant related to the level of confidence and the function type to be assessed[2]. A typical value for C is 1,000. The aim of more advanced simulation methods is to reduce the size of the sample necessary for assessing failure probability.

These simulation methods have been developed over recent years and are today, used instead of, or as well as, the methods approached here. The need for a combined usage has been realized in cases where it has become important to check the exactitude of the approached methods, such as when studying failure with multiple modes, or analyzing the reliability of many elements.

Since it is very difficult to manipulate level III methods, this means that we often use approximations, hence the existence of level II methods, which seek to approach the failure probability.

3.5.4. *Basler-Cornell index*

The principle of level II methods is to systematically replace the limit state by a surface approximation. The most intuitive is the hyperplane (in the case of limit state

---

1 Level I relates to the semi-probabilistic format.
2 The sought failure probabilities are generally in the order of 10^{-6}.

functions introducing many variables). Let us therefore consider a limit state function $g(z_1,\ldots,z_n)$ with n random variables. The limit state function is replaced by its first order Taylor expansion at $\{z^*\} = (z_1^*,\ldots,z_n^*)$:

$$g(\{z\}) \approx g(\{z^*\}) + \sum_{i=1}^{n} \left.\frac{\partial g(\{z\})}{\partial z_i}\right|_{\{z^*\}} (z_i - z_i^*) \qquad [3.42]$$

The mean and variance values of variable $M = g(Z_1,\ldots,Z_n)$ are calculated on the one hand from equation [3.42], and on the other hand, by performing a linearization at $\{z^*\}$:

$$\mathbb{E}(M) \approx g(\{z^*\}) + \sum_{i=1}^{n} \left.\frac{\partial g(\{z\})}{\partial z_i}\right|_{\{z^*\}} \left(\mathbb{E}(Z_i) - z_i^*\right) \qquad [3.43]$$

$$\mathbb{V}(Z) \approx \sum_{i=1}^{p}\sum_{j=1}^{p} \left.\frac{\partial g}{\partial x_i}\right|_{\{z^*\}} \left.\frac{\partial g}{\partial x_j}\right|_{\{z^*\}} \mathbb{COV}(Z_i,Z_j) \qquad [3.44]$$

The Basler-Cornell reliability index is, then, defined by:

$$\beta_c = \frac{\mathbb{E}(M)}{\sqrt{\mathbb{V}(M)}} = \frac{g(\{z^*\}) + \sum_{i=1}^{n} \left.\dfrac{\partial g(\{z\})}{\partial z_i}\right|_{\{z^*\}} \left(\mathbb{E}(Z_i) - z_i^*\right)}{\sqrt{\sum_{i=1}^{p}\sum_{j=1}^{p} \left.\dfrac{\partial g}{\partial x_i}\right|_{\{z^*\}} \left.\dfrac{\partial g}{\partial x_j}\right|_{\{z^*\}} \mathbb{COV}(Z_i,Z_j)}} \qquad [3.45]$$

The point with mean values $(\mathbb{E}(Z_1),\ldots,\mathbb{E}(Z_n))$ is one of the most intuitive linearization points, thus giving:

$$\beta_c(\{\mathbb{E}(Z)\}) = \frac{g(\mathbb{E}(\{Z\}))}{\sqrt{\sum_{i=1}^{p}\sum_{j=1}^{p} \left.\dfrac{\partial g}{\partial x_i}\right|_{\mathbb{E}(Z)} \left.\dfrac{\partial g}{\partial x_j}\right|_{\mathbb{E}(Z)} \mathbb{COV}(Z_i,Z_j)}} \qquad [3.46]$$

This approach is called the MVFOSM method (*mean value first order second moment*).

EXAMPLE 3.6.– Let us consider the case of a reinforced concrete beam subjected to a bending moment M_a. The ultimate resisting moment is given by the expression:

$$\mathfrak{M} = A\,f_y\left(d - k\,\frac{A\,f_y}{\sigma_b\,b}\right)$$

where A, d, b, f_y, σ_b respectively are the steel section, the beam height, the section's width, the steel strength and the concrete compressive strength. k is a coefficient which depends on the shape of the stress distribution on the compressed concrete section. These variables are taken to be independent. The limit state function is:

$$M = g(A, f_y, k, \sigma_b, b, M_a) = \mathfrak{M} - M_a = A\,f_y\left(d - k\,\frac{A\,f_y}{\sigma_b\,b}\right) - M_a$$

By applying equation [3.45] and taking into account the independence of the variables, we obtain:

$$\beta_c = \frac{\mathbb{E}(A)\,\mathbb{E}(f_y)\left(\mathbb{E}(d) - \mathbb{E}(k)\,\dfrac{\mathbb{E}(A)\,\mathbb{E}(f_y)}{\mathbb{E}(\sigma_b)\,\mathbb{E}(b)}\right) - \mathbb{E}(M_a)}{\sqrt{\mathbb{V}(M)}}$$

with:

$$\mathbb{V}(M) \approx \mathbb{E}(A)^2\,\mathbb{E}(f_y)^2\,\mathbb{V}(d) + \left(\mathbb{E}(A)\,\mathbb{E}(d) - 2\,\mathbb{E}(k)\,\frac{\mathbb{E}(A)^2\,\mathbb{E}(f_y)}{\mathbb{E}(\sigma_b)\,\mathbb{E}(b)}\right)^2\,\mathbb{V}(f_y)$$

$$+ \left(\mathbb{E}(f_y)\,\mathbb{E}(d) - 2\,\mathbb{E}(k)\,\frac{\mathbb{E}(f_y)^2\,\mathbb{E}(A)}{\mathbb{E}(\sigma_b)\,\mathbb{E}(b)}\right)^2\,\mathbb{V}(A)$$

$$+ \left(\mathbb{E}(k)\,\frac{\mathbb{E}(f_y)^2\,\mathbb{E}(A)^2}{\mathbb{E}(\sigma_b)\,\mathbb{E}(b)}\right)^2\left(\frac{\mathbb{V}(b)}{\mathbb{E}(b)^2} + \frac{\mathbb{V}(\sigma_b)}{\mathbb{E}(\sigma_b)^2}\right) + \mathbb{V}(M_a) \qquad \square$$

EXAMPLE 3.7.– Fatigue damage of a joint is assessed by the Miner law, $D = \sum_{i=1}^{N} \dfrac{n_i}{N_i}$,

where n_i and N_i respectively are the number of stress cycles and the number of failure cycles for the stress variation S_i. The number of failure cycles is given by the mean Wöhler curve: $N_i = A\, S^{-3}$. We can write the Miner's damage as follows:

$$D = \sum_{i=1}^{N}\frac{n_i}{N_i} = \sum_{i=1}^{N}\frac{n_i}{A\,S_i^{-3}} = \frac{1}{A}\sum_{i=1}^{N}\frac{\gamma_i N}{S_i^{-3}} = \frac{N}{A}\sum_{i=1}^{N}\gamma_i\,S_i^3 = \frac{N}{A}S_e^3$$

S_e is called the equivalent extended strain and $N = \sum_i n_i$ is the total number of stress cycles. Fatigue assessment consists of ensuring that $D = \dfrac{N}{A}S_e^3 \geq \Delta$ where Δ is a maximum damage threshold (generally close to 1).

The Basler-Cornell index for this limit state is given by:

$$\beta_c = \frac{\mathbb{E}(\Delta) - \dfrac{\mathbb{E}(N)}{\mathbb{E}(A)}\mathbb{E}(S_e)^3}{\sqrt{\mathbb{V}(\Delta)+\left(\dfrac{\mathbb{E}(S_e)^3}{\mathbb{E}(A)}\right)^2\mathbb{V}(N)+\left(\dfrac{3\mathbb{E}(S_e)^2\mathbb{E}(N)}{\mathbb{E}(A)}\right)^2\mathbb{V}(S_e)+\left(\dfrac{\mathbb{E}(S_e)^3\mathbb{E}(N)}{\mathbb{E}(A)^2}\right)^2\mathbb{V}(A)}}$$

We assume that $S_e = 53$ MPa , that A and Δ have respective means and standard deviations of 2.51×10^{11} MPa and 1.0, and 7.53×10^{10} MPa and 0.3, and that the total number of cycles is related to the year t by the relationship $N = 30{,}660\,t$. The Cornell index is simplified as:

$$\beta_c = \frac{\mathbb{E}(\Delta) - \dfrac{\mathbb{E}(N)}{\mathbb{E}(A)}\mathbb{E}(S_e)^3}{\sqrt{\mathbb{V}(\Delta)+\left(\dfrac{\mathbb{E}(S_e)^3\mathbb{E}(N)}{\mathbb{E}(A)^2}\right)^2\mathbb{V}(A)}}$$

$$= \frac{1 - \dfrac{30{,}660\,t}{2.51\,10^{11}} 53^3}{\sqrt{0.3^2 + \left(\dfrac{30{,}660\,t}{\left(2.51\,10^{11}\right)^2} 53^3\right)^2 \left(7.53\,10^{10}\right)^2}}$$

$$= \frac{1 - 0.018\,t}{\sqrt{0.09 + 3.03\,10^{-5}\,t^2}}$$

By accepting that the variables are lognormal (note that in the Cornell approach, this knowledge is exempt), it is more advisable to work on the following safety margin:

$$\hat{M} = \ln(\Delta) - \ln\left(\frac{N}{A} S_e^3\right) = \ln(\Delta) - \ln(N) - 3\ln(S_e) + \ln(A)$$

This limit state is linear and only introduces normal (for $\ln(\Delta)$ and $\ln(A)$), and deterministic variables (for $\ln(S_e)$ and $\ln(N)$). The reliability index is then simply given by equation [3.31]:

$$\beta = \frac{\mathbb{E}(\ln \Delta) - \mathbb{E}(\ln N) - 3\mathbb{E}(\ln S_e) + \mathbb{E}(\ln A)}{\sqrt{\mathbb{V}(\ln \Delta) + \mathbb{V}(\ln A)}}$$

The mean values and standard deviations of $\ln(\Delta)$ and $\ln(A)$ are given by the following properties (see Chapter 1):

$$\mathbb{E}(\ln X) = \ln\left(\frac{\mathbb{E}(X)}{\sqrt{1 + \dfrac{\mathbb{V}(X)}{\mathbb{E}(X)^2}}}\right); \quad \mathbb{V}(\ln X) = \ln\left(1 + \frac{\mathbb{V}(X)}{\mathbb{E}(X)^2}\right)$$

which gives $\mathbb{E}(\ln A) = 26.206$, $\mathbb{V}(\ln A) = 0.086$, $\mathbb{E}(\ln \Delta) = -0.043$, $\mathbb{V}(\ln \Delta) = 0.086$. The reliability index is written as:

$$\beta = \frac{-0.043 - \ln(30{,}660) - \ln t - 3\ln 53 + 26.206}{\sqrt{0.086 + 0.086}} = \frac{3.915 - \ln t}{0.415}$$

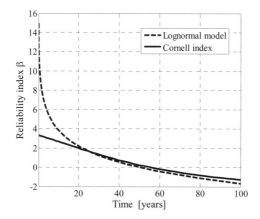

The above figure compares the differences between the two calculated reliability indexes. We note here that the Cornell index is bounded for the first few years, and underestimates the reliability in relation to the index calculated by the limit state logarithm. Conversely, it overestimates the indexes for the later years. We will see in example 3.11 that the Cornell index gives erroneous results.   □

Moreover, the Cornell index is not, unfortunately, an unvarying safety measure. The following example illustrates this.

EXAMPLE 3.8.– In fact, let us consider the limit state $g(\{Z\})^3 = 0$. The Cornell index for this limit state is therefore:

$$\beta_{g^3}\left(\mathbb{E}(\{Z\})\right) = \frac{g\left(\mathbb{E}(\{Z\})\right)^3}{\sqrt{\sum_{i=1}^{p}\sum_{j=1}^{p}\left.\frac{\partial g^3}{\partial x_i}\right|_{\mathbb{E}(Z)}\left.\frac{\partial g^3}{\partial x_j}\right|_{\mathbb{E}(Z)}\mathrm{COV}(Z_i,Z_j)}}$$

$$= \frac{g\left(\mathbb{E}(\{Z\})\right)^3}{\sqrt{\sum_{i=1}^{p}\sum_{j=1}^{p}3g(\mathbb{E}(\{Z\}))^2\left.\frac{\partial g}{\partial x_i}\right|_{\mathbb{E}(Z)}3g(\mathbb{E}(\{Z\}))^2\left.\frac{\partial g}{\partial x_j}\right|_{\mathbb{E}(Z)}\mathrm{COV}(Z_i,Z_j)}}$$

$$= \frac{1}{3}\frac{g\left(\mathbb{E}(\{Z\})\right)}{\sqrt{\sum_{i=1}^{p}\sum_{j=1}^{p}\left.\frac{\partial g}{\partial x_i}\right|_{\mathbb{E}(Z)}\left.\frac{\partial g}{\partial x_j}\right|_{\mathbb{E}(Z)}\mathrm{COV}(Z_i,Z_j)}} = \frac{1}{3}\beta_g\left(\mathbb{E}(\{Z\})\right)$$

Following the expression choice for the limit state, the obtained reliability indexes can, therefore, differ. This lack of invariance does not only come from the point of linearization. The factor $1/3$ exists independently of this problem.   □

Another choice consists of choosing a point on the failure surface. The Cornell index is then written:

$$\beta_c\{z^*\} = \frac{\displaystyle\sum_{i=1}^{n}\frac{\partial g(\{z\})}{\partial z_i}\bigg|_{\{z^*\}}\left(\mathbb{E}(Z_i)-z_i^*\right)}{\sqrt{\displaystyle\sum_{i=1}^{p}\sum_{j=1}^{p}\frac{\partial g}{\partial x_i}\bigg|_{\{z^*\}}\frac{\partial g}{\partial x_j}\bigg|_{\{z^*\}}\mathbb{COV}(Z_i,Z_j)}} \qquad [3.47]$$

The question, now, is knowing which point $\{z^*\}$ from the limit state to use.

EXAMPLE 3.9.– For a linearization point chosen on the limit state, it is easy to demonstrate that the previous index verifies the invariance condition:

$$\beta_{g^3}\left(\{z^*\}\right) = \frac{\displaystyle\sum_{i=1}^{n}\frac{\partial g^3}{\partial z_i}\bigg|_{\{z^*\}}\left(\mathbb{E}(Z_i)-z_i^*\right)}{\sqrt{\displaystyle\sum_{i=1}^{p}\sum_{j=1}^{p}\frac{\partial g^3}{\partial x_i}\bigg|_{\{z^*\}}\frac{\partial g^3}{\partial x_j}\bigg|_{\{z^*\}}\mathbb{COV}(Z_i,Z_j)}}$$

$$= \frac{\displaystyle\sum_{i=1}^{n}3g\left(\{z^*\}\right)^2\frac{\partial g}{\partial z_i}\bigg|_{\{z^*\}}\left(\mathbb{E}(Z_i)-z_i^*\right)}{\sqrt{\displaystyle\sum_{i=1}^{p}\sum_{j=1}^{p}3g\left(\{z^*\}\right)^2\frac{\partial g}{\partial x_i}\bigg|_{\mathbb{E}(Z)}3g\left(\{z^*\}\right)^2\frac{\partial g}{\partial x_j}\bigg|_{\mathbb{E}(Z)}\mathbb{COV}(Z_i,Z_j)}}$$

$$= \frac{\displaystyle\sum_{i=1}^{n}\frac{\partial g}{\partial z_i}\bigg|_{\{z^*\}}\left(\mathbb{E}(Z_i)-z_i^*\right)}{\sqrt{\displaystyle\sum_{i=1}^{p}\sum_{j=1}^{p}\frac{\partial g}{\partial x_i}\bigg|_{\{z^*\}}\frac{\partial g}{\partial x_j}\bigg|_{\{z^*\}}\mathbb{COV}(Z_i,Z_j)}} = \beta_g\left(\{z^*\}\right) \qquad \square$$

3.5.5. Hasofer-Lind index

In section 3.2 we saw that the reliability index, for the simple case of a linear limit state with normal independent variables, was the distance to the origin of a limit state expressed in a standardized space of standard normal variables. This principle was used by Hasofer and Lind in 1974 [HAS 74] for complex limit states in order to define an invariant reliability index.

The procedure consists of redefining the problem in terms of random independent standard normal variables U_i which define a U-space with dimension n. The transformation linking $\{U\}$ to $\{Z\}$, written $\{U\} = T(\{Z\})$ is called the *Rosenblatt transform* (see section 3.5.8). It is generated in such as way so that the failure probabilities are not modified. If the new failure surface is $g_U(\{U\}) = 0$, then we have:

$$P_f = P\big(g(\{Z\}) \leq 0\big) = P\big(g_U(\{U\}) \leq 0\big)$$ [3.48]

The point U^* from the surface closest to the origin $g_U(\{U\}) = 0$ is called the *design point*. In truth, this point is the point on the limit state with the maximum probability density value. Its distance to the origin is the reliability index, known as the *Hasofer-Lind index* (Figure 3.9).

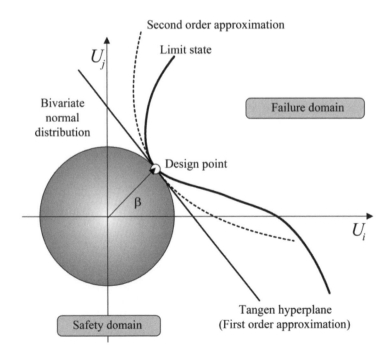

Figure 3.9. *Limit state – design point*

In the first order reliability method (FORM) the failure surface $g_U\left(\{U\}\right)=0$ is approximated by a hyperplane which is tangent to the design point (Figure 3.9). The equation of this first order approximation is given by the expression:

$$\tilde{g}_U\left(\{u\}\right)=g_U\left(\{u^*\}\right)+\sum_{i=1}^{n}\left.\frac{\partial g_U\left(\{u\}\right)}{\partial u_i}\right|_{\{u^*\}}\left(u_i-u_i^*\right)$$

$$=\sum_{i=1}^{n}\left.\frac{\partial g_U\left(\{u\}\right)}{\partial u_i}\right|_{\{u^*\}}\left(u_i-u_i^*\right)=0$$

[3.49]

The distance to the origin from this hyperplane is given by:

$$\beta=\frac{-\displaystyle\sum_{i=1}^{n}\left.\frac{\partial g_U\left(\{u\}\right)}{\partial u_i}\right|_{\{u^*\}}u_i^*}{\sqrt{\displaystyle\sum_{i=1}^{p}\left(\left.\frac{\partial g_U}{\partial u_i}\right|_{\{u^*\}}\right)^2}}$$

[3.50]

The Cornell index applied to the limit state $g_U\left(\{U\}\right)$ with linearization point $\{u^*\}$ gives:

$$\beta_c=\frac{\displaystyle\sum_{i=1}^{n}\left.\frac{\partial g_U\left(\{u\}\right)}{\partial u_i}\right|_{\{u^*\}}\left(\underbrace{\mathbb{E}(U_i)}_{=0}-u_i^*\right)}{\sqrt{\displaystyle\sum_{i=1}^{p}\sum_{j=1}^{p}\left.\frac{\partial g_U}{\partial u_i}\right|_{\{u^*\}}\left.\frac{\partial g_U}{\partial u_j}\right|_{\{u^*\}}\underbrace{\mathbb{COV}(Z_i,Z_j)}_{=0\ \text{si}\ i\neq j}}}$$

$$=\frac{-\displaystyle\sum_{i=1}^{n}\left.\frac{\partial g_U\left(\{u\}\right)}{\partial u_i}\right|_{\{u^*\}}u_i^*}{\sqrt{\displaystyle\sum_{i=1}^{p}\left(\left.\frac{\partial g_U}{\partial u_i}\right|_{\{u^*\}}\right)^2\underbrace{\sigma_{U_i}^2}_{=1}}}=\frac{-\displaystyle\sum_{i=1}^{n}\left.\frac{\partial g_U\left(\{u\}\right)}{\partial u_i}\right|_{\{u^*\}}u_i^*}{\sqrt{\displaystyle\sum_{i=1}^{p}\left(\left.\frac{\partial g_U}{\partial u_i}\right|_{\{u^*\}}\right)^2}}=\beta$$

[3.51]

The Cornell and Hasofer-Lind indexes are therefore identical if the point of linearization is effectively taken at the design point in the standardized normal space. If we set out:

$$\frac{\left.\dfrac{\partial g_U\left(\{u\}\right)}{\partial u_i}\right|_{\{u^*\}}}{\sqrt{\displaystyle\sum_{i=1}^{p}\left(\left.\dfrac{\partial g_U}{\partial u_i}\right|_{\{u^*\}}\right)^2}} = \alpha_i \qquad [3.52]$$

then according to [3.51]:

$$\{u^*\} = -\beta\{\alpha\} \qquad [3.53]$$

Here, the coefficients α_i have identical functions as the coefficients identified for linear limit states.

3.5.6. *Rackwitz-Fiessler algorithm*

We can use many algorithms to solve this minimization problem with constraints. The Rackwitz-Fiessler algorithm [RAC 78] is by far the most commonly used in reliability studies, not only because of its very simple mode of expression, but also for its good results. However, we must pay attention to the fact that there is no guarantee of convergence for all situations. There are, also, other techniques and we lead our readers to the reference [LEM 09] which presents these other methods.

The algorithm is implemented in the reduced variable space. We are moving away from an unspecified point $\{u^0\}$, e.g. the origin which is also the mean point in the standardized normal space, and we are linearizing the limit state function $g_U\left(\{u\}\right)$ at this point. We are then looking for the point $\{u^1\}$ of this surface which is closest to the origin, and the procedure is then repeated at point $\{u^1\}$.

The design point $\{u^*\}$ is therefore determined as the limit of a series of points $\{u^0\}, \{u^1\} \cdots \{u^k\}$. The orthonormal vector $\{\alpha^k\}$ of the direction cosines at the k-th step is given by the following expression:

$$\frac{\left.\dfrac{\partial g_U\left(\{u\}\right)}{\partial u_i}\right|_{\{u^k\}}}{\sqrt{\displaystyle\sum_{i=1}^{p}\left(\left.\dfrac{\partial g_U\left(\{u\}\right)}{\partial u_i}\right|_{\{u^k\}}\right)^2}} = \alpha_i^k \qquad [3.54]$$

The distance to the origin from the tangent hyperplane of $g_U\left(\{u\}\right)$ at $\{u^k\}$ is given by:

$$\beta^k = \frac{g_U\left(\{u\}^k\right) - \displaystyle\sum_{i=1}^{n}\left.\dfrac{\partial g_U\left(\{u\}\right)}{\partial u_i}\right|_{\{u^k\}} u_i^k}{\sqrt{\displaystyle\sum_{i=1}^{p}\left(\left.\dfrac{\partial g_U}{\partial u_i}\right|_{\{u^k\}}\right)^2}} \qquad [3.55]$$

which allows us to simply determine the point $\{u^{k+1}\}$:

$$\{u^{k+1}\} = -\beta^k \{\alpha^k\} \qquad [3.56]$$

The iterations are stopped as soon as some criteria (difference between two successive reliability indexes, design points etc.) are fulfilled.

3.5.7. *Isoprobability transformations*

The previous algorithm requires us to know the function $g_U\left(\{u\}\right)$ and therefore to be able to transform the set of basic/physical variables into independent standard normal variables. Obtaining the transformation $\{U\} = T\left(\{Z\}\right)$ is not direct, particularly if the variables are correlated. Out of all the existing transformations, the Nataf transform is particularly well used. It only requires us to know the first and second moments and the correlations. The marginal densities of the variables and the correlation matrix ρ_{ij} must therefore be known *a priori*. The principle of this method consists of considering a series of standard normal variables $\{Y\} = (Y_1,...,Y_n)^t$ issued from the marginal transformations [3.57], where Φ represents the cumulative probability function of the standard normal distribution:

$$y_i = \Phi^{-1}(F_{Z_i}(z_i)) \qquad [3.57]$$

It must be noted that the intermediate variables $\{Y\}$ are not *uncorrelated*. Their correlation is *a priori* different from the correlation of physical variables $\{Z\}$. The correlations of these intermediate variables ρ_{ij}^* are solutions of the integral equation [3.58], where ϕ_2 represents the density of the bivariate normal distribution:

$$\rho_{ij} = \int_{-\infty}^{+\infty} \int_{-\infty}^{+\infty} \frac{z_i - \mathbb{E}[Z_i]}{\sigma_{Z_i}} \frac{z_j - \mathbb{E}[Z_j]}{\sigma_{Z_j}} \phi_2(y_i, y_j, \rho_{ij}^*) \, dy_i \, dy_j \qquad [3.58]$$

In practice, empirical relationships providing acceptable estimates of the correlations for intermediate variables are used. The Choleski decomposition is thus performed on the correlation matrix of the intermediate variables. The coordinates of the physical variables in the standardized space can, then, be determined [LEM 09, MEL 99].

We will limit this chapter to the single case of independent random variables, which represents a large majority of cases. In this case, only the marginal transformations [3.57] are useful and necessary:

$$\Phi(u) = F_Z(z) = \Phi(T(z)) \qquad [3.59]$$

i.e.:

$$u = T(z) = \Phi^{-1}(F_Z(z)) \qquad [3.60]$$

Table 3.2 gives a few transformations of the most used random variables in reliability calculations.

Variable	Transformation
Normal of mean μ and standard deviation σ	$X = \mu + \sigma U$
Lognormal of mean μ and standard deviation σ	$X = \dfrac{\mu}{\sqrt{1 + \dfrac{\sigma^2}{\mu^2}}} \exp\left(U \sqrt{\ln\left[1 + \dfrac{\sigma^2}{\mu^2} \right]} \right)$
Exponential of parameter λ	$X = -\dfrac{1}{\lambda} \ln[\Phi(-U)]$
Gumbel-max distribution with parameters u and α	$X = u - \alpha \ln[-\ln[U]]$

U is a standard normal variable represented by a cumulative probability function noted Φ

Table 3.2. *Examples of isoprobability transformations*

EXAMPLE 3.10.– Let us consider a linear limit state with independent lognormal variables. We obtain (according to Table 3.2):

$$g_U(U_R, U_S) = \frac{\mathbb{E}(R)}{\sqrt{1 + \dfrac{\mathbb{V}(R)}{\mathbb{E}(R)^2}}} \exp\left(U_\Delta \sqrt{\ln\left[1 + \frac{\mathbb{V}(R)}{\mathbb{E}(R)^2}\right]}\right)$$

$$- \frac{\mathbb{E}(S)}{\sqrt{1 + \dfrac{\mathbb{V}(S)}{\mathbb{E}(S)^2}}} \exp\left(U_S \sqrt{\ln\left[1 + \frac{\mathbb{V}(S)}{\mathbb{E}(S)^2}\right]}\right)$$

By setting out $\mathbb{CDV}_R = \sqrt{\dfrac{\mathbb{V}(R)}{\mathbb{E}(R)^2}}, \mathbb{CDV}_S = \sqrt{\dfrac{\mathbb{V}(S)}{\mathbb{E}(S)^2}}$, we obtain:

$$g_U(U_R, U_S) = \frac{\mathbb{E}(R)}{\sqrt{1 + \mathbb{CDV}_R^2}} \exp\left(U_\Delta \sqrt{\ln\left[1 + \mathbb{CDV}_R^2\right]}\right)$$

$$- \frac{\mathbb{E}(S)}{\sqrt{1 + \mathbb{CDV}_S^2}} \exp\left(U_S \sqrt{\ln\left[1 + \mathbb{CDV}_S^2\right]}\right)$$

$$= \exp\left(U_\Delta \sqrt{\ln\left[1 + \mathbb{CDV}_R^2\right]} + \ln\left[\frac{\mathbb{E}(R)}{\sqrt{1 + \mathbb{CDV}_R^2}}\right]\right)$$

$$- \exp\left(U_S \sqrt{\ln\left[1 + \mathbb{CDV}_S^2\right]} + \ln\left[\frac{\mathbb{E}(S)}{\sqrt{1 + \mathbb{CDV}_S^2}}\right]\right)$$

The previous limit state may be replaced by the linear limit state:

$$G(U_R, U_S) = U_\Delta \sqrt{\ln\left[1 + \mathbb{CDV}_R^2\right]} + \ln\left[\frac{\mathbb{E}(R)}{\sqrt{1 + \mathbb{CDV}_R^2}}\right]$$

$$- U_S \sqrt{\ln\left[1 + \mathbb{CDV}_S^2\right]} - \ln\left[\frac{\mathbb{E}(S)}{\sqrt{1 + \mathbb{CDV}_S^2}}\right]$$

which makes it possible to apply the results for a linear limit state with standard normal variables:

$$\beta = \frac{\ln\left[\dfrac{\mathbb{E}(R)}{\sqrt{1+\mathbb{CDV}_R^2}}\right] - \ln\left[\dfrac{\mathbb{E}(S)}{\sqrt{1+\mathbb{CDV}_S^2}}\right]}{\sqrt{\ln\left[1+\mathbb{CDV}_R^2\right] + \ln\left[1+\mathbb{CDV}_S^2\right]}}$$

which gives for a lognormal distribution:

$$\mathbb{E}(\ln X) = \mathbb{E}(X)\exp(-\frac{\mathbb{V}(\ln X)}{2})$$
$$\mathbb{V}(\ln X) = \ln\left(\mathbb{CDV}(X)^2 + 1\right) \approx \mathbb{CDV}(X)^2 \quad \text{for} \quad \mathbb{CDV}(X) \ll 0,3$$

The reliability index then simplifies into:

$$\beta = \frac{\ln\left[\dfrac{\mathbb{E}(R)}{\sqrt{1+\mathbb{CDV}_R^2}}\right] - \ln\left[\dfrac{\mathbb{E}(S)}{\sqrt{1+\mathbb{CDV}_S^2}}\right]}{\sqrt{\ln\left[1+\mathbb{CDV}_R^2\right] + \ln\left[1+\mathbb{CDV}_S^2\right]}}$$

$$= \frac{\mathbb{E}(\ln R) - \mathbb{E}(\ln S)}{\sqrt{\sigma_{\ln R}^2 + \sigma_{\ln S}^2}} \approx \frac{\mathbb{E}(\ln R) - \mathbb{E}(\ln S)}{\sqrt{\mathbb{CDV}_R^2 + \mathbb{CDV}_S^2}}$$

We must note here that this result can be obtained without using an isoprobabilistic transformation, but instead by using the safety margin $\hat{M} = \ln R - \ln S$ which is, then, a linear limit state introducing two normal variables $\ln R$ and $\ln S$. □

EXAMPLE 3.11.– Let us consider again example 3.7 and let us express the limit state in the standardized normal space. We obtain (according to Table 3.2):

$$g_U(U_\Delta, U_A) = \frac{\mathbb{E}(\Delta)}{\sqrt{1+\dfrac{\mathbb{V}(\Delta)}{\mathbb{E}(\Delta)^2}}}\exp\left(U_\Delta\sqrt{\ln\left[1+\dfrac{\mathbb{V}(\Delta)}{\mathbb{E}(\Delta)^2}\right]}\right)$$

$$- \frac{N\,S_e^3}{\dfrac{\mathbb{E}(A)}{\sqrt{1+\dfrac{\mathbb{V}(A)}{\mathbb{E}(A)^2}}}\exp\left(U_A\sqrt{\ln\left[1+\dfrac{\mathbb{V}(A)}{\mathbb{E}(A)^2}\right]}\right)}$$

$$= 0.958\exp\left(0.294\,U_\Delta\right) - 0.019\,t\,\exp\left(-0.294\,U_A\right)$$

The partial derivatives are:

$$\frac{\partial g_U}{\partial U_A} = 0.019 \times 0.294\, t \exp(-0.294\, U_A) = 0.0056\, t\, \exp(-0.294\, U_A)$$

$$\frac{\partial g_U}{\partial U_\Delta} = 0.958 \times 0.294 \exp(0.294\, U_A) = 0.282\, \exp(0.294\, U_\Delta)$$

The tangent hyperplane to g_U at any point and the orthonormal vector are given by:

$$L_U\left(U_\Delta, U_A\right) = g_U\left(U_\Delta^*, U_A^*\right) + \left.\frac{\partial g_U}{\partial U_\Delta}\right|_{U_\Delta^*}\left(U_\Delta - U_\Delta^*\right) + \left.\frac{\partial g_U}{\partial U_A}\right|_{U_A^*}\left(U_A - U_A^*\right)$$

$$\alpha_{U_\Delta}^* = \frac{\left.\dfrac{\partial g_U}{\partial U_\Delta}\right|_{U_\Delta^*}}{\sqrt{\left(\left.\dfrac{\partial g_U}{\partial U_\Delta}\right|_{U_\Delta^*}\right)^2 + \left(\left.\dfrac{\partial g_U}{\partial U_A}\right|_{U_A^*}\right)^2}}\ ; \alpha_{U_A}^* = \frac{\left.\dfrac{\partial g_U}{\partial U_A}\right|_{U_A^*}}{\sqrt{\left(\left.\dfrac{\partial g_U}{\partial U_\Delta}\right|_{U_\Delta^*}\right)^2 + \left(\left.\dfrac{\partial g_U}{\partial U_A}\right|_{U_A^*}\right)^2}}$$

The reliability index is given by:

$$\beta^* = \frac{g_U\left(U_\Delta^*, U_A^*\right) - \left.\dfrac{\partial g_U}{\partial U_\Delta}\right|_{U_\Delta^*} U_\Delta^* - \left.\dfrac{\partial g_U}{\partial U_A}\right|_{U_A^*} U_A^*}{\sqrt{\left(\left.\dfrac{\partial g_U}{\partial U_\Delta}\right|_{U_\Delta^*}\right)^2 + \left(\left.\dfrac{\partial g_U}{\partial U_A}\right|_{U_A^*}\right)^2}}$$

The following table describes the different calculation stages for $t = 10$ years.

Iteration	Variable	U_X^k	$\alpha_{U_X}^k$	β^k	U_X^{k+1}
1	U_Δ	0	0.981	2.67	-2.620
	U_A	0	0.194		-0.520
2	U_Δ	-2.620	0.872	4.21	-3.671
	U_A	-0.520	0.490		-2.064
3	U_Δ	-3.671	0.683	3.83	-2.614
	U_A	-2.064	0.731		-2.798

4	U_Δ	-2.614	0.716	3.89	-2.792
	U_A	-2.798	0.697		-2.719
5	U_Δ	-2.792	0.716	3.89	-2.792
	U_A	-2.719	0.697		-2.719

The following figure compares the results from example 3.7 with those given by the Hasofer-Lind index, coupled with the Rackwitz solution method. We observe that the lognormal model is superposed on the Hasofer-Lind indexes. This proves that the Cornell index can induce sizeable errors on the reliability assessment.

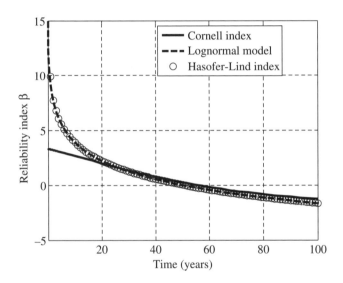

EXAMPLE 3.12.– Consider a bi-clamped beam subjected to a force P at mid-span. The serviceability criterion to be respected, is that the maximum displacement must not exceed 1/100 of the span, i.e.: $f_{max} = \dfrac{1}{192} \dfrac{P L^3}{EI} \geq \dfrac{1}{100} L$. The variables P, I and E are normal variables whose means and standard deviations are respectively:

$$m_P = 4 \text{ kN}, \ \sigma_P = 1 \text{ kN}; \ m_I = 167.4\,10^7 \text{ m}^4, \ \sigma_I = 16.74\,10^7 \text{ m}^4$$
$$m_E = 2\,10^7 \text{ kN/m}^2, \ \sigma_E = 0.5\,10^7 \text{ kN/m}^2$$

The beam is 6 m long (deterministic). The failure surface is then brought back to:

$$g(z_E, z_I, z_P) = 6\, z_e\, z_i - 112.5\, z_p = 0$$

The following standard normal variables may, therefore, be formed:

$$U_P = \frac{P - m_P}{\sigma_P}; \; U_E = \frac{E - m_E}{\sigma_E}; \; U_I = \frac{I - m_I}{\sigma_I}$$

The failure surface in the standardized space is:

$$g_U(u_E, u_I, u_P) = 1004.4 \; (2.0 + 0.5 \, u_E)(1.0 + 0.1 \, u_I) - 112.5 \, (4.0 + u_P) = 0$$

Applying the Rackwitz-Fiessler algorithm to determine the Hasofer-Lind index gives the results in the following table:

Iteration	1	2	3	4	5
β	2.8104	3.0223	3.0001	3.0000	3.0000
α_E	− 0.9086	− 0.9580	− 0.9668	− 0.9677	− 0.9677
α_I	− 0.3634	− 0.1543	− 0.1120	− 0.1101	− 0.1098
α_P	0.2059	0.2418	0.2298	0.2269	0.2268

Applying the Cornell index with the MVFOSM method is more direct and gives:

$$\beta_c = \frac{6 \, m_E \, m_I - 112.5 \, m_P}{\sqrt{\left(6 \, m_I \, \sigma_E\right)^2 + \left(6 \, m_E \, \sigma_I\right)^2 + \left(112.5\right)^2}}$$

$$\approx \frac{6 \times 2 \times 167.4 - 450.0}{\sqrt{36 \times 167.4^2 \times 0.5^2 + 36 \times 4 \times 16.74^2 + 112.5}}$$

$$\approx 2.82$$

highlighting once again the error which may occur when using the Cornell index. □

3.5.8. *Calculation of the failure probability*

For the first order approximation $\tilde{g}_U(\{u\}) = 0$, the probability of failure is approximated by:

$$P_f \approx \Phi(-\beta) \tag{3.61}$$

where $\Phi(.)$ is the cumulative probability function of a standard normal variable (see Chapter 1). This approach is called the FORM because it consists of replacing the limit state by a tangent hyperplane. In a second order approach (*second order reliability method* – SORM), the failure surface is approached by a paraboloid approximation (which goes through the design point and which has the same curvature at this point). The geometric representation of these two approaches is illustrated by Figure 3.9. The second order approximation must consider the second order terms omitted from equation [3.49]:

$$\tilde{g}_U(\{u\}) \approx \sum_{i=1}^{n} \frac{\partial g_U(\{u\})}{\partial u_i}\bigg|_{\{u^*\}} (u_i - u_i^*) + \frac{1}{2}(\{u\} - \{u^*\})' [H] (\{u\} - \{u^*\}) \qquad [3.62]$$

The Hessian matrix $[H] = \nabla^2 g_U(\{u^*\})$ must be determined then diagonalized so that the main curvatures κ_i can be calculated. These curvatures provide additional information contained in the SORM formulation, compared to FORM. In other words, a quadratic SORM approximation is possible if the curvatures can be calculated. Solving what can be expressed as an eigenvalue problem is generally fastidious, especially in cases of small gradient zones. In these cases, it is then necessary to use a third order approximation. Presuming the gradients are not equal to zero, the computation of the curvatures can be carried out by calculating the matrix:

$$[B] = \frac{\nabla^2 g_U(\{u^*\})}{2\|\nabla g_U(\{u^*\})\|} \qquad [3.63]$$

Then, a coordinate rotation is performed so as to orientate one of the axes from the new system of coordinates using the direction cosine vector:

$$\{v^*\} = \begin{Bmatrix} 0 \\ 0 \\ \vdots \\ \beta \end{Bmatrix} = [\mathcal{R}]\{u^*\} \qquad [3.64]$$

The eigenvalues λ_i of the matrix:

$$[A] = [\mathcal{R}_{n-1}] \frac{\nabla^2 g_U(\{u^*\})}{2\|\nabla g_U(\{u^*\})\|} {}^t[\mathcal{R}_{n-1}] \qquad [3.65]$$

are determined, where $[\mathcal{R}_{n-1}]$ represents the $n-1$ lines of $[\mathcal{R}]$. The main curvatures κ_i are deduced from the eigenvalues by the relationship $\kappa_i = 2\lambda_i$.

The SORM approximation at the design point is therefore expressed by:

$$\tilde{g}_U(u) = u_n - \beta - \frac{1}{2}\sum_{i=1}^{n-1}\kappa_i u_i^2 \qquad [3.66]$$

With such a second-order function, the failure probability can be calculated by many approaches. The failure probability, approached by Breitung [BRE 84] is given by:

$$P_f \approx \Phi(-\beta)\prod_{i=1}^{n-1}(1+\kappa_i\,\beta)^{\frac{-1}{2}} \qquad [3.67]$$

Hohenbichler and Rackwitz [HOH 88] propose a more general failure probability approximation in relation to the Breitung approximation. It is given by:

$$P_f \approx \Phi(-\beta)\prod_{i=1}^{n-1}\left(1+\kappa_i\,\frac{\varphi\,(\beta)}{\Phi(-\beta)}\right)^{-\frac{1}{2}} \qquad [3.68]$$

In this approach, the term $\dfrac{\varphi\,(\beta)}{\Phi(-\beta)}$ is obtained by the following expression:

$$\frac{\varphi\,(\beta)}{\Phi(-\beta)} = \beta + \beta^{-1} - 2\,\beta^{-3} + 10\,\beta^{-5} - 74\,\beta^{-7} + 706\,\beta^{-9} + \cdots \qquad [3.69]$$

The Breitung approximation may be obtained by the first term of equation [3.69]. Let us note that the SORM method may be brought back to a first order approach by defining a hyperplane which is tangent to the design point, but at a distance β' defined as the following [CRE 02]:

$$\beta' \equiv -\Phi^{-1}(P_f) \qquad [3.70]$$

β' is called the *generalized reliability index*, and therefore differs from the Hasofer-Lind index. This book does not go into any more detail on this calculation process, and the reader could turn to references [MEL 99] and [LEM 09], particularly for calculating curvatures, which is one of the most fastidious aspects of the method.

It is important for us to note that applying the FORM or SORM methods, in the case of an implicit limit state, requires either the generation of a limit state function approximation around the design point, or the numerical assessment of Hessian gradients and matrices by deterministic calculations. Finally, it is appropriate to specify that the potential difference between the failure probabilities obtained by FORM and SORM may be related to non-linearities or strong curvatures, etc. Calculation time is independent of the order of magnitude for the failure probability. However, in the FORM method, it varies linearly with the space dimension n, and the extra term in SORM is evaluated in n^2 [DEV 96].

EXAMPLE 3.13.– To illustrate the importance of the curvature when calculating failure probability in certain cases, we will consider the following simple limit state [LEM 09]:

$$g(x, y) = 0, 1\, x^2 - \frac{\sqrt{2}}{2} y + 2,5 = 0$$

with x and y as two standard normal variables with zero means and respective variances equal to 2. Applying the FORM method leads to the introduction of standardized variables u and v:

$$\hat{g}(x, y) = 0.2\, u^2 - v + 2.5 = 0, \quad \text{or } v = 2.5 + 0.2\, u^2$$

and thus to study a limit state in the standardized normal space. To determine the design point and the reliability index, we must minimize $\left(u^2 + v^2\right)$ under the constraint $0.2\, u^2 - v + 2.5 = 0$, simply meaning $\left(u^2 + \left[2.5 + 0.2u^2\right]\right)$. We can verify easily that the only solution is $u^* \equiv 0$, and therefore $v^* = 2.5$. The design point in the basic space is then $x^* = 0$ and $y^* = \sqrt{2} \times 2.5 = 3.536$. The reliability index is therefore $\beta = \sqrt{0^2 + 2.5^2} = 2.5$ which gives $P_f = \Phi(-\beta) = 0.0062$ as the failure probability.

To calculate the second order failure probability, we must have curvatures in the standardized space. For a problem with two variables $v = f(u)$, the curvature at a point u^* is given by the expression:

$$\kappa\big|_{u^*} = \frac{f''(u^*)}{\left[1 + \left(f'(u^*)\right)^2\right]^{3/2}}$$

When applied to our problem, this gives:

$$\kappa\big|_0 = \frac{f''(0)}{\left[1+\underbrace{\left(f'(0)\right)^2}_{=0}\right]^{3/2}} = f''(0) = 0.4$$

By using the Breitung and Hohenbichler expressions, the corrected failure probability can be calculated. For these two approximations, the probability is: 0.0044 and 0.0043 respectively. The generalized reliability indexes are therefore 2.62 and 2.63. Calculations per integration give 0.0041 and 2.64 for the failure probabilities and the generalized reliability index. The error produced by the first order approach (FORM) is – in this example – very simple: close to 50%!   □

3.5.9. Monte-Carlo methods

Simulation methods, or *Monte-Carlo* methods, consist of reproducing the real world based on hypotheses and models. This process has greatly risen with the development of computers and their persistent greater powers of calculation. In engineering, simulation may be used to predict or study a system's performance. For a set of values representing the system's parameters, this performance may be calculated and repeated as many times as the sets of values are introduced. By these simulations, the sensitivity of the performance to the system's parameters may be assessed. In the reliability theory framework, we can very quickly see the interest of what such a procedure may contribute for calculating failure probability. It is sufficient to generate sets of values (or samples) for the variables of a limit state and to count the number of times that the limit state is violated, i.e. the number of times the limit state function takes a negative value. Brought back to the number of value sets used, this number enables us to estimate the probability of violating the limit state, which is none other than the failure probability. At this stage, it would appear intuitive that the higher the number of value sets, the more accurate the failure probability will be. This convergence in probability will be all the more difficult to obtain when the sought probabilities are low, requiring a very large number of value sets. However, reduction techniques do exist [MEL 99].

The problem of generating sets of values is equally as troubling. On the one hand, in the case of large samples, random number generators may present a bias, since generators present some cyclic features after a large number of values. On the other hand, the usual generators are restricted to uniform variables between 0 and 1, or to standard normal variables. The transition to other distributions can, however, be easily achieved by using the isoprobability transformations introduced in section 3.5.8. However, dealing with joint distributions, in the case of multiple correlations between variables, is more difficult and requires use of the Rosenblatt or Nataf transformation.

In this chapter, we will not tackle all the theoretical concepts and elements relating to the Monte-Carlo simulation techniques. Those curious to know more should see reference [MEL 99]. However, it would be appropriate to specify a few basic, useful concepts for the engineer or researcher wanting to understand the aim of certain techniques.

The Monte-Carlo simulation method aims at generating a failure probability estimation. Random trials are carried out and an assessment of the structure's response is done for each set of data. There are many Monte-Carlo methods, aiming to optimize the sampling strategies for reducing the calculating costs essentially related to the number of trials. First of all, we shall define the indicator function of the failure domain $1_{D_R}(.)$. Equation [3.40] may also be expressed under the form:

$$P_f = \mathcal{P}\big(g(\{Z\}) \le 0\big) = \int_{\mathbb{R}^n} 1_{D_R}(z_1,....,z_n) f_Z(z_1,....,z_n)\, dz_1 \cdots dz_n \qquad [3.71]$$

Written in another way, we obtain the equation:

$$P_f = \mathbb{E}\big[1_{D_R}(\{Z\})\big] \qquad [3.72]$$

It is therefore appropriate to call upon the empirical estimator:

$$P_f \approx \frac{1}{N} \sum_{i=1}^{n} 1_{D_R}(\{Z_i\}) \qquad [3.73]$$

We note that N assessments of the limit state function are necessary for generating a good estimate of the failure probability. There are many empirical ratios which enable us to predict the number of trials N. A ratio which is often used is given by:

$$N = \frac{C}{P_f} \qquad [3.74]$$

where the constant C represents the level of confidence with which we wish to know P_f. A typical value for C is 1,000. In order to reduce the number of trials, many strategies have been developed. In the following, we propose to present the principle of one of these strategies, namely importance and directional sampling.

In the method known as *importance sampling* [MEL 90], equation [3.71] may be put in the following form, where $h(.)$ is a probability density function *a priori* unknown, which must be chosen in such a way that the sampling is carried out in the area which participates the most in integral [3.71]:

$$P_f = \int_{\mathbb{R}^n} 1_{D_R}\left(z_1,....,z_n\right) \frac{f_Z\left(z_1,....,z_n\right)}{h\left(z_1,....,z_n\right)}\, h\left(z_1,....,z_n\right) dz_1 \cdots dz_n \qquad [3.75]$$

As previously stated, it is possible (when approaching this integral) to use the estimator [3.73] where $\{v_i\}$ are obtained randomly according to $h(.)$:

$$P_f \approx \frac{1}{N} \sum_{i=1}^{n} 1_{D_R}\left(\{v_i\}\right) \frac{f_Z\left(\{v_i\}\right)}{h\left(\{v_i\}\right)} \qquad [3.76]$$

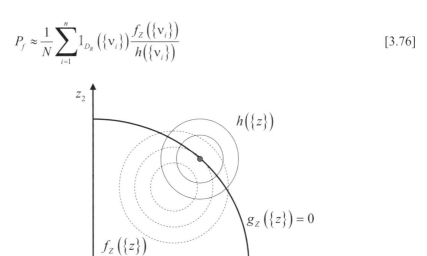

Figure 3.10. *Importance sampling*

In a reliability calculation, this technique may be broken down into many stages. As for the level II methods, the basic problem is transformed in a standardized normal space. In this space, an approximation of the design point (Figure 3.10) is then determined by generating a few samples, and only keeping the point belonging to the failure domain which is closest to the origin of the standardized normal space. Finally, the density of the importance sampling $h(.)$ can be estimated and centered in this point (generally, this involves a standard normal distribution). This is how a precise trial can be performed. In fact, the number of trials necessary for a good estimate of the failure probability primarily increases with the dimension. Of course, this is to be linked with the fact that, for a fixed number of trials, when the physical space dimension increases, the sampling density drops. In other words, the sampling densities reflect the amount of information contained in their response.

Directional sampling [DIT 88] can be found within the concept of conditional probability. In fact, equation [3.71] can be put in the following form, where Ω_n, $\{\Gamma\}$ and R respectively represent the unitary hypersphere centered in the origin of the space with dimension n, a random unitary vector whose realizations are uniformly chosen from the center of Ω_n towards the exterior, and the radius r whose realization is the solution of equation $g_{\Gamma,R}(\gamma\, r) = 0$:

$$P_f = \int_{\gamma \in \Omega_n} P\Big(g_{\Gamma,R}(\Gamma R) \leq 0 \big| \Gamma = \gamma\Big)\, f_\Gamma\big(\{\gamma\}\big)\, d\gamma \qquad [3.77]$$

Thus the failure probability can be approached by the estimator:

$$P_f \approx \frac{1}{N}\sum_{i=1}^{n} P\Big(g_{\Gamma,R}(\Gamma\, R) \leq 0 \big| \Gamma = \gamma_i\Big)$$

$$= \frac{1}{N}\sum_{i=1}^{n} P(R \geq r_i) = \frac{1}{N}\sum_{i=1}^{n}\Big[1 - \chi_n^2(r_i^2)\Big] \qquad [3.78]$$

because R^2 follows a chi-square distribution with n degrees of freedom. Figure 3.11 illustrates the principle of this method [LEM 09].

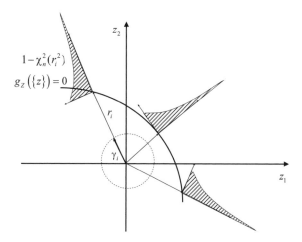

Figure 3.11. *Directional sampling on a unitary sphere*

3.5.10. *Response surfaces*

In the case of complex structures, the performance functions are not directly explicit, and they are assessed by numerical calculations (finite differences, finite elements, etc.). It is still possible to use the previous methods but they present certain characteristics which often limit their use (large number of trials, derivative or Hessian calculations, etc.). An alternative method simply consists of *artificially* generating the limit state function using a polynomial adjusted to the results from a limited number of calculations using finite elements [BUC 90, RAJ 93]. This type of approach is known as the *quadratic response surface method*. One of the aims of response surfaces is to obtain an explicit mathematic model:

– which represents the failure surface of complex structures for particular components (failure modes) and for the structure (failure/collapse mechanisms);

– which represents a good approximation of the behavior of the model studied;

– which is a good approximation technique enabling us to decrease calculation times and to carry out a reliable study of complex structures;

– for which we can perform a structural-reliability coupling between a deterministic finite-elements code and reliability techniques.

3.5.10.1. *Basic concepts*

A structural response is generally represented by a limit state function $g(\{Z\})$ which characterizes a failure mode. In the case of complex structures, the implicit function $g(\{Z\})$ can only be assessed in a discrete way by the sample values $\{z\}^k$, $k = 1, \ldots, m$. The initial idea of the response surface methods is to replace function $g(\{Z\})$ which is an *a priori* unknown function, with an equivalent explicit function $\hat{g}(\{Z\})$. The response surface methods therefore seek out a function, generally a polynomial surface, whose coefficients are determined in a way so as to minimize the approximation error in the area around the design point (limit state point with the highest density value). Assessing these coefficients requires a series of numerical experiments which correspond to numerical calculations with input parameters chosen in conformity with an experiment design. The choice of the polynomial terms to be considered, as well as the definition of the experiment designs to be performed, are difficult operations which we will now describe in more detail.

3.5.10.2. *Choice of a polynomial degree*

The choice of degree for the polynomial surface, its characteristics and the design of the experiment are closely linked. The number of coefficients to be determined increases (in the same way as the size of the experiment design) with the degree of the polynomial function. Usually, we use linear or quadratic polynomial surfaces. Three response surfaces are often used: linear response surface, quadratic response surface and quadratic response surface with mixed terms. The performance function $\hat{g}(\{z\})$ is then in the form:

$$\hat{g}(\{z\}) = A + {}^t\{z\}\{B\} + {}^t\{z\}[C]\{z\} \tag{3.79}$$

with $\{B\} = \begin{Bmatrix} b_1 \\ \vdots \\ b_n \end{Bmatrix}$; $[C] = \begin{bmatrix} c_{11} & \cdots & c_{1n} \\ \vdots & \ddots & \vdots \\ c_{1n} & \cdots & c_{nn} \end{bmatrix}$.

If $[C] \equiv [0]$, then the response surface is *linear*; if $[C]$ is a diagonal matrix, then the response surface is said to be *quadratic without mixed terms*. Finally, if $[C]$ is a full matrix, then the response surface is said to be *quadratic with mixed terms*.

3.5.10.3. *Experimental design*

Experimental design is the definition of the data for numerical calculations needed to build the approximation of the response surface. There is no specific guide for selecting an experimental design. The number of experimental points in the design experiment is often chosen so as to have as many equations as unknown coefficients of the polynomial surface. Some approaches introduce a larger number of experimental points, and it is therefore necessary to call upon regression analyses in order to assess the polynomial surface coefficients. The simplest experimental design is the *star* design (Figure 3.12a). It includes $2n + 1$ points and is simple to generate because it does not require many experiments. Each variable may take on three distinct values, in addition to the center point. We choose a central value $\bar{z}_i$, and a difference, often a multiple of the standard deviation σ_i, of each random variable [BUC 90]: $z_i = \bar{z}_i \pm h\sigma_i$. The value of h is often arbitrary and ranges between 1 and 2 [MOH 07]. This type of experimental design enables us to study the behavior of each variable according to an axis, but it does not provide any indication of the interaction between variables. The star design can be described in a matrix:

$$[\mathcal{Z}] = \begin{bmatrix} \bar{z}_1 + h\sigma_1 & \bar{z}_2 & \bar{z}_3 & \cdots & \bar{z}_n \\ \bar{z}_1 - h\sigma_1 & \bar{z}_2 & \bar{z}_3 & \cdots & \bar{z}_n \\ \bar{z}_1 & \bar{z}_2 + h\sigma_2 & \bar{z}_3 & \cdots & \bar{z}_n \\ \bar{z}_1 & \bar{z}_2 - h\sigma_2 & \bar{z}_3 & \cdots & \bar{z}_n \\ \bar{z}_1 & \bar{z}_2 & \bar{z}_3 + h\sigma_3 & \cdots & \bar{z}_n \\ \bar{z}_1 & \bar{z}_2 & \bar{z}_3 - h\sigma_3 & \cdots & \bar{z}_n \\ \vdots & \vdots & \vdots & \vdots & \vdots \\ \bar{z}_1 & \bar{z}_2 & \bar{z}_3 & \cdots & \bar{z}_n + h\sigma_n \\ \bar{z}_1 & \bar{z}_2 & \bar{z}_3 & \cdots & \bar{z}_n - h\sigma_n \\ \bar{z}_1 & \bar{z}_2 & \bar{z}_3 & \cdots & \bar{z}_n \end{bmatrix} \qquad [3.80]$$

Another type of experimental design used is the *factorial* design (Figure 3.12b). It includes 2^n points. It enables us to take into account the interaction between variables but requires a larger number of points than the star design. Additionally, one of the best performing experimental designs in terms of representing data, is the central composite design. It includes $2^n + 2n+1$ points [MYE 02]. It, then, involves 2^n factorial points, increased by $2n + 1$ points with coordinates $-\lambda\, h\, \sigma, \lambda\, h\, \sigma$ along the axes. The choice of λ is very important because it ensures stability. Usually, a value of λ between 1 and $\sqrt{n}$ is chosen (Figure 3.12c). However, it requires a higher number of calculation points in the case of a large number of variables.

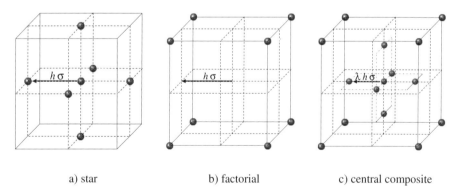

a) star   b) factorial   c) central composite

Figure 3.12. *Experiment design examples for three variables*

Generally, the experimental design matrix is written as:

$$[Z] = \begin{bmatrix} z_{11} & \cdots & z_{1i} & \cdots & z_{1n} \\ \vdots & \cdots & \vdots & \cdots & \vdots \\ z_{m1} & \cdots & z_{mi} & \cdots & z_{mn} \end{bmatrix} \qquad [3.81]$$

with n as the number of random variables and m the number of sample values performed for each variable (we presume that $m > n$). In this matrix, each line represents a set of data and each column is a sample.

3.5.10.4. Determination of response surface coefficients

The response surface functions can be formally written by:

$$\hat{g}(\{z\}) = {}^t\{G(\{z\})\}\{D\} \qquad [3.82]$$

with:

$$\{D\} = \{A \ b_1 \ \cdots \ b_n \ c_{11} \ \cdots \ c_{nn} \ c_{12} \ \cdots \ c_{1n} \ \cdots \ c_{n1} \ \cdots \ c_{n\,n-1}\} \qquad [3.83]$$

$${}^t\{G(\{z\})\} = \{1 \ z_1 \ \cdots \ z_n \ z_1^2 \ \cdots \ z_n^2 \ z_1 z_2 \ \cdots \ z_1 z_n \ \cdots \ z_n z_1 \ \cdots \ z_n z_{n-1}\} \qquad [3.84]$$

When we have m sets of data from an experimental design, it is sufficient to solve the system using a least squares approach:

$$\begin{Bmatrix} \hat{g}(\{z\}^1) \\ \vdots \\ \hat{g}(\{z\}^m) \end{Bmatrix} = \begin{bmatrix} {}^t\{G(\{z\}^1)\} \\ \vdots \\ {}^t\{G(\{z\}^m)\} \end{bmatrix} \{D\} \qquad [3.85]$$

i.e.

$$\{D\} = \left(\begin{bmatrix} {}^t\{G(\{z\}^1)\} \\ \vdots \\ {}^t\{G(\{z\}^m)\} \end{bmatrix} \begin{bmatrix} {}^t\{G(\{z\}^1)\} \\ \vdots \\ {}^t\{G(\{z\}^m)\} \end{bmatrix} \right)^{-1} \begin{bmatrix} {}^t\{G(\{z\}^1)\} \\ \vdots \\ {}^t\{G(\{z\}^m)\} \end{bmatrix} \begin{Bmatrix} \hat{g}(\{z\}^1) \\ \vdots \\ \hat{g}(\{z\}^m) \end{Bmatrix} \qquad [3.86]$$

3.5.10.5. *Algorithm for computing response surfaces*

The fundamental objective in researching a response surface function is based on estimating the failure probability. This particularly involves focusing on the area where the probabilities are the highest in the failure domain, a domain which is found around the design point, with this point not being known *a priori*. An iterative technique is used initially to estimate this design point.

With a lack of specific information, the starting values for calculating the polynomial $\hat{g}(\{z\})$ coefficients can be chosen so as to form a set of data centered around a point $\{\bar{z}\}_1$, fixed on the mean of variables $\mathbb{E}[\{Z\}]$, by using one of the aforementioned experimental designs.

By using these points, the coefficients which determine the first approximation of the polynomial function $\hat{g}_1(\{z\})$ can be determined. It is then possible to assess the design point $\{z_d\}_1$ on the surface $\hat{g}_1(\{z\}) = 0$ using the Rackwitz-Fiessler algorithm, for instance, or any other optimization technique. A new center $\{\bar{z}\}_2$ is used as a starting point for a new iteration.

In this case, the new center point may be obtained by linear interpolation between the design point and the previous center point [BUC 90].

$$\{\bar{z}\}_2 = \{\bar{z}\}_1 + \left(\{z_d\}_1 - \{\bar{z}\}_1\right)\frac{g\left(\{\bar{z}\}_1\right)}{\left[g\left(\{\bar{z}\}_1\right) - g\left(\{z_d\}_1\right)\right]} \qquad [3.87]$$

In the same way as before, the experimental design is built on this new center $\{\bar{z}\}_2$. By using the points from the experimental design, the second approximation of the polynomial function $\hat{g}_2(\{z\})$ can therefore be determined. It is then possible to assess the design point $\{z_d\}_2$ on the polynomial surface $\hat{g}_2(\{z\}) = 0$. At this iteration, the distance is defined as being between the design point $\{z_d\}_2$ and the new center $\{\bar{z}\}_2$. The convergence is guaranteed by several criteria related to the convergence of the design point, the reliability index, etc.

This criterion does not guarantee, however, that there will always be convergence towards the real limit state. It is therefore appropriate to double the verification that the design point is on the true limit state. The last iteration provides the reliability index, the design point's coordinates and the direction cosines for each variable.

The question regarding the work space (physical or standardized normal) is obviously asked. The previous iterative schema may lead to physically unrealizable points. In effect, it works very well if the variable support is real, and less well if the support is bounded or asymmetrical. This explains why it is preferable to generate the response surface in the standardized normal space rather than in the physical space. However, the two spaces may not be related by bijective relationships.

3.5.10.6. Second order calculation of the failure probability

The response surface method provides an explicit approximation of the limit state function $\hat{g}\left(\{z\}\right)$ as much in the physical variable space as in the standardized normal space. As the response surface is quadratic, it would be intuitive to compare it to the SORM approach whose limit surface is approached by a paraboloid in the standardized space.

In the classic SORM approach, the second order surfaces are defined by adjusting its main curvatures to those of the limit state at the design point. These main curvatures are obtained by solving an eigenvalue problem with the Hessian matrix. The calculated curvatures enable us to calculate a second order approximation of the failure probability, by means of expressions like the Breitung or Hohenbichler expressions (section 3.5.9).

However, calculating the Hessian matrix is difficult when there is a large number of variables, or when some of them have very small direction cosines. This difficulty can be avoided in the case of the response surface method, because this (if adjusted in the normal space) avoids gradient and Hessian matrix calculations. In fact, let us assume the convergence: at the last iteration, the response surface is also constructed in the standardized space, and not only in the physical space. Let $\hat{g}_U\left(\{U\}\right)$ be the limit state function approached in the normal space, and let us presume that it is a quadratic with no mixed terms:

$$\hat{g}_U\left(\{U\}\right) = A + {}'\{U\}\{B\} + {}'\{U\}[C]\{U\} \qquad [3.88]$$

where $[C]$ is diagonal. The Hessian gradient calculation $[H]$ is direct:

$$[H] = \frac{2}{\|\{G\}\|}[C] \qquad [3.89]$$

By only keeping the $n-1$ terms λ_i of the diagonal matrix $[H]$, we can then extract the curvatures and then apply any SORM approximation of the failure probability. This approach is called *point-fitting SORM* (PFSORM), because the curvatures are calculated by adjusting a quadratic function with no mixed terms on the last experimental design.

In summary, the response surface approach not only enables us to clarify a limit state, but also to efficiently calculate an approximation of the second order failure probability whilst avoiding any numerical instability.

EXAMPLE 3.14.– Let us take example 3.13 again, and let us try to build an incomplete quadratic response surface based on a star design ($h = 1$). The central starting point in the standardized space is $(0,0)$. The experimental designs in this space and in the physical space, as well as the performance function values, are therefore:

$$[Z_U] = \begin{bmatrix} -1 & 0 \\ +1 & 0 \\ 0 & -1 \\ 0 & +1 \\ 0 & 0 \end{bmatrix}; [Z_X] = \begin{bmatrix} -\sqrt{2} & 0 \\ +\sqrt{2} & 0 \\ 0 & -\sqrt{2} \\ 0 & +\sqrt{2} \\ 0 & 0 \end{bmatrix};$$

$$g\left([Z_X]\right) = {}^t\{2.7 \quad 2.7 \quad 3.5 \quad 1.5 \quad 2.5\}$$

Solving equation [3.86] then gives:

$$\{D\} = \left({}^t\begin{bmatrix} 1 & -\sqrt{2} & 0 & 2 & 0 \\ 1 & \sqrt{2} & 0 & 2 & 0 \\ 1 & 0 & -\sqrt{2} & 0 & 2 \\ 1 & 0 & \sqrt{2} & 0 & 2 \\ 1 & 0 & 0 & 0 & 0 \end{bmatrix}\begin{bmatrix} 1 & -\sqrt{2} & 0 & 2 & 0 \\ 1 & \sqrt{2} & 0 & 2 & 0 \\ 1 & 0 & -\sqrt{2} & 0 & 2 \\ 1 & 0 & \sqrt{2} & 0 & 2 \\ 1 & 0 & 0 & 0 & 0 \end{bmatrix} \right)^{-1} {}^t\begin{bmatrix} 1 & -\sqrt{2} & 0 & 2 & 0 \\ 1 & \sqrt{2} & 0 & 2 & 0 \\ 1 & 0 & -\sqrt{2} & 0 & 2 \\ 1 & 0 & \sqrt{2} & 0 & 2 \\ 1 & 0 & 0 & 0 & 0 \end{bmatrix}\begin{Bmatrix} 2.7 \\ 2.7 \\ 3.5 \\ 1.5 \\ 2.5 \end{Bmatrix}$$

$$= \left(\begin{bmatrix} 5 & 0 & 0 & 4 & 4 \\ 0 & 4 & 0 & 0 & 0 \\ 0 & 0 & 4 & 0 & 0 \\ 4 & 0 & 0 & 8 & 0 \\ 4 & 0 & 0 & 0 & 8 \end{bmatrix}\right)^{-1}\begin{Bmatrix} 2.7 - \sqrt{2} \\ 2.7 + \sqrt{2} \\ 2.7 - 3.5\sqrt{2} \\ 2.7 + 1.5\sqrt{2} \\ 2.7 \end{Bmatrix}$$

$$= \begin{bmatrix} 1 & 0 & 0 & -0.5 & 0.5 \\ 0 & 0.25 & 0 & 0 & 0 \\ 0 & 0 & 0.25 & 0 & 0 \\ -0.5 & 0 & 0 & 0.375 & 0.25 \\ -0.5 & 0 & 0 & 0.25 & 0.375 \end{bmatrix} \begin{Bmatrix} 2.7-\sqrt{2} \\ 2.7+\sqrt{2} \\ 2.7-3.5\sqrt{2} \\ 2.7+1,5\sqrt{2} \\ 2.7 \end{Bmatrix}$$

$$= \begin{Bmatrix} 2.5 \\ 0 \\ -\dfrac{\sqrt{2}}{2} \\ 0.1 \\ 0 \end{Bmatrix}$$

From the first iteration, we find the limit state function coefficients:

$$\hat{g}(x,y) = 2.5+0\times x - \frac{\sqrt{2}}{2}\times y + 0.1\times x^2 + 0\times y^2 = 2.5+0.1\,x^2 - \frac{\sqrt{2}}{2}y$$

with $\left(0 \quad 2.5\sqrt{2}\right)$ as the design point.

In the normal space, the same adjustment gives:

$$\hat{g}_U(u,v) = 2.5+0.2\,u^2 - v$$

with $\left(u^* \quad v^*\right) = \left(0 \quad 2.5\right)$ as the design point. The gradient is $\nabla g_U(u^*,v^*) = \left(0 \quad -1\right)$ and the Hessian, which gives $\kappa_1 = 0.4$. Therefore, we find again the curvature calculated in example 3.11. The failure probability calculation (SORM) is identical. □

3.5.11. Sensitivity measures

The results from a reliability analysis are not just reduced to the reliability index, the failure probability and the direction cosines. Other factors, called *sensitivity measures*, enable us to assess the influence of variables on the failure probability (or the reliability index). These various factors are therefore particularly useful for judging their role in reliability. In other words, the variables which present influencing factors deserve to be particularly well known, to the detriment of less influencing variables. This importance analysis therefore aims to identify significant variables [LEM 09].

The sensitivity of a variable in a reliability calculation can be analyzed in two different ways. The first, a rather classic way, consists of studying its influence on the performance function. This influence is generally measured by the standardized gradient of this function in relation to a variable called the *elasticity measure*:

$$S_{elas}^{i} = \frac{z_i}{g(\{z\})} \left. \frac{\partial g(\{z\})}{\partial z_i} \right|_{\{z\}_r}$$

[3.90]

$\{z\}_r$ is a representative value (often the mean value). $S_{méca}^{i}$ cannot, however, be calculated at the design point. The fact that the sensitivity is "standardized" enables us to compare the weight of different variables on the performance function. Only influencing variables deserve to be considered as random. It is advisable, however, to remain vigilant over a section of variables on the only elasticity measures, because the presence of strong linearities, the choice of the value representing the elasticity calculation, the probabilistic dispersion, the type of distribution or the correlations intervening between variables may also greatly influence the results.

This explains why studying the importance of variables must be coupled with other studies of sensitivity. The first factor is directly provided by the reliability calculation and involves the *sensitivity* of the reliability index to the *standardized variables*:

$$\left. \frac{\partial \beta}{\partial u_i} \right|_{\{u^*\}} = -\alpha_i$$

[3.91]

The direction cosines therefore represent the sensitivities of the reliability index in the normal space, with regard to all the standardized variables. We can generalize this sensitivity to physical variables. However, for independent variables, the two concepts are joined, hence systematically resorting to direction cosines in order to represent the sensitivity of the reliability index to physical or standardized variables.

Also of importance is the *sensitivity* of the reliability index to the *distribution parameters* used for all the variables. Generally, to be able to compare the weight of different variables, these are the *elasticity measures with respect to the distribution parameters*:

$$S_{rel}^{p_{Z_i,k}} = \frac{p_{Z_i,k}}{\beta} \frac{\partial \beta}{\partial p_{Z_i,k}}\Bigg|_{\{u^*\}}$$ [3.92]

with p_i being one of the parameters of variable Z_i. Calculating these parameters uses the Nataf transformation for correlated variables. In the case of independent variables, they can be calculated easily:

$$S_{rel}^{p_{Z_i,k}} = -\alpha_i \frac{p_{Z_i,k}}{\beta} \frac{1}{\varphi(u_i^*)} \frac{\partial F_{Z_i}(x_i^*)}{\partial p_{Z_i,k}}$$ [3.93]

Thus, for a normal variable, we obtain:

$$S_{rel}^{\mathbb{E}[Z]} = \frac{\alpha_Z}{\beta} \frac{\mathbb{E}[Z]}{\sigma[Z]}; \quad S_{rel}^{\sigma[Z]} = -\alpha_Z^2$$ [3.94]

Another sensitivity measure, called the *omission factor*, also plays a very important role in reliability calculations:

$$\mathcal{O}_i = \frac{\beta(Z_i \equiv z_{r,i})}{\beta}$$ [3.95]

where $\beta(Z_i \equiv z_{r,i})$ is the reliability index calculated when presuming that the variable Z_i is deterministic and equal to $z_{r,i}$. This factor expresses the relative error made on the reliability index when a variable is replaced by a deterministic variable. A value close to 1 indicates that the variable may be kept deterministic. Calculating omission factors involves $n + 1$ reliability calculations, which may prove to be numerically prohibitive. In the case of a linear limit state or weak non-linearity (whose first order approximation sufficiently represents the limit state in the standardized normal space), we can used the following expressions (exact for a linear limit state):

$$\mathcal{O}_i = \frac{1 + \dfrac{\alpha_i \, u_{r,i}}{\beta}}{\sqrt{1 - \alpha_i^2}}$$ [3.96]

Let us note here that if the representative value $z_{r,i} = \mathbb{E}[Z_i]$, then the previous expression can be simplified as:

$$\mathcal{O}_i = \frac{1 + \dfrac{\alpha_i \times 0}{\beta}}{\sqrt{1 - \alpha_i^2}} = \frac{1}{\sqrt{1 - \alpha_i^2}}$$ [3.97]

EXAMPLE 3.15.– Again let us take example 3.7 and calculate the different importance factors. According to this exercise, the reliability index is given by the expression:

$$\beta = \frac{\mathbb{E}(\ln \Delta) - \ln N(t) - 3 \ln S_e + \mathbb{E}(\ln A)}{\sqrt{\mathbb{V}(\ln \Delta) + \mathbb{V}(\ln A)}}$$

relatively to the safety margin:

$$\hat{M} = \ln(\Delta) - \ln\left(\frac{N}{A} S_e^3\right) = \ln(\Delta) - \ln(N) - 3\ln(S_e) + \ln(A)$$

The elasticity measure at the mean point is thus:

$$S_{elas}^{\ln \Delta} = \frac{z_i}{g(\{z\})} \frac{\partial g(\{z\})}{\partial z_i}\bigg|_{\mathbb{E}[\ln \Delta]} = \frac{\mathbb{E}[\ln \Delta]}{\mathbb{E}(\ln \Delta) - \ln N(t) - 3 \ln S_e + \mathbb{E}(\ln A)}$$

$$S_{elas}^{\ln A} = \frac{\mathbb{E}[\ln A]}{\mathbb{E}(\ln \Delta) - \ln N(t) - 3 \ln S_e + \mathbb{E}(\ln A)}$$

The elasticity measures with respect the distribution parameters are given by:

$$S_{rel}^{\mathbb{E}[\ln \Delta]} = \frac{\mathbb{E}[\ln \Delta]}{\beta} \frac{\partial \beta}{\partial \mathbb{E}[\ln \Delta]} = \frac{\alpha_{\ln(\Delta)}}{\beta} \frac{\mathbb{E}[\ln(\Delta)]}{\sigma[\ln(\Delta)]} = \frac{\alpha_{\ln(\Delta)}}{\beta} \mathbb{COV}[\ln \Delta]$$

$$S_{rel}^{\mathbb{E}[\ln A]} = \frac{\mathbb{E}[\ln A]}{\beta} \frac{\partial \beta}{\partial \mathbb{E}[\ln A]} = \frac{\alpha_{\ln(A)}}{\beta} \mathbb{COV}[\ln A]$$

$$S_{rel}^{\mathbb{V}[\ln \Delta]} = \frac{\mathbb{V}[\ln \Delta]}{\beta} \frac{\partial \beta}{\partial \mathbb{V}[\ln \Delta]} = -\alpha_{\ln(\Delta)}^2 ; \quad S_{rel}^{\mathbb{V}[\ln A]} = \frac{\mathbb{V}[\ln A]}{\beta} \frac{\partial \beta}{\partial \mathbb{V}[\ln A]} = -\alpha_{\ln(A)}^2$$

$$S_{rel}^{\ln S_e} = \frac{\ln S_e}{\beta} \frac{\partial \beta}{\partial \ln S_e} = -3 \frac{\ln S_e}{\beta} ; \quad S_{rel}^{\ln N(t)} = \frac{\ln N(t)}{\beta} \frac{\partial \beta}{\partial \ln N(t)} = -\frac{\ln N(t)}{\beta}$$

The direction cosines of the two variables are worth, respectively:

$$\alpha_{\ln \Delta} = \frac{\sigma[\ln(\Delta)]}{\sqrt{\mathbb{V}(\ln \Delta) + \mathbb{V}(\ln A)}}; \; \alpha_{\ln A} = \frac{\sigma[\ln(A)]}{\sqrt{\mathbb{V}(\ln \Delta) + \mathbb{V}(\ln A)}}$$

They are therefore independent of time. This implies that the elasticities $S_{rel}^{\mathbb{V}[\ln \Delta]}$ and $S_{rel}^{\mathbb{V}[\ln A]}$ do not vary according to time. However, $S_{rel}^{\mathbb{E}[\ln \Delta]}$ and $S_{rel}^{\mathbb{E}[\ln A]}$ vary in opposite proportions to the reliability index, however their relative ratio does not vary. We can also remark that:

$$
\begin{aligned}
S_{rel}^{\mathbb{E}[\ln \Delta]} &= \frac{\sigma[\ln(\Delta)]}{\sqrt{\mathbb{V}(\ln \Delta) + \mathbb{V}(\ln A)}} \frac{1}{\beta} \frac{\mathbb{E}[\ln(\Delta)]}{\sigma[\ln(\Delta)]} = \frac{\mathbb{E}[\ln(\Delta)]}{\sqrt{\mathbb{V}(\ln \Delta) + \mathbb{V}(\ln A)}} \frac{1}{\beta} \\
&= \frac{\sqrt{\mathbb{V}(\ln \Delta) + \mathbb{V}(\ln A)}}{\mathbb{E}(\ln \Delta) - \ln N(t) - 3\ln S_e + \mathbb{E}(\ln A)} \frac{\mathbb{E}[\ln(\Delta)]}{\sqrt{\mathbb{V}(\ln \Delta) + \mathbb{V}(\ln A)}} \\
&= \frac{\mathbb{E}[\ln(\Delta)]}{\mathbb{E}(\ln \Delta) - \ln N(t) - 3\ln S_e + \mathbb{E}(\ln A)} = S_{elas}^{\ln \Delta}
\end{aligned}
$$

The numerical application gives:

$$S_{elas}^{\ln \Delta} = \frac{-0.043}{3.915 - \ln(t)}; \; S_{elas}^{\ln A} = \frac{26.206}{3.915 - \ln(t)}$$

$$S_{rel}^{\mathbb{E}[\ln \Delta]} = \frac{-0.043}{3.915 - \ln(t)}; \; S_{rel}^{\mathbb{E}[\ln A]} = \frac{26.206}{3.915 - \ln(t)}$$

$$S_{rel}^{\mathbb{V}[\ln \Delta]} = -0.042; \; S_{rel}^{\mathbb{V}[\ln A]} = -0.042$$

$$S_{rel}^{\ln S_e} = -\frac{4.943}{3.915 - \ln(t)}; \; S_{rel}^{\ln N(t)} = -\frac{4.287 - 0.415 \ln t}{3.915 - \ln(t)} \qquad \square$$

3.6. System reliability

The problems raised in the previous sections consider failure modes, each one described by a limit state. The design and assessment of structures generally introduces many failure modes. For example, a structural element such as a beam can fail either by buckling, or by flexure; a structure will fail when a sub-set made of other components fails. In these two examples, failure is conditioned by the appearance of many failure modes. These two problems are very different. Thus, failure of a structural element will be conditioned by the occurrence of one failure mode (series system) or by many failure modes (parallel system). In the case of the structure, occurrences of failure or the different paths which lead to failure can be dealt with by identifying the order in which all the elementary failure modes appear. *Failure trees* are intended for identifying elementary failure successions (failure modes).

In order to tackle the problems concerning system reliability, we will firstly consider an example which allows us to introduce certain concepts from system reliability.

EXAMPLE 3.16.– Let us consider a two-span beam carrying two loads. The live load moment at any cross-section is the maximum bending moment induced by these two loads with amplitudes equal to $P = 896$ kN. Sections 1, 2 and 3 are the cross-sections where the bending moments reach their maximum absolute values.

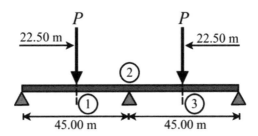

Variable	Section 1	Section 2	Section 3	Distribution
Resisting bending moment $\mathfrak{M}_1, \mathfrak{M}_2, \mathfrak{M}_3$				Normal
Mean (kN.m)	9,500	10,500	9,500	
Coefficient of variation (%)	5.0	5.0	5.0	
Live load bending moment $\mathfrak{M}_{e1}, \mathfrak{M}_{e2}, \mathfrak{M}_{e3}$				Normal
Mean (kN.m)	6,290	7,580	6,290	
Coefficient of variation (%)	10.0	10.0	10.0	

The reliability indexes for each section are easily deduced:

$$\beta = \frac{9,500 - 6,290}{\sqrt{9,500^2 \times 0.05^2 + 6,290^2 \times 0.10^2}} \approx 4.07 \quad \text{Section 1}$$

$$\beta = \frac{10,500 - 7,580}{\sqrt{10,500^2 \times 0.05^2 + 7,580^2 \times 0.10^2}} \approx 3.17 \quad \text{Section 2}$$

$$\beta \approx 4.07 \qquad\qquad\qquad\qquad\qquad\qquad\qquad \text{Section 3}$$

which corresponds to the failure probabilities:

$$P_f = \Phi(-4.07) \approx 2.33\,10^{-5} \quad \text{Section 1}$$
$$P_f = \Phi(-2.58) \approx 7.62\,10^{-4} \quad \text{Section 2}$$
$$P_f = \Phi(-4.07) \approx 2.33\,10^{-5} \quad \text{Section 3}$$

In general, when the reliability analysis of a structure is only based on the reliability of the elements, the lowest reliability index is used to describe the structure's reliability. This is equivalent to considering the weakest link of the set of failure components, and therefore modeling the structure as a *series system* (the system's failure is reached when one of the elements is failing). In the present case, by default this probability will be at least equal to the smallest of the failure probabilities, i.e. $2.33\,10^{-5}$.

This hypothesis is frequently used, although it is incorrect. In fact, the reliability of such a system not only depends on the reliability of its components but also on the way in which these components are combined to form the structural failure: we are referring to the *failure* (or *collapse*) *mechanism*. The failure mechanisms of the example are given on the following figure. The structure's failure is therefore conditioned by the occurrence of one of these two mechanisms: it is a series system, not in relation to the failure components, but in relation to the failure mechanisms.

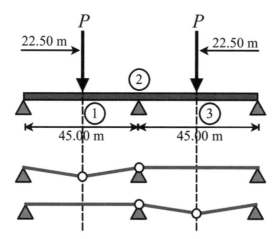

By applying the limit analysis principles, the limit states of these ruin mechanisms may be established:

$$\begin{cases} M_1 = \mathfrak{M}_1 - \mathfrak{M}_2 + P \times 22.50 & \text{Mechanism 1} \\ M_2 = \mathfrak{M}_3 - \mathfrak{M}_2 + P \times 22.50 & \text{Mechanism 2} \end{cases}$$

These identical mechanisms have the following for reliability indexes:

$$\beta = \frac{9,500 - 10,500 + 22.50 \times 896}{\sqrt{(9,500 \times 0.05)^2 + (10,500 \times 0.05)^2 + (22.50 \times 896 \times 0.10)^2}} \approx 8.96$$

i.e. as failure probabilities: $1.57\,10^{-19}$ which is much lower than the value of $2.33\,10^{-5}$.   □

We may draw a certain number of conclusions from these calculations. Firstly, the reliability indexes relative to the *failure mechanisms* are higher than the reliability indexes of the *failure modes*. This observation confirms that using the lowest reliability index in the three critical sections to characterize the structure's reliability level is not the most suitable: in fact, it may largely under estimate the structure's reliability.

Secondly, the failure mechanisms introduce common variables (the resisting bending moment on the middle bearing point, the live load point-load), which implies that these two mechanisms are statistically correlated, meaning that they are not independent. Since the beam can be seen as a series system, with each failure component being a failure mechanism, the reliability index of a bridge will be lower than the reliability indexes of each mechanism. The following sections specify a few of these concepts, and a few of the methods of the reliability theory for systems.

3.6.1. *Mathematical concepts*

In example 3.16, we introduced the concept of a series system. This type of system is, with *parallel systems*, one of the major concepts in the theory of system reliability. Before presenting the methods for quantifying the reliability of these systems, it is important to define the mathematical concepts which enable us to describe a *structural model*: this involves the concepts of *structure function*, the *system representation* and *redundancy*.

3.6.1.1. *Structure function*

A structure function ϕ is expressed in the following way:

– the structure can be in a single state: functioning or failed;

– the structure can be broken down into components (or failure modes) which may be functioning or failed.

The state of a system, then, depends on the state of its failure components. There are two fundamental types of system: *series systems* and *parallel systems*.

The series system, or weakest link system, is a system where the failure of one single component leads to the failure of the entire system (Figure 3.13a). In a parallel system (Figure 3.13b), the system's failure is conditioned by the failure of all its components.

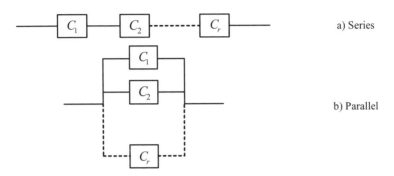

a) Series

b) Parallel

Figure 3.13. *Series and parallel systems*

Consider a system S made of r components C_i and let us define for each component a variable F_i indicating its functioning state:

$$F_i = \begin{cases} 1 & \text{the component is functioning} \\ 0 & \text{the component has failed} \end{cases} \qquad [3.98]$$

In the same way, F_S is the variable indicating the system's functioning state:

$$F_S = \begin{cases} 1 & \text{the system is functioning} \\ 0 & \text{the system has failed} \end{cases} \qquad [3.99]$$

Clearly, the value of F_S will depend on the value of F_i by the structural function ϕ:

$$F_S = \phi\left(F_1, F_2, \cdots, F_r\right) \qquad [3.100]$$

For a series system, this structure function is simply expressed as the product of Boolean variables:

$$F_{S,series} = \prod_{i=1}^{r} F_i \qquad [3.101]$$

For a parallel system, it is given by:

$$F_{S,parallel} = 1 - \prod_{i=1}^{r}(1 - F_i)$$

[3.102]

3.6.1.2. *System representations*

Any system can always be formally represented as a series system with parallel sub-systems. Two different approaches can be used for this, based on *path vectors* and *cut vectors* [MAD 86][3]. The concept of a failure mechanism often substitutes the cut vector concept. A path vector L of S is a sub-set of components, so much so that the system continues to function if all the elements of L are functioning, and those elements which do not belong to L have failed.

$$F_i = \begin{cases} 1 & i \in L \\ 0 & i \notin L \end{cases} \quad \text{and} \quad F_S = 1$$

[3.103]

If there are no path vectors L' included in L, then the path vector L is said to be minimal. A minimal path vector L is therefore a series system, since it fails if one of its components is no longer functioning. The system S is therefore described by a parallel system made of series sub-systems which are nothing other than the minimum path vectors.

A cut vector C of S is a sub-set of failure components, insofar as the system has failed if all the elements of C have failed, and those not belonging to C are functioning.

$$F_i = \begin{cases} 0 & i \in C \\ 1 & i \notin C \end{cases} \quad \text{and} \quad F_S = 0$$

[3.104]

A cut vector C is related to the concept of *failure mechanisms*. If there is no cut vector C' in C, then the cut vector C is said to be minimal. Or in a different way, the *mechanism C* is said to be *fundamental*. A cut vector C is therefore a parallel system, since it is functioning if one of its components functions. The system S is therefore described by a series system made of parallel sub-systems which are nothing other than the minimum cut vectors (or fundamental mechanisms).

---

3 Some authors use the terms of link and cut sets instead of path and cut vectors.

EXAMPLE 3.17.– Consider a two-span composite two-girder bridge.

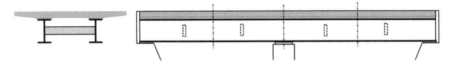

This structure has crossed-beams at $\frac{1}{3}$ and $\frac{2}{3}$ of the span, and we presume that the critical failure modes are at mid-span and on bearings. These failure modes are noted P1, P2, and P3 for the first beam P, and Q1, Q2, and Q3 for the beam Q. The failure modes of the cross-beams are noted E1, E2, E3 and E4.

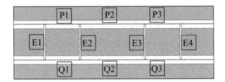

The problem can be represented in the following way:

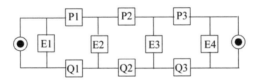

This representation is called the "reliability network" representation. We will go into more detail in the following section.

When all the critical sections P1, P2, P3, Q1, Q2 and Q3 are functioning, and if the loads are uniformly distributed, then the structural participation of the cross-beams is neglected. We presume that the different failure modes are independent and all have the same probability, equal to 0.90. When the two girders are functioning, the structure function and the reliability are respectively:

$$\phi(P1, P2, P3, Q1, Q2, Q3, E1, E2, E3, E4) = 1 - (1 - P1 \bullet P2 \bullet P3)(1 - Q1 \bullet Q2 \bullet Q3)$$

$$\mathcal{F}(P1, P2, P3, Q1, Q2, Q3, E1, E2, E3, E4) = 1 - (1 - 0.90 \times 0.90 \times 0.90)^2 = 0.93$$

According to the previous diagram, (E1,P1,E2,Q2,Q3) is a path vector, but is not minimal, because by removing E1, then (P1,E2,Q2,Q3) will still continue to function. This path vector exists thanks to the cross-beam E2. If such a cross-beam is absent, the structure would fail. We notice that cross-beams E2 and E4 are not essential elements in relation to the network representation, because they do not belong to minimal path vectors (if P1 and Q1 have failed, having E1 functioning does not change anything). This does not mean that, mechanically, these components do not function if P1 or Q1 are not functioning (P3 or Q3 respectively), because they will redistribute the loading. This explains that they can contribute to the path vectors but are not essential for ensuring that the whole system is functioning. This is not, however, the case for cross-beams E2 and E3 which ensure, in the case of simultaneous failure on beams P and Q, load redistribution.

Let us imagine a failure in Q1; the path vectors (P1,E2,Q2,Q3), (P1,E2,Q2,E3,P3), (P1,P2,P3), (P1,P2,E3,Q3) are minimal, which gives the following structure function:

$$\phi_{Q1} = 1 - (1 - P1 \bullet P2 \bullet P3) \times (1 - P1 \bullet E2 \bullet Q2 \bullet Q3)$$
$$\times (1 - P1 \bullet E2 \bullet Q2 \bullet E3 \bullet P3)$$
$$\times (1 - P1 \bullet P2 \bullet E3 \bullet Q3)$$

which gives:

$$\phi_{Q1} = 1 - (1 - 0.9^3) \times (1 - 0.9^3 \times 0.2) \times (1 - 0.9^3 \times 0.2^2) \times (1 - 0.9^3 \times 0.2) \approx 0.81$$

under the hypothesis that the cross-beams reliability is less than for the beams (here, a probability of 0.20).

In the same way, (P1,E2,Q2) is a minimal cut vector, because even if the other components are functioning, the dysfunction of cross-beam E2 will prevent the load redistribution from one beam to another.   □

EXAMPLE 3.18.– A simple spans viaduct made of three beams is supported by two large piers, each supported by two groups of four piles:

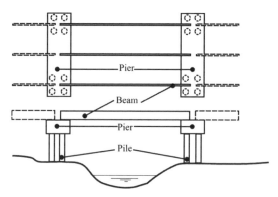

During an earthquake, the structure may collapse with many failure modes:

– the beams may be dislodged by sliding from the left pier (A) or the right pier (B);

– the failure of two pile groups on a pier (G1, G2 G3 and G4 denote the failure of the four pile groups).

The system representating the span failure is given by:

$$E = (A \cup B) \cup (G1 \cap G2) \cup (G3 \cap G4)$$

or graphically by:

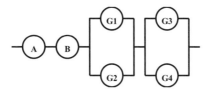

□

The representation in example 3.18 is called a *block diagram representation (BDR)*. The change from a block diagram representation into a fault tree representation (FTR) is possible and is even quite natural (Figure 3.14).

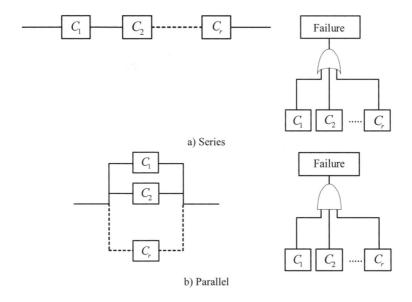

Figure 3.14. *Relationship between block diagram and fault tree representations*

EXAMPLE 3.19.– Let us consider again example 3.18. The fault tree representation is then:

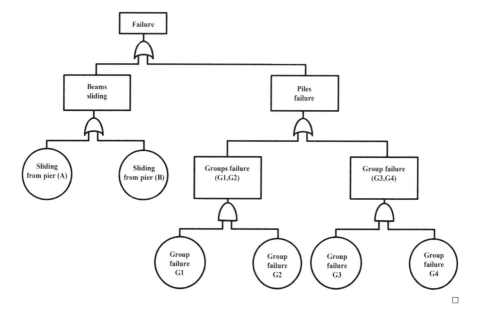

3.6.1.3. *Reliability network, redundancy*

Another representation, which is less common, is the *reliability network representation* or *graph representation*. Example 3.15 illustrates this. This representation [THA 87], which is much simpler, enables us to assess the way in which a structure functions and, in particular, to highlight the *load redistribution* capacity and *redundancy*.

Figure 3.15 illustrates this network representation for the case of simple and continuous span bridges. The failure components are the critical cross-sections (Figure 3.15b). Figure 3.15c shows that, for a simple span bridge, the reliability network representation is a parallel system where each component is characterized by the section's failure. For the continuous span bridge (Figure 3.15d), the representation is a parallel system of series sub-systems. These series sub-systems relate to failure modes of sections 1, 2 and 3, the parallel system being formed by the structure's five beams. Figure 3.15e shows the case of the continuous structure with diaphragms. The reliability network representation enables us to highlight that the cross-beams ensure the load redistribution.

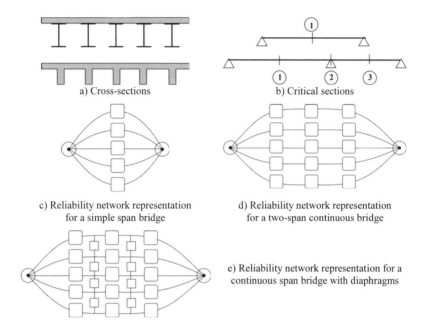

a) Cross-sections

b) Critical sections

c) Reliability network representation
for a simple span bridge

d) Reliability network representation
for a two-span continuous bridge

e) Reliability network representation for a
continuous span bridge with diaphragms

Figure 3.15. *Reliability network representations for simple and continuous span bridges*

EXAMPLE 3.20.– A suspension cable may be considered as a system made from a set of N strands laid out in parallel. A strand itself consists of a set of M twisted wires. Studying the behavior of a cable may be considered as a multi-scale study, where we can distinguish the scale of the wire, the strand and the cable. On the strand's scale, we must distinguish the strand segment characterized by the recovery length of the wires, and the strand itself. The following figure gives an outline of these different scales.

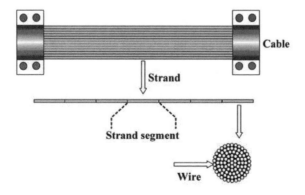

When the wires are twisted, any broken wire has the capacity to recover the imposed load over a given length, called the recovery length (from 1.0 to 2.5 times the wire pitch) and which will define the strand segment's length. The behavior of a strand will be closely related to the weakest strand segment, as it is a series system. Finally, with the strands being laid out in parallel, the cable strength will depend on their individual resistance and the load distribution between them. The reliability network representation of the suspension cable between two clamps may therefore be expressed by the following reliability network:

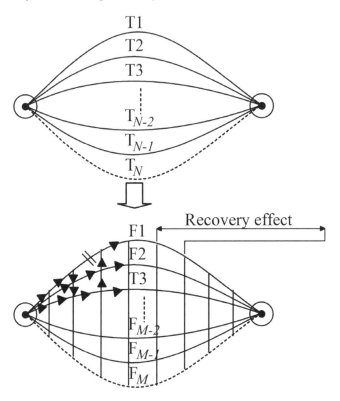

In this representation, the cable failure occurs when the set of N strands between the two clamps fails. With each strand consisting of M wires, there is a redistribution effect between adjacent wires, beyond the wires' recovery length. This results in a redundancy effect, because there is no complete wire failure when a local rupture occurs. The arrows in the previous figure indicate this redistribution effect for a local failure, indicated by the double lines. The cable between two clamps is described as a parallel system; the strand between two clamps also behaves as a complex series/parallel system by identifying the cut and/or path vectors. The reliability network representation is able to formalize this redistribution effect for a structure. However, it is more difficult to apprehend in a system representation.   □

3.6.1.4. *Redundancy*

The concept of *redundancy* is used to describe the capacity of the components or the sub-systems in a system to replace other components or sub-systems if failure occurs [THA 87]. Any structure, represented differently from a series system, may be considered as redundant; because there may well be failed components without the system which represents the structure is also failing.

Active redundancy represents a case where components are functioning simultaneously without it being essential for the system. This is particularly the case for a post-tensioned cross-section in a structure, consisting of many strands which are not essential, but which are active. *Passive redundancy* consists of components or sub-systems in *reserve*, which are not intended to function in normal service conditions. This is the case for certain bridge cross-beams or diaphragms.

3.6.2. *Calculation of the system probability of failure*

3.6.2.1. *Failure probability of a series system*

By definition, a series system fails as soon as one of its failure component has failed. Let a series system S allow m failure components. The failure mode i can be described by a safety margin M_i. The probability $P(M_i \leq 0)$ is the failure probability for this mode. Following the definition of a series system, its failure probability is expressed as the union of events $E_i = (M_i \leq 0)$:

$$P_f^S = \mathcal{P}\left(\bigcup_{i=1}^m (M_i \leq 0)\right) = \mathcal{P}\left(\bigcup_{i=1}^m E_i\right) \qquad [3.105]$$

We can show that this failure probability [MEL 99] may be bounded by:

$$\max_{i \in (1,m)} P_f^i \leq P_f^S \leq \sum_{i=1}^m P_f^i \qquad [3.106]$$

The term on the left relates to a case of perfect correlation (equal to 1) between the system's components. The term on the right is an approximation (for small probabilities) of the case when the elements are statistically independent. This clearly shows that the failure probability of a series system increases, on the one hand, with the number of components it possesses, and on the other hand, is largely conditioned by the weakest element, i.e. allowing the larger failure probability. These bounds are called *simple bounds* for a series system.

EXAMPLE 3.21.– Let us return to example 3.16. The beam is therefore described as a series system where each component is a failure mechanism. By applying the simple bounds of a series system, it is possible to get:

$$\max\left(5.93\,10^{-6};3.24\,10^{-6}\right) \le P_f^s \le \left(5.93\,10^{-6} + 3.24\,10^{-6}\right)$$

i.e. $5.93\,10^{-6} \le P_f^s \le 9.17\,10^{-6}$.   □

The simple bounds may be improved by taking into account the joint probabilities between the safety margins. Ditlevsen [DIT 79] thus proposes improved bounds called Ditlevsen bounds:

$$P_f^s \ge P_f^1 + \sum_{i=2}^{m}\left[\max\left\{0,\ P_f^i - \sum_{j=1}^{i-1}\mathcal{P}\left((g_i(Z)<0)\cap(g_j(\{Z\})<0)\right)\right\}\right]$$

$$P_f^s \le \sum_{i=1}^{m}P_f^i - \sum_{i=2}^{m}\max_{j<i}\ \mathcal{P}\left((g_i(\{Z\})<0)\cap(g_j(\{Z\})<0)\right)$$

[3.107]

The difficulty of calculating these bounds lies in assessing the joint probabilities $\left(g_i(\{Z\})\cap g_j(\{Z\})<0\right)$. This can be estimated by *simulations*. A different strategy consists of linearizing the limit states around the design points, then calculating the joint probabilities in the standardized normal space. Such a calculation uses the bivariate normal distribution. This can be calculated by Hermite polynomials.

$$H_0(x) = 1$$
$$H_1(x) = x$$
$$H_2(x) = x^2 - 1$$ [3.108]
$$H_n(x) = x\,H_{n-1}(x) - (n-1)\,H_{n-2}(x)$$
$$\Phi_2(x_1,x_2,\rho) = \frac{1}{2\pi}\exp\left(-\frac{x_1^2}{2} - \frac{x_2^2}{2}\right)\sum_{i=1}^{\infty}\frac{\rho^i}{i!}H_{i-1}(x_1)H_{i-1}(x_2)$$

This method provides good results. However, we also propose bounds, enabling an easier implementation [DIT 79]. If $\sum_{k}\alpha_{i,k}U_k + \beta_i = 0$ represents the first order approximation of the limit states, we obtain:

$$\Phi(-\beta_i)\,\Phi\left(-\frac{\beta_j-\rho_{ij}\beta_i}{\sqrt{1-\rho_{ij}^2}}\right)+\Phi(-\beta_j)\,\Phi\left(-\frac{\beta_i-\rho_{ij}\beta_j}{\sqrt{1-\rho_{ij}^2}}\right)\geq \mathcal{P}_{ij}$$

$$\max\left(\Phi(-\beta_i)\,\Phi\left(-\frac{\beta_j-\rho_{ij}\beta_i}{\sqrt{1-\rho_{ij}^2}}\right),\Phi(-\beta_j)\,\Phi\left(-\frac{\beta_i-\rho_{ij}\beta_j}{\sqrt{1-\rho_{ij}^2}}\right)\right)\leq \mathcal{P}_{ij}$$

[3.109]

with $\mathcal{P}_{ij}=\mathcal{P}\big((g_i(\{Z\})<0)\cap(g_j(\{Z\})<0)\big)$ et $\rho_{ij}=\displaystyle\sum_k \alpha_{i,k}\,\alpha_{j,k}$.

EXAMPLE 3.22.– The two limit states from example 3.16 give the following for direction cosines:

$$\alpha_{R_1}^1 = \frac{3{,}141}{\sqrt{3{,}141^2+3{,}391^2+4{,}919^2}} \approx 0.465$$

$$\alpha_{R_1}^2 = 0.000$$

$$\alpha_{R_2}^1 = \frac{-3{,}391}{\sqrt{3{,}141^2+3{,}391^2+4{,}919^2}} \approx -0.502$$

$$\alpha_{R_2}^2 = \frac{-3{,}391}{\sqrt{6{,}674^2+3{,}391^2+6{,}379^2}} \approx -0.345$$

$$\alpha_{R_3}^1 = 0.000;\quad \alpha_{R_3}^2 = \frac{6{,}674}{\sqrt{6{,}674^2+3{,}391^2+6{,}379^2}} \approx 0.679$$

$$\alpha_P^1 = \frac{4{,}919}{\sqrt{3{,}141^2+3{,}391^2+4{,}919^2}} \approx 0.728$$

$$\alpha_P^2 = \frac{6{,}379}{\sqrt{6{,}674^2+3{,}391^2+6{,}379^2}} \approx 0.649$$

With the other variables being deterministic, the direction cosines are zero.

$$\rho_{12}=\alpha_{R_1}^1\times\alpha_{R_1}^2+\alpha_{R_2}^1\times\alpha_{R_2}^2+\alpha_{R_3}^1\times\alpha_{R_3}^2+\alpha_P^1\times\alpha_P^2$$
$$=0+0.502\times0.345+0+0.728\times0.649\approx0.645$$

We therefore get the following bounds from the joint probability between the two safety margins:

$$\Phi(-4.38)\,\Phi\left(-\frac{4.51-0.645\times4.38}{\sqrt{1-0.645^2}}\right)+\Phi(-4.51)\,\Phi\left(-\frac{4.38-0.645\times4.51}{\sqrt{1-0.645^2}}\right)\geq \mathcal{P}_{12}$$

$$\max\left(\Phi(-4.38)\,\Phi\left(-\frac{4.51-0.645\times4.38}{\sqrt{1-0.645^2}}\right),\Phi(-4.51)\,\Phi\left(-\frac{4.38-0.645\times4.51}{\sqrt{1-0.645^2}}\right)\right)\leq \mathcal{P}_{12}$$

i.e. $8.79\ 10^{-8} \leq \mathcal{P}_{12} \leq 1.67\ 10^{-7}$.

Using the Ditlevsen bounds, the failure probability is bounded by:

$$P_f^1 + P_f^2 - \mathcal{P}_{12} \leq P_f^s \leq P_f^1 + P_f^2 - \mathcal{P}_{12}$$

i.e. by taking the extreme bounds:

$$\underbrace{5.93\ 10^{-6} + 3.24\ 10^{-6} - 1.67\ 10^{-7}}_{9.00\ 10^{-6}} \leq P_f^s \leq \underbrace{5.93\ 10^{-6} + 3.24\ 10^{-6} - 8.79\ 10^{-8}}_{9.08\ 10^{-6}}$$

which are much better than simple bounds. □

3.6.2.2. *Failure probability for a parallel system*

By definition, a parallel system fails as soon as all its components fail. Let there be a parallel system S allowing m failure components. The probability of failure is expressed by:

$$P_f^S = \mathcal{P}\left(\bigcap_{i=1}^{m}(M_i \leq 0)\right) \qquad\qquad [3.110]$$

In particular, it is also possible to give simple bounds for the failure probability of a parallel system:

$$\prod_{i=1}^{m} P_f^i \leq P_f^S \leq \min_{i\in(1,m)} P_f^i \qquad\qquad [3.111]$$

The term on the right corresponds to the case of perfect correlation (equal to 1) between the system's components. The term on the left corresponds to the case where the elements are statistically independent.

The probability failure framings of a parallel system are often useless. It is, then, preferable to use a calculation which uses the *multivariate normal distribution*, but this distribution is difficult to implement numerically (see [CRE 97]). In this case, the failure probability is calculated by linearizing the limit states around their design point. If $\sum\limits_{k} \alpha_{i,k} U_k + \beta_i = 0$ represents these first order approximations of the limit states in the standardized space (grouping together all the variables intervening in the different limit states), then we obtain:

$$
\begin{aligned}
P_f^S &= \mathcal{P}\left(\bigcap_{i=1}^{m} (M_i \leq 0) \right) \\[2mm]
&= \mathcal{P}\left(\bigcap_{i=1}^{m} \left[\sum_k \alpha_{i,k} U_k + \beta_i \leq 0 \right] \right) = \mathcal{P}\left(\bigcap_{i=1}^{m} \left[\sum_k \alpha_{i,k} U_k \leq -\beta_i \right] \right) \quad [3.112] \\[2mm]
&= \Phi_m \left(-\left\{ \begin{matrix} \vdots \\ \beta_i \\ \vdots \end{matrix} \right\}; \left[\sum_k \alpha_{i,k}\, \alpha_{j,k} \right] \right)
\end{aligned}
$$

$\left[\sum\limits_{k} \alpha_{i,k}\, \alpha_{j,k} \right]$ is the correlation matrix between the various safety margins.

EXAMPLE 3.23.– Let us take example 3.18; the system is therefore described as a series system with two parallel sub-systems. We will calculate the probabilities of these two sub-systems. Each group of piles supports ¼ of the load, and each pile supports ¼ of the load transferred onto the group of piles. We presume that the load's coefficient of variation is 0.30, and that the mean resistance of a pile is $\bar{S}/8$, with a 20% coefficient of variation ($\bar{S}$ is the load's mean value).

We also allow that the failure of a pile group induces a conditional failure probability of 0.90 for the second pile. The failure of one pile is not independent of the other piles, since load transfer between piles takes place when the other piles are failing. For a given group of piles, there are then 4! failure paths. Thus, if the piles are numbered 1 to 4, then 4-2-1-3, 3-4-2-1, 1-4-2-3, constitute the failure paths. Let us formally consider one of these paths that we will call *a-b-c-d*. The failure probability of pile *a* is given by a standard reliability calculation:

$$P_f(a) = \Phi\left(-\frac{\bar{S}/8 - \bar{S}/16}{\sqrt{\left(\bar{S}/8 \times 0.2\right)^2 + \left(\bar{S}/16 \times 0.3\right)^2}}\right)$$

$$= \Phi\left(-\frac{1/8 - 1/16}{\sqrt{\left(1/8 \times 0.2\right)^2 + \left(1/16 \times 0.3\right)^2}}\right) = 0.022$$

When pile a is failing, the load is transferred onto the other piles; the failure of pile b is therefore given by:

$$P_f(b/a) = \Phi\left(-\frac{\bar{S}/8 - \bar{S}/12}{\sqrt{\left(\bar{S}/8 \times 0.2\right)^2 + \left(\bar{S}/12 \times 0.3\right)^2}}\right) = 0.119$$

In a similar way, we obtain:

$$P_f(c/a,b) = \Phi\left(-\frac{\bar{S}/8 - \bar{S}/8}{\sqrt{\left(\bar{S}/8 \times 0.2\right)^2 + \left(\bar{S}/8 \times 0.3\right)^2}}\right) = 0.500$$

$$P_f(d/a,b,c) = \Phi\left(-\frac{\bar{S}/8 - \bar{S}/4}{\sqrt{\left(\bar{S}/8 \times 0.2\right)^2 + \left(\bar{S}/4 \times 0.3\right)^2}}\right) = 0.943$$

The probability of the failure option a-b-c-d occurring is therefore:

$$P_f(a-b-c-d) = P_f(a) \times P_f(b/a) \times P_f(c/a,b) \times P_f(d/a,b,c)$$

$$\approx 0.0227 \times 0.119 \times 0.500 \times 0.943 = 0.0012$$

The failure probability for a group of piles is calculated by the 4! possible combinations:

$$P(G1) = P(G2) = P(G3) = P(G4) = 4! \times 0.0012 \approx 0.0288$$

The failure probability of the two sub-systems is given by:

$$P(G1 \cap G2) = P(G3 \cap G4) = P(G1) \times P(G2/G1) = 0.9 \times P(G1) \approx 0.0011 \qquad \square$$

3.6.3. *Robustness and vulnerability*

In the context of risk analysis, vulnerability implies the capacity of a structure to sustain effects due to hazards, i.e. its reaction when faced with the impact of a certain phenomenon. In the case of a performance analysis, we must be assured that there will be no dysfunction. Vulnerability corresponds, then, to the tendency to dysfunction. In a certain sense, vulnerability may be confused with the concept of failure probability. A structure's robustness, according to its definition given in EN1991-1-7 [EN 06], is based on the assessment of a structure's capacity to sustain events such as fire, explosions, consequences linked to human errors, etc., without being damaged disproportionately in relation to its cause. The consideration of redundancy properties and non-linear behaviors is necessary if we wish to provide a credible assessment of the structural performance in damaged situations. This is because, in fact, robustness involves checking the post-failure behavior. Table 3.3 gives a few examples of the characteristics related to hazards, vulnerability and robustness.

	Characteristic	Indicators	Consequences
Exposure (hazard)	Impact Explosion/fire Earthquake, wind Traffic CO_2,Cl Vandalism Error, etc.	Amplitude Duration Variation Kinetics, etc.	
Vulnerability	Yielding Rupture Crack growth Delamination Corrosion, etc.	Regulations Standards/codes Age Materials Quality Condition Protection, etc.	Repair costs Fatalities Injuries Pollution, etc.
Robustness	Partial failure Global failure	Standards/codes Ductility Redundancy Segmentation Monitoring Emergency procedures, etc.	Reconstruction costs Demolition costs Evacuation costs Fatalities Injuries Socio-economic loss Pollution, etc.

Table 3.3. *Hazards, vulnerability and robustness*

Figure 3.16 gives an outline of the events leading to failure, from the occurrence of a hazard to the system's failure. This diagram may help to understand the meaning given to vulnerability and robustness, and to define a robustness index:

$$\mathbb{I}_{Ro} = \frac{R_{E,A}}{R_{E,A} + R_{F,E,A}} = \frac{C_D \times P(E \mid A)}{C_D \times P(E \mid A) + C_{IND} \times P_f(F \mid E, A)} \qquad [3.113]$$

$R_{E,A}$ and $R_{F,E,A}$ respectively represent the risk related to damage E, if hazard A occurs, and the risk related to the failure of system F conditioned by the damage E and hazard A. C_D and C_{IND} are then the costs linked to damage E and the system's failure. $P(E \mid A)$ and $P_f(F \mid E, A)$ are the probabilities of damage E and the system's failure F. If $\mathbb{I}_{Ro} = 1$, then the system is said to be fully robust; if $\mathbb{I}_{Ro} \to 0$ then the risk of failure is very high, and therefore the system is not very robust [BAK 08].

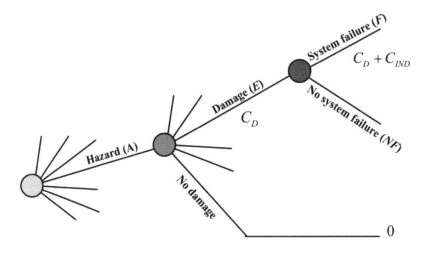

Figure 3.16. *Failure event tree*

EXAMPLE 3.24.– Let us consider n strands in an unbounded prestressed concrete beam. They are supporting a load T with mean 277 kN. The mean rupture strength R_i of one strand is 173 kN with a 15% coefficient of variation. The strands are assumed to independently carry the total load. The performance or limit state function for one strand is given by:

$$M_i = R_i - \frac{T}{n}, \qquad i = 1, \cdots, n$$

The probability of failure for each strand is then (assuming that the variables are independent and normal):

$$P_f = \Phi\left(-\frac{173 - \dfrac{277}{n}}{\sqrt{(173 \times 0.15)^2 + \left(\dfrac{277 \times \text{COV}[T]}{n}\right)^2}}\right)$$

Assuming that the strand behavior is brittle, if one strand fails, the probability of failure of a second strand is:

$$P_f = \Phi\left(-\frac{173 - \dfrac{277}{n-1}}{\sqrt{(173 \times 0.15)^2 + \left(\dfrac{277 \times \text{COV}[T]}{n-1}\right)^2}}\right)$$

The global failure starting from the rupture of one strand is therefore expressed in terms of successive strand failures:

$$P_f^S = (n-1)! \times \Phi\left(-\frac{173 - \dfrac{277}{n}}{\sqrt{(173 \times 0.15)^2 + \left(\dfrac{277 \times \text{COV}[T]}{n}\right)^2}}\right)$$

$$\times \prod_{i=1}^{n-1} \Phi\left(-\frac{173 - \dfrac{277}{n-i}}{\sqrt{(173 \times 0.15)^2 + \left(\dfrac{277 \times \text{COV}[T]}{n-i}\right)^2}}\right)$$

since from one failure there are $(n-1)!$ possible paths to global failure.

Let us fix for indirect costs $C_{IND} = 10C_D, = 50C_D, = 100C_D$, and for the coefficient of variation $\text{COV}[T] = 5\%, 15\%, 25\%$. The evolution of the robustness index versus the number of strands is given by the following figure. The robustness clearly increases with the number of components and the reduction of the coefficient of variation of the load. Conversely, it decreases if the indirect costs increase.

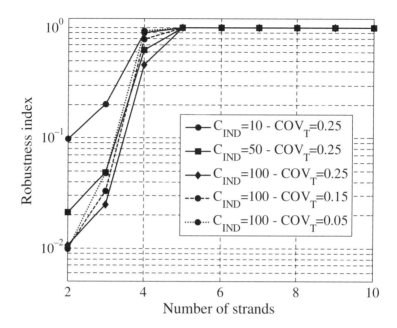

3.7. Determination of collapse/failure mechanisms

For analyzing robustness, the necessity to study the consequences of a local damage on the overall reliability of the structure has been highlighted. This implies identifying the most significant collapse mechanisms branched on the local damage. For simple structures, potential failure/collapse mechanisms can be derived from the principle of virtual work (see example 3.16). For large structures, it is difficult in practice to determine the collapse mechanisms and methods for identifying significant failure mechanisms are welcome. Automatic generation of safety margins has been developed to determine the formation of a mechanism.

3.7.1. *Generation of safety margins for truss structures*

The procedure which is presented in this section is based on Murotsu's work [MUR 80]. It can be summarized in two stages:

– the yielding condition occurs at the member ends and is approximated by a linear surface,

– the safety margin is expressed in terms of linear combinations of nodal resistances and loads.

This formulation helps to simplify the reliability analysis, but relies on several hypotheses that it is necessary to detail. Firstly, the structures are truss structures. Secondly the following properties are assumed:

– members are homogeneous, and critical sections where plastic hinges may occur are the locations at which the concentrated loads are applied. The structural analysis is made such as the members' ends are located at these places;

– yielding occurs when one of the limit state is reached;

– materials are perfectly elasto-plastic or elasto-brittle.

The failure criterion for a member occurs when the applied loading exceed the member's strength. If $\{f_i\}$ is the vector of the nodal forces applied to the ith structural member with nodes (k,l), expressed in the local coordinate system (Figure 3.17), yielding occurs at one or both of the member's end. The safety margin is given by:

$$\begin{cases} \mathbf{M}_{i,k} = R_{i,k} - {}^t\{\Lambda_{i,k}\}\{f_i\} \\ \mathbf{M}_{i,l} = R_{i,l} - {}^t\{\Lambda_{i,l}\}\{f_i\} \end{cases}$$ [3.114]

For space structures, the yielding criterion introduces combined load effects.

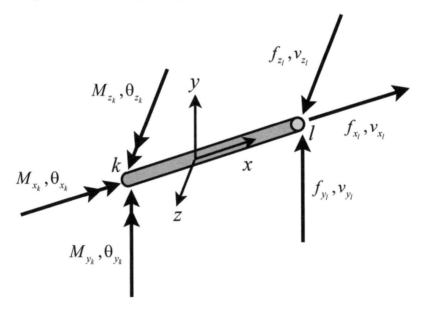

Figure 3.17. *Nodal forces and nodal displacements of a beam element*

Let us consider $\{\delta_i\}$ the vector of nodal displacements in the local coordinate system; the relation between the nodal forces and the nodal displacements is written:

$$\{\delta_i\} = [k_i]\{f_i\} \qquad\qquad [3.115]$$

with:

$$\{f\}_i = \left\{f_{x_k}, f_{y_k}, f_{z_k}, M_{x_k}, M_{y_k}, M_{z_k}, f_{x_l}, f_{y_l}, f_{z_l}, M_{x_l}, M_{y_l}, M_{z_l}\right\}^t$$

$$\{\delta\}_i = \left\{v_{x_k}, v_{y_k}, v_{z_k}, \theta_{x_k}, \theta_{y_k}, \theta_{z_k}, v_{x_l}, v_{y_l}, v_{z_l}, \theta_{x_l}, \theta_{y_l}, \theta_{z_l}\right\}^t \qquad [3.116]$$

$[k_i]$ is the elastic member stiffness matrix. The nodal forces and displacements are related to the global coordinate system by the transformation matrix:

$$\begin{cases} \{\delta_i\} = [P_i]\{d_i\} \\ \{f\} = [P_i]\{F_i\} \end{cases} \qquad\qquad [3.117]$$

where $\{F_i\}, \{d_i\}$ are the nodal force and displacement vectors referred to the global coordinate system. The stiffness matrix in this global system is given by:

$$[K_i] = {}^t[P_i][k_i][P_i] \qquad\qquad [3.118]$$

After rearranging the nodal displacement and force vectors into the global displacement and force vectors $\{d\}, \{F\}$, it comes:

$$\{d\} = [K]\{F\} \qquad\qquad [3.119]$$

with:

$$[K] = \sum_{i=1}^{n}[K_i] = \sum_{i=1}^{n}{}^t[P_i][k_i][P_i] \qquad\qquad [3.120]$$

The collapse of the structure occurs when a collapse/failure mechanism exists. This mechanism is formed as follows. When one hinges occurs, the loads are distributed over the other surviving components and the next hinge is determined. Repeating the process q times, the hinges $(r_i)_{1 \le i \le q}$ are formed. If the determinant of the stiffness matrix $[K]_q$, with q hinges, is null, then a collapse mechanism is formed.

To build this stiffness matrix, it is necessary to modify the relation between force and displacement vectors taking into account that a section of the member has yielded. The total displacement vector $\{\delta_i\}$ is therefore decomposed into an elastic displacement $\{\delta_i^e\}$ and a plastic displacement $\{\delta_i^p\}$:

$$\{\delta_i\} = \{\delta_i^e\} + \{\delta_i^p\} = \{\delta_i^e\} + \{\delta_{i,k}^p\} + \{\delta_{i,l}^p\} \qquad [3.121]$$

Based on the plastic deformation theory, the plastic deformations are expressed by [THO 86]:

$$\begin{cases} \{\delta_{i,k}^p\} = \lambda_{i\,k} \dfrac{\partial \mathbf{M}_{i,k}}{\partial \{f_i\}} = -\lambda_k \{\Lambda_{i,k}\} \\[3mm] \{\delta_{i,l}^p\} = \lambda_{i,l} \dfrac{\partial \mathbf{M}_{i,l}}{\partial \{f_i\}} = -\lambda_l \{\Lambda_{i,l}\} \end{cases} \qquad [3.122]$$

where $\lambda_{i,k}, \lambda_{i,l}$ are factors that indicate the magnitude of the plastic deformation.

The nodal force is given by:

$$\begin{aligned} \{f_i\} &= [k_i]\{\delta_i^e\} = [k_i]\left(\{\delta_i\} - \{\delta_i^p\}\right) \\ &= [k_i]\{\delta_i\} + \lambda_{i,k}[k_i]\{\Lambda_{i,k}\} + \lambda_{i,l}[k_i]\{\Lambda_{i,l}\} \end{aligned} \qquad [3.123]$$

leading to:

$$\begin{cases} R_{i,k} - {}^t\{\Lambda_{i,k}\}\left([k_i]\{\delta_i\} + \lambda_{i,k}[k_i]\{\Lambda_{i\,k}\} + \lambda_{i,l}[k_i]\{\Lambda_{i\,l}\}\right) = 0 \\[2mm] R_{i,l} - {}^t\{\Lambda_{i,l}\}\left([k_i]\{\delta_i\} + \lambda_{i,k}[k_i]\{\Lambda_{i\,k}\} + \lambda_{i,l}[k_i]\{\Lambda_{i\,l}\}\right) = 0 \end{cases} \qquad [3.124a]$$

From equation [3.124a], the factors $\lambda_{i,k}$ and $\lambda_{i,l}$ can be deduced and the following equation can be obtained:

$$\{f_i\} - \{\hat{f}_i\} = \left[\hat{k}_i\right]\{\delta_i\} \qquad [3.124b]$$

The explicit forms of $\{\hat{f}_i\}, \left[\hat{k}_i\right], \lambda_{i,k}$ and $\lambda_{i,l}$ are given in Table 3.4.

Elastic member	$\lambda_{i,k} = \lambda_{i,l} = 0$ $\left[\hat{k}_i\right] = \left[k_i\right]; \left\{\hat{f}_i\right\} = \{0\}$
Failure at left end	$\lambda_{i,k} = \dfrac{R_{i,k} - {}^t\left\{\Lambda_{i,k}\right\}\left[k_i\right]\left\{\delta_i\right\}}{{}^t\left\{\Lambda_{i,k}\right\}\left[k_i\right]\left\{\Lambda_{i,k}\right\}} ; \lambda_{i,l} = 0$ $\left[\hat{k}_i\right] = \begin{cases} \left[k_i\right] - \dfrac{\left[k_i\right]\left\{\Lambda_{i,k}\right\} {}^t\left\{\Lambda_{i,k}\right\}\left[k_i\right]\left\{\delta_i\right\}}{{}^t\left\{\Lambda_{i,k}\right\}\left[k_i\right]\left\{\Lambda_{i,k}\right\}} & \text{ductile} \\ [0] & \text{brittle} \end{cases}$ $\left\{\hat{f}_i\right\} = \begin{cases} \dfrac{R_{i,k}\left[k_i\right]\left\{\Lambda_{i,k}\right\}}{{}^t\left\{\Lambda_{i,k}\right\}\left[k_i\right]\left\{\Lambda_{i,k}\right\}} & \text{ductile} \\ [0] & \text{brittle} \end{cases}$
Failure at right end	$\lambda_{i,k} = 0; \lambda_{i,l} = \dfrac{R_{i,l} - {}^t\left\{\Lambda_{i,l}\right\}\left[k_i\right]\left\{\delta_i\right\}}{{}^t\left\{\Lambda_{i,l}\right\}\left[k_i\right]\left\{\Lambda_{i,l}\right\}}$ $\left[\hat{k}_i\right] = \begin{cases} \left[k_i\right] - \dfrac{\left[k_i\right]\left\{\Lambda_{i,l}\right\} {}^t\left\{\Lambda_{i,l}\right\}\left[k_i\right]\left\{\delta_i\right\}}{{}^t\left\{\Lambda_{i,l}\right\}\left[k_i\right]\left\{\Lambda_{i,l}\right\}} & \text{ductile} \\ [0] & \text{brittle} \end{cases}$ $\left\{\hat{f}_i\right\} = \begin{cases} \dfrac{R_{i,l}\left[k_i\right]\left\{\Lambda_{i,l}\right\}}{{}^t\left\{\Lambda_{i,l}\right\}\left[k_i\right]\left\{\Lambda_{i,l}\right\}} & \text{ductile} \\ [0] & \text{brittle} \end{cases}$
Failure at both ends	$\begin{Bmatrix} \lambda_{i,k} \\ \lambda_{i,l} \end{Bmatrix} = -\left[G_{i,k,l}\right]^{-1}\left[H_{i,k,l}\right]\left\{\delta_i\right\} + \left[G_{i,k,l}\right]^{-1}\begin{Bmatrix} R_{i,k} \\ R_{i,l} \end{Bmatrix}$ $\left[G_{i,k,l}\right] = \begin{bmatrix} {}^t\left\{\Lambda_{i,k}\right\}\left[k_i\right]\left\{\Lambda_{i,k}\right\} & {}^t\left\{\Lambda_{i,k}\right\}\left[k_i\right]\left\{\Lambda_{i,l}\right\} \\ {}^t\left\{\Lambda_{i,l}\right\}\left[k_i\right]\left\{\Lambda_{i,k}\right\} & {}^t\left\{\Lambda_{i,l}\right\}\left[k_i\right]\left\{\Lambda_{i,l}\right\} \end{bmatrix} ; \left[H_{i,k,l}\right] = \begin{bmatrix} {}^t\left\{\Lambda_{i,k}\right\}\left[k_i\right] \\ {}^t\left\{\Lambda_{i,l}\right\}\left[k_i\right] \end{bmatrix}$ $\left[\hat{k}_i\right] = \begin{cases} \left[k_i\right] - {}^t\left[H_{i,k,l}\right]\left[G_{i,k,l}\right]^{-1}\left[H_{i,k,l}\right] & \text{ductile} \\ [0] & \text{brittle} \end{cases}$ $\left\{\hat{f}_i\right\} = \begin{cases} {}^t\left[H_{i,k,l}\right]\left[G_{i,k,l}\right]^{-1}\begin{Bmatrix} R_{i,k} \\ R_{i,l} \end{Bmatrix} & \text{ductile} \\ [0] & \text{brittle} \end{cases}$

Table 3.4. *Expressions of* $\left\{\hat{f}_i\right\}, \left[\hat{k}_i\right], \lambda_{i,k}, \lambda_{i,l}$

The collapse of the structure is defined by the production of a failure mechanism. Let us assume the critical sections (hinges) $\left(r_i\right)_{1\le i\le q-1}$. The stiffness matrix is therefore:

$$\left[\hat{K}^{(q)}\right]\{d\} = \{F\} + \left\{F^{(q)}\right\}$$ [3.125]

introducing the modified stiffness matrix and force vector:

$$\left[\hat{K}^{(q)}\right] = \sum_{i=1}^{n}{}^{t}[P_i]\left[k_i^{(q)}\right][P_i]$$

$$\left\{F^{(q)}\right\} = -\sum_{i=1}^{n}{}^{t}[P_i]\left\{f_i^{(q)}\right\}$$ [3.126]

The collapse occurs when the stiffness matrix becomes singular.

3.7.2. β-unzipping method

Based on the automatic generation of safety margins for truss structures, there are several methods for determining the reliability of structures through the identification of significant collapse mechanisms. In this section, a very simple technique, by Thoft-Christensen and Sorensen [THO 83], is presented. Another technique, called the branch and bound method, is also very well known [THO 86] but is, in the end, less general than the β-unzipping method (in particular for different failure mode definitions).

The reliability of a structure is based on a multi-level analysis. At level 0, the reliability is estimated on the basis of failure of a single component (failure mode): the structure's reliability is equal to the reliability of this single element. Such a reliability analysis is not a system analysis but a component analysis. If a structure consists of n failure modes (components) characterized by their reliability indexes $\left(\beta_i\right)_{1\le i\le n}$, then the system reliability index in a level 0 analysis is given by:

$$\beta_S = \min_{i=1,\cdots,n} \beta_i$$ [3.127]

A more satisfactory approach of the structure's reliability can be performed at level 1 where the possibility of failure of any failure mode is taken into account by modeling the structural system as a series system. The probability of failure takes into account the correlations between the different safety margins. At level 1, the system failure can be calculated by means of the n-multivariate normal distribution but may be reduced to the most significant failure modes. Such failure modes can be selected by choosing the β-values ranging from β_{min} to $\beta_{min} + \Delta\beta_1$. $\Delta\beta_1$ must be chosen in an appropriate way. The selected failure modes are called critical failure modes.

The level 2 analysis is based on the analysis of a series system composed of parallel sub-systems. to obtain the so-called *critical pairs of failure modes*, the structure is modified by assuming failure in the critical failure components and adding fictitious loads corresponding to the load redistribution. Assuming the failure of component i, and modifying the stiffness matrix and the nodal forces, new safety margins and reliability indexes can be determined on the surviving failure components. The components with the lowest values are combined with the failure component for defining the critical pairs of failure modes. The procedure can be continued as far as possible and level k can be modeled as a series system of parallel sub-systems made of k failure components. Figure 3.18 gives an illustration of the system model at levels 0, 1 and 2.

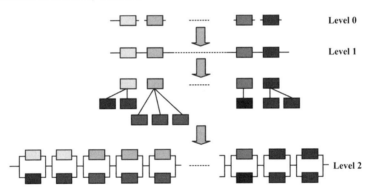

Figure 3.18. *System model at levels 0, 1 and 2*

EXAMPLE 3.25.– Let us consider the following simple frame loaded with two independent loads P and H.

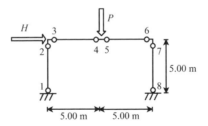

The properties of the resistances and loads are given in the table below.

Variable	Mean	Coefficient of variation
R_i	135 kN.m	10%
L	5 m	/
P	45 kN	10%
H	55 kN	10%

A finite element analysis of this structure provides the coefficient of influence and the safety margins for each critical section:

$$\begin{cases} M_1 = R_1 - 1.567H + 0.4973P \\ M_2 = R_2 - 0.9387H + 0.9991P \\ M_3 = R_3 + 0.9387H + 0.9991P \\ M_4 = R_4 - 0.00151H - 1.501P \\ M_5 = R_5 + 0.9356H + 0.9991P \\ M_6 = R_6 - 0.9356H - 0.9991P \\ M_7 = R_7 - 1.558H - 0.4973P \end{cases}$$

A reliability analysis for each critical component gives the reliability indexes. For instance:

$$\beta_1 = \frac{m_{R_1} - 1.567\, m_H + 0.4973\, m_P}{\sqrt{\sigma_{R_1}^2 + (1.567)^2\, \sigma_H^2 + (0.4973)^2\, \sigma_P^2}}$$

$$= \frac{135 - 1.567 \times 55 + 0.4973 \times 45}{\sqrt{13.5^2 + (1.567)^2 \times 5.5^2 + (0.4973)^2 \times 4.5^2}} \approx 4.40$$

The following table gives the reliability indexes for the 8 potential hinges.

Component	Reliability index
1	4.40
2	8.48
3	8.48
4	4.46
5	4.46
6	2.55
7	2.55
8	1.67

Let us take $\Delta\beta_1 = 3.0$. The critical failure modes are therefore 8, 7, 6, 1 and 4. The correlation matrix between the five safety margins can be calculated using the relation:

$$\rho_{ij} = \frac{a_{i,H}\,a_{j,H}\,\mathbb{V}[H] + a_{i,P}\,a_{j,P}\,\mathbb{V}[P]}{\sqrt{\left(a_{i,R_i}^2\,\mathbb{V}[R_i] + a_{i,H}^2\,\mathbb{V}[H] + a_{i,P}^2\,\mathbb{V}[P]\right)\left(a_{j,R_j}^2\,\mathbb{V}[R_j] + a_{j,H}^2\,\mathbb{V}[H] + a_{j,P}^2\,\mathbb{V}[P]\right)}}$$

if the safety margins are formally written $M_i = a_{i,R_i}R_i + a_{i,H}H + a_{i,P}P$. This gives:

$$[\rho] = \begin{bmatrix} 1.00 & 0.97 & 0.22 & 0.96 & 0.06 \\ 0.97 & 1.00 & 0.20 & 0.88 & 0.13 \\ 0.22 & 0.20 & 1.00 & 0.14 & 0.93 \\ 0.96 & 0.88 & 0.14 & 1.00 & -0.06 \\ 0.06 & 0.13 & 0.93 & -0.06 & 1.00 \end{bmatrix}$$

The probability of failure can be calculated using the multivariate normal distribution, with simple or Ditlevsen bounds. This leads to the generalized reliability index $\beta_S^1 \approx 1.62$ at the level 1 analysis.

The level 2 analysis implies to form parallel sub-systems from each critical failure mode; this results in critical pairs of failure modes. Let us consider the critical component 8 and analyze the modified frame including a critical hinge:

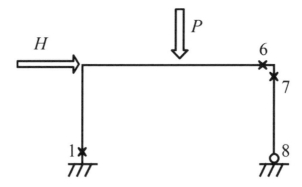

Fixing $\Delta\beta_2 = 1.0$, the level 2 analysis implies modifying the stiffness matrix and determining new safety margins. Using the automatic generation of safety margins, the following critical pairs of failure modes are identified:

(8,7), (8,6), (8,1), (7,8), (6,8), (1,8), (1,7), (1,6), (4,1)

The analysis of this series system of parallel sub-systems provides the generalized reliability index $\beta_S^2 \approx 2.45$. Following the same procedure, the level 3 analysis requires the introduction of two hinges. With $\Delta\beta_3 = 1.0$, the critical triples of failure modes are:

(8,7,1), (8,6,1), (8,6,4),(8,1,7), (8,1,6), (7,8,1), (6,8,1), (6,8,4), (1,8,7), (1,8,6), (1,7,8), (1,6,8), (1,6,4), (4,1,2), (4,1,8)

The reliability analysis gives the generalized reliability index $\beta_S^3 \approx 3.10$. This value can be compared the estimate of the system reliability at the failure mechanism level. Four collapse mechanisms can be identified for this simple frame: (8,7,1,2), (8,7,1,4), (8,6,1,4), (6,4,3).

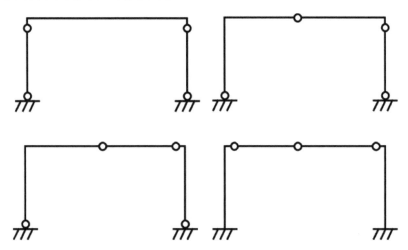

The safety margins are given using the principle of virtual works. The probability of failure of the structure is obtained as a series system of the four collapse mechanisms, giving $\beta_S \approx 4.19$, which is far from the level 3 analysis.

3.8. Calibration of partial factors

At the beginning of this chapter, we presented the different ways of introducing safety into design rules. Today, the most classical approach is based on partial (safety) factors: this is the approach most often encountered, particularly in Eurocodes [GUL 02]. The idea is to replace the single safety coefficient of the allowable stress design principle with a series of factors – said to be *partial* – on the different limit state variables. In the case of a limit state with a resistance variable R and a load variable S, this means having partial factors for R and S, written γ_R and γ_S. In other words, provided that each of these variables is a characterized by an appropriate value (known as a *characteristic*), design checking consists of ensuring that:

$$\gamma_S \ S_k \le \frac{R_k}{\gamma_R} \qquad [3.128]$$

The probabilistic approach to structural reliability enables us to establish such factors. Take the case of a linear limit state with two normal variables; we have seen that the failure probability was expressed by:

$$P_f = \Phi \left(-\frac{m_R - m_S}{\sqrt{\sigma_R^2 + \sigma_S^2}} \right) \qquad [3.129]$$

We will consider that the design rule tries to ensure a minimum safety level by the acceptable failure probability P_f^0. This is the same as imposing:

$$P_f = \Phi \left(-\frac{m_R - m_S}{\sqrt{\sigma_R^2 + \sigma_S^2}} \right) \le P_f^0 \qquad [3.130]$$

or differently:

$$\beta = \frac{m_R - m_S}{\sqrt{\sigma_R^2 + \sigma_S^2}} \ge \beta_0 = -\Phi^{-1}\left(P_f^0 \right) \qquad [3.131]$$

The index β_0 is often known as the *target reliability index*. It corresponds to the acceptable failure probability P_f^0.

The condition [3.117] may be expressed differently using direction cosines α_R and α_S defined by equation [3.25]:

$$m_R - m_S \ge \beta_0 \sqrt{\sigma_R^2 + \sigma_S^2} = \beta_0 \frac{1}{\sqrt{\sigma_R^2 + \sigma_S^2}} \left(\sigma_R^2 + \sigma_S^2 \right)$$

$$[3.132]$$

$$\ge \beta_0 \underbrace{\frac{\sigma_R}{\sqrt{\sigma_R^2 + \sigma_S^2}}}_{\alpha_R} \sigma_R + \beta_0 \underbrace{\frac{\sigma_S}{\sqrt{\sigma_R^2 + \sigma_S^2}}}_{-\alpha_S} \sigma_S$$

meaning:

$$m_R - \beta_0 \ \alpha_R \ \sigma_R \geq m_S - \beta_0 \ \alpha_S \ \sigma_S \qquad [3.133]$$

Values $R_d = m_R - \beta_0 \ \alpha_R \ \sigma_R$ and $S_d = m_S - \beta_0 \ \alpha_S \ \sigma_S$ are called *design values*. In fact, these are the coordinates of the design point in the physical space. Equation [3.133] results in the criterion.

$$R_d \geq S_d \qquad [3.134]$$

We will define the partial factors as the ratios between the representative values and the design values:

$$\gamma_R = \frac{R_k}{R_d}; \ \gamma_S = \frac{S_d}{S_k} \qquad [3.135]$$

thus the design rule [3.134] becomes:

$$\frac{R_k}{\gamma_R} \geq \gamma_S \ S_k \qquad [3.136]$$

Most of the time, the representative values are the *characteristic values*, i.e. the fractiles of variables R and S. In this simple case, the partial factors are defined by:

$$\gamma_R = \frac{R_k}{m_R} \frac{1}{1 - \beta_0 \ \alpha_R \ COV_R}; \ \gamma_S = \frac{S_d}{S_k} = \frac{m_S}{S_k}(1 - \beta_0 \ \alpha_S \ COV_S) \qquad [3.137]$$

The previous approach for a linear limit state with two normal variables can be generalized to a non-linear limit state introducing unspecific variables. For this, we will consider the design point in the standardized space $\{U^*\}$. The design point in the physical space is directly deduced from $\{U^*\}$ by the relationship $\{Z^*\} = T^{-1}(\{U^*\})$. $\{U^*\}$ is deduced from the reliability index by the relationship [3.53]. If we allow this design point to be found at distance β_0 (meaning, that the failure probability is the acceptable failure probability), we then obtain:

$$\{U^*\} = -\{\alpha\} \ \beta_0 \qquad [3.138]$$

Let us separate the variables into resistance variables (in numbers of m) and into load variables (in numbers of $n\text{-}m$). For the resistance variables $1 \le i \le m$, we will introduce a fractile $R_{k,i}$:

$$Z_i^* = T^{-1}\left(-\alpha_i\,\beta_0\right) = \frac{T^{-1}\left(-\alpha_i\,\beta_0\right)}{Z_{k,i}} Z_{k,i} = \frac{Z_{k,i}}{\gamma_{Ri}} \qquad [3.139]$$

and for the load variables $m+1 \le j \le n$:

$$Z_i^* = T^{-1}\left(-\alpha_i\,\beta_0\right) = \frac{T^{-1}\left(-\alpha_i\,\beta_0\right)}{Z_{k,i}} Z_{k,i} = \gamma_{Si}\, Z_{k,i} \qquad [3.140]$$

These partial factors are approximate values. In fact, they consist of using an initial design which provides the direction cosines α_i, and relocating (into the standardized normal space) the initial design point U_i^* towards a point found at distance β_0. This point is not necessarily the design point U^* which is sought after[4]. This problem will be tackled in Chapter 4 because the non-unique nature of partial factors enables us to adjust some of them, whilst retaining old design values and therefore avoiding the introduction of those which could be totally different to older practices (which could be induced by expressions [3.139] and [3.140]).

This problem is linked to the problem of invariance mentioned in section 3.3. In fact, by taking equation [3.136] again, there is nothing preventing us from multiplying the two members of inequality by a coefficient η, whilst also retaining the characteristic values:

$$\frac{R_k}{\left(\gamma_R \times \dfrac{1}{\eta}\right)} \ge \left(\gamma_S \times \eta\right)\, S_k \qquad [3.141]$$

which leads us to work with two new partial factors:

$$\hat{\gamma}_R = \left(\gamma_R \times \frac{1}{\eta}\right);\; \hat{\gamma}_S = \left(\gamma_S \times \eta\right) \qquad [3.142]$$

---

4 The reader may consult reference [GAY 04] which offers a summary of the approaches possible for calibrating partial factors.

In fact, we will note here that the important point is to preserve the product $\gamma_R \times \gamma_S = \hat{\gamma}_R \times \hat{\gamma}_S$.

EXAMPLE 3.26.– If the limit state is linear, we can go a bit further in determining partial factors by using the following approximation $\sqrt{1 + \dfrac{\sigma_S^2}{\sigma_R^2}} \approx a + b\dfrac{\sigma_S}{\sigma_R}$. Many expressions can be used: the following figure compares the following functions as an illustration $\sqrt{1 + x^2}$, $0.8 + 0.7\,x$ and $0.5(1 + x)$.

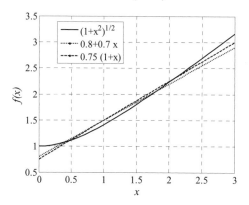

We can then write:

$$m_R - m_S \geq \beta_0 \sqrt{\sigma_R^2 + \sigma_S^2} \approx \beta_0\,\sigma_R \left(a + b\frac{\sigma_S}{\sigma_R} \right)$$

i.e.:

$$m_R - a\,\beta_0\,\sigma_R \geq m_S + b\,\beta_0\,\sigma_S$$

The design values R_d and S_d can therefore be approached for a linear limit state with normal variables by the following expressions:

$$R_d = m_R - a\,\beta_0\,\sigma_R$$
$$S_d = m_S + b\,\beta_0\,\sigma_S$$

We should recall that by definition, $\sqrt{\alpha_R^2 + \alpha_S^2} = 1$. The approximation choice is equivalent to writing $\alpha_R \approx a$ and $\alpha_S \approx -b$, and therefore checking that $\sqrt{a^2 + b^2} \approx 1$ (for the proposed approximations, this value is worth 1.2025 and

1.125 respectively). Eurocode EN1990 proposes retaining $a = 0.8$ and $b = 0.7$ when the ratio σ_S/σ_R is between 0.16 and 7.6. Outside this range, we recommend to implementing the value 1.0 on the parameter a showing the largest standard deviation. The approximation $\alpha_R = -\alpha_S = 0.75$ has the advantage of not favoring one variable over another. It is particularly used in LRFD (*load and resistance factor design*) format, in Canadian and American regulations.

By setting out $\gamma_R = \dfrac{R_k}{R_d}$, $\gamma_S = \dfrac{S_d}{S_k}$, the partial factors are deduced from characteristic values by the following ratios:

$$\gamma_R = \frac{R_k}{m_R - a\,\beta_0\,\sigma_R} = \frac{R_k}{m_R}\left(\frac{1}{1 - a\,\beta_0\,\text{COV}_R}\right)$$

$$\gamma_S = \frac{m_S + b\,\beta_0\,\sigma_R}{S_k} = \frac{m_S}{S_k}\left(1 + b\,\beta_0\,\text{COV}_R\right)$$

□

EXAMPLE 3.27.– Let us consider a design rule which recommends verifying the following inequality:

$$\frac{R_k}{\gamma_R} \geq \gamma_Q\,Q_k + \gamma_G\,G_k$$

where R, Q and G represent the variables for resistance, live load and permanent load. We assume that these variables are normally distributed and that the bias between characteristic values and mean values are written: v_R, v_G, v_Q. Let us presume that the target index is β_0. We are therefore looking to have:

$$\beta = \frac{m_R - m_Q - m_G}{\sqrt{\sigma_R^2 + \sigma_Q^2 + \sigma_G^2}} = \beta_0$$

We therefore obtain: $\dfrac{\dfrac{R_k}{v_R} - \dfrac{Q_k}{v_Q} - \dfrac{G_k}{v_G}}{\sqrt{\left(\dfrac{R_k}{v_R}\text{COV}_R\right)^2 + \left(\dfrac{Q_k}{v_Q}\text{COV}_Q\right)^2 + \left(\dfrac{G_k}{v_G}\text{COV}_G\right)^2}} = \beta_0$. This

equation gives a ratio relating $\dfrac{R_k}{v_R}$ to $\dfrac{Q_k}{v_Q}$ and $\dfrac{G_k}{v_G}$:

$$\left[\left(\frac{R_k}{\nu_R}\right)^2 - \beta_0^2\left(\frac{R_k}{\nu_R}COV_R\right)^2\right] - 2\left(\frac{Q_k}{\nu_Q} + \frac{G_k}{\nu_G}\right)\frac{R_k}{\nu_R} + \cdots$$

$$\cdots\left[\left(\frac{Q_k}{\nu_Q} + \frac{G_k}{\nu_G}\right)^2 - \beta_0^2\left(\frac{Q_k}{\nu_Q}COV_Q\right)^2 - \beta_0^2\left(\frac{G_k}{\nu_G}COV_G\right)^2\right] = 0$$

which conditions the design (which will act on R_k) ensuring the performance criterion β_0. The direction cosines are then given by the following ratios:

$$\alpha_R = \frac{\dfrac{R_k}{\nu_R}COV_R}{\sqrt{\left(\dfrac{R_k}{\nu_R}COV_R\right)^2 + \left(\dfrac{Q_k}{\nu_Q}COV_Q\right)^2 + \left(\dfrac{G_k}{\nu_G}COV_G\right)^2}}$$

$$\alpha_Q = -\frac{\dfrac{Q_k}{\nu_Q}COV_Q}{\sqrt{\left(\dfrac{R_k}{\nu_R}COV_R\right)^2 + \left(\dfrac{Q_k}{\nu_Q}COV_Q\right)^2 + \left(\dfrac{G_k}{\nu_G}COV_G\right)^2}}$$

$$\alpha_G = -\frac{\dfrac{G_k}{\nu_G}COV_G}{\sqrt{\left(\dfrac{R_k}{\nu_R}COV_R\right)^2 + \left(\dfrac{Q_k}{\nu_Q}COV_Q\right)^2 + \left(\dfrac{G_k}{\nu_G}COV_G\right)^2}}$$

According to equations [3.139] and [3.140], we obtain:

$$\frac{m_R}{R_k}\left(1 - \alpha_R\,\beta_0\,COV_R\right) = \frac{1}{\gamma_R}$$

$$\frac{m_Q}{Q_k}\left(1 - \alpha_Q\,\beta_0\,COV_R\right) = \gamma_Q$$

$$\frac{m_G}{G_k}\left(1 - \alpha_G\,\beta_0\,COV_R\right) = \gamma_G \qquad\qquad \square$$

EXAMPLE 3.28.– Let us take example 3.27 and assume that $m_Q/m_G = 2.0$. In this case, the relationship between the resistance and the loads is written by taking

$v_R = 0.95$; $v_Q = 1.18$; $v_G = 0.95$; $COV_R = 0.11$; $COV_Q = 0.25$; $COV_G = 0.10$ and
$\beta_0 = 3.5$:

$$m_R^2 - 7.04\, m_R\, m_G + 8.62\, m_G^2 = 0$$

which gives us the more severe solution:

$$m_R = \left(\frac{7.04 + 3.88}{2}\right) m_G \approx 5.46\, m_G$$

The direction cosines are easily deduced:

$$\alpha_R = \frac{5.46 \times 0.11}{\sqrt{(5.46 \times 0.11)^2 + (2 \times 0.25)^2 + (0.10)^2}} \approx 0.7623$$

$$\alpha_Q = -\frac{0.5}{0.7879} \approx -0.6346$$

$$\alpha_G = -\frac{0.1}{0.7879} \approx -0.1319$$

which means that we can determine the partial coefficients:

$$\frac{1}{\gamma_R} = 0.95 \times (1 - 0.7623 \times 3.5 \times 0.11) \approx 0.67$$
$$\gamma_Q = 1.18 \times (1 + 0.6346 \times 3.5 \times 0.25) \approx 1.83$$
$$\gamma_G = 0.95 \times (1 + 0.1319 \times 3.5 \times 0.10) \approx 0.99$$

This gives the design condition to be verified: $\dfrac{R_k}{1.49} \geq 1.83\, Q_k + 0.99\, G_k$.   □

3.9. Nature of a probabilistic calculation

The technical community is often perplexed when faced with failure probability calculations. In fact, what meaning can we give to a result which generally borders 10^{-6}? One of the main difficulties for a good understanding of the results from a reliability analysis is that it is not possible to translate a failure probability of 10^{-6} into a comprehensible measurement. In Chapter 1, we mentioned the different forms taken by the concept of probability. The engineer most often sees probability as a frequency measure, i.e. the number of outcomes stated out of the total number of possible outcomes. When we refer to an annual failure probability of 10^{-6}, unfortunately we should not conceive it as the annual frequency of one structure (from a certain type) failing out of 1,000,000! What this probability actually tells us in fact has only relatively little meaning in the absolute. It will only be useful to us compared to the probability calculated for another structure with the same analysis methods and the same probability distributions for the variables. As the failure probability must be analyzed relatively, it is important to do so in a coherent framework. In fact, the reliability indexes are very sensitive to the choice of distributions used. A low reliability index (or a high failure probability) does not necessarily indicate imminent failure in relation to a given failure mode. However, it clearly indicates that maintenance, repair, restoration or reinforcement is necessary to avoid failure in a more or less close time delay. This not only involves connecting the condition of a structure or one of its elements with structural performance, but also allocating budgets which ensure a sufficient level of performance at a minimum total maintenance cost over the service lifetime.

3.10. Failure probabilities and acceptable risks

3.10.1. *Acceptable failure probabilities*

Calibrating the partial factors of a semi-probabilistic approach for performance assessment is based on the definition of an *acceptable failure probability*. It would be more practical to speak about the *probability of acceptable performance loss*, because the concept covers structural safety (ultimate limit state) as well as serviceability (service limit states). The choice of such an acceptable probability was already approached in Chapter 2 (section 2.3.2). In theory, provided that human errors are not taken into consideration, this acceptable probability could be obtained by optimizing total costs, including failure costs. This optimization is quite difficult when it is a matter of calibrating an acceptable probability and we often use other approaches.

One of the traditional approaches is, in fact, calibrating the acceptable failure probability from existing standards. This is relevant if the standard rule is said to be optimal, meaning that it offers a good compromise between economy and safety. This hypothesis is far from being verified, since some dimensions are known as being very conservative, and therefore not optimal on the economic front. Another problem for calibrating an acceptable reliability index lies in the absence of data on uncertainties linked to resistances and loads effects. In fact, this is rarely mentioned in standards which are generally deterministic. The approach therefore consists of designing a set of structures or structural elements according to existing rules. Next, the uncertainties with resistances, models and loads may be chosen on the basis of data taken from literature. The reliability indexes for each component (for a given limit state) can be calculated from these hypotheses. Of course, each failure component receives a different reliability index. For a return procedure on the initial hypotheses, we are looking to obtain a set of indexes ranging within a given interval. From this interval, a *target reliability index* $\beta_0 = -\Phi\left(P_f^0\right)$ can be chosen.

We must also underline that a certain differentiation in the acceptable probabilities is needed. In fact, a limitation of reliability in relation to a serviceability limit state can be less severe in relation to an ultimate limit state, where the consequences in terms of safety are very different! In the same way, the consequences of failure, as well as the type of rupture (brittle or ductile) must play a role in the calibration of acceptable probabilities. Finally, the reference period (a year or many years) must be introduced. The annual failure probabilities are, in fact, lower than the failure probabilities over the service period.

In this way, Eurocode [EN 02] defines three categories of failure consequences (in terms of losses of human lives, or economic and environmental losses):

– CC1: low consequences;

– CC2: medium consequences;

– CC3: high consequences.

These categories of failure consequences are affected differently according to the structure's usage (see Table 3.5).

Usage frequency	Failure consequence		
	Low	Medium	High
Low	CC1	CC2	CC3
Medium	CC2	CC2	CC3
High	-	CC3	CC3

Table 3.5. *Consequence matrix (according to [EN 02])*

These categories of consequences relate to reliability classes on which failure probabilities are assigned, which differ according to the considered limit state. Table 3.6 indicates the values retained for these acceptable failure probabilities in Eurocodes. Eurocode recognized that the risks today associated with structural failures can be considered as CC2. In this way, the Eurocode stipulates an annual minimal failure probability P_f^0 of $1.5 \ 10^{-6}$ ($\beta_0 = 4.7$) and of $7.2 \ 10^{-5}$ ($\beta_0 = 3.8$, $T = 50$ years) for the ultimate limit states, and $1.4 \ 10^{-3}$ ($\beta_0 = 3.0$) and $7.0 \ 10^{-2}$ ($\beta_0 = 1.5$, $T = 50$ years) for the serviceability limit states.

Reliability classes	ULS		Fatigue		SLS	
	1 year	50 years	1 year	50 years	1 year	50 years
RC3	5.2	4.3	-	-	-	-
RC2	4.7	3.8	-	1.5-3.8	2.9	1.5
RC1	4.2	3.3	-	-	-	-

Table 3.6. *Target reliability indexes (according to [EN 02])*

If P_{fa} is the annual failure probability, an approximation of the failure probability P_{fT} over the design working period T is given by

$$P_{fT} = 1 - \left(1 - P_{fa}\right)^T \approx \sum_{k=1}^{T} P_{fa} = T \, P_{fa}.$$ An expression which can also be used in terms

of the reliability index is: $\Phi(\beta_T) = \left[\Phi(\beta_a)\right]^T$.

These acceptable probabilities are acceptable for new structures, meaning they are valid during the design phase[5]. For existing structures, we must take into account the consequences of failure and inspection quality, in a way which is more effective than for calibrating design rules. This modulation of target reliability indexes is, in fact, hidden in some design standards for new structures, since Eurocode 3 [EN 05] proposes to modify the partial factor for fatigue, according to whether welded joints are inspected or not. This implies that the subjacent reliability index varies according to the quality of the inspection. Moreover, they are not adapted to other limit states than those mentioned in the Eurocode. Thus, in the case of limit states for "material" durability (carbonation, chloride penetration, leaching, etc.) the given target indexes cannot be directly applied.

EXAMPLE 3.29.– Let us consider a temporary structure giving access to a site, built for an intended service period of five years. The failure consequences and the usage frequency are judged as medium. According to Table 3.6, the reliability class proposed by Eurocode is RC2. The acceptable failure probabilities for SLS and ULS are given by:

$$\mathcal{P}_{f,5}^{ULS} \leq 5 \times \Phi(-4.7) = 6.5 \; 10^{-6} \qquad \beta_5^{ULS} \geq 4.36$$
$$\mathcal{P}_{f,5}^{SLS} \leq 5 \times \Phi(-2.9) = 9.3 \; 10^{-3}, \text{ or } \beta_5^{SLS} \geq 2.35 \qquad\qquad \square$$

---

5 Due to the limit states that they cover (Chapter 2, section 2.6.2) it is not *a priori* recommended to use them as limit states which are relative to durability. In fact, the target indexes for the SLS may prove to be very conservative since they have been calibrated for structural serviceability limit states.

There are many standards which introduce this variation in assessing existing structures. The Scandinavian standards [NOR 87] also indicate acceptable annual failure probabilities for ultimate limit states (Table 3.7). The *Joint Committee of Structural Safety* [JCS 00] proposed probabilistic rules for evaluating existing structures (Table 3.8) for ultimate limit states with an annual reference period. For serviceability limit states, acceptable index values from 1 to 2 are recommended.

Failure consequence (Safety class)	Failure: ductile failure with remaining capacity	Failure: ductile failure without remaining capacity	Failure: brittle failure
Less serious (Low safety class)	$P_f \leq 10^{-3}$, $\beta \geq 3.1$	$P_f \leq 10^{-4}$, $\beta \geq 3.71$	$P_f \leq 10^{-5}$, $\beta \geq 4.26$
Serious (medium safety class)	$P_f \leq 10^{-4}$, $\beta \geq 3.71$	$P_f \leq 10^{-5}$, $\beta \geq 4.26$	$P_f \leq 10^{-6}$, $\beta \geq 4.75$
Very serious (high safety class)	$P_f \leq 10^{-5}$, $\beta \geq 4.26$	$P_f \leq 10^{-6}$, $\beta \geq 4.75$	$P_f \leq 10^{-7}$, $\beta \geq 5.20$

Table 3.7. *Acceptable annual failure probabilities according to NKB*

Consequences (seriousness)	Safety measure costs		
	High	Normal	Low
Minor	3.1	3.7	4.2
Moderate	3.4	4.2	4.4
Critical	3.7	4.4	4.7

Table 3.8. *Acceptable annual failure probabilities according to JCSS*

A rather advanced version of this reliability index modulation can be found in the Canadian regulation CAN/CSA-S6-00. The basic target index (new structure) is modified with the rise (increased acceptability level) or with the fall (reduced acceptability level) following various parameters. For new structures, the code CSA-S6 recommends a reliability index of 3.50 for 75 years design working life, i.e. an annual reliability index of 3.7 (Table 3.9).

The Canadian philosophy for assessing existing bridges is to maintain a consistent level of risk for the loss of human lives for each element of a structure. Those structures receiving regular inspections, with alarms in case of failure and allowing load redistribution, are less likely to induce human loss than a structure without such attributes. This is why the code CSA-S6 allows higher acceptable failure probabilities if the risk of losing human lives is reduced in the case of failure.

To assess the structural elements of existing constructions, the behavior of these structural elements is divided into three categories:

– E1: failure with no alarm;

– E2: failure with no alarm but with residual bearing capacity;

– E3: progressive failure with alarm.

The behavior in a structure's system is also divided into three categories:

– S1: element failure leading to global failure;

– S2: element failure leading to global failure but with a low probability;

– S3: element failure only leads to local failure.

Finally, the level of inspection is distributed into three categories:

– I1: the element cannot be inspected, but observing nearby elements may give indications of its performance under service conditions;

– I2: the element is inspected regularly, the results are recorded (routine inspection);

– I3: the element is inspected in detail by a qualified inspector (detailed inspection).

According to these various categories and classes, and for two traffic conditions (normal traffic/authorized traffic "PA/PB and PS", and controlled authorized traffic PC), annual reliability indexes are given (Table 3.9). An entire set of partial coefficients for the loads has been calibrated from these acceptable indexes.

Normal and authorized traffic PA, PB, PS					PC Traffic				
Behavior		Inspection			Behavior		Inspection		
System	Element	I1	I2	I3	System	Element	I1	I2	I3
S1	E1	4.00	3.75	3.75	S1	E1	3.50	3.25	3.25
	E2	3.75	3.50	3.50		E2	3.25	3.00	2.75
	E3	3.50	3.25	3.00		E3	3.00	2.75	2.50
S2	E1	3.75	3.50	3.50	S2	E1	3.25	3.00	2.75
	E2	3.50	3.25	3.00		E2	3.00	2.75	2.50
	E3	3.25	3.00	2.75		E3	2.75	2.50	2.25
S3	E1	3.50	3.25	3.00	S3	E1	3.00	2.75	2.50
	E2	3.25	3.00	2.75		E2	2.75	2.50	2.25
	E3	3.00	2.75	2.50		E3	2.50	2.25	200

Table 3.9. *Annual reliability indexes for assessing bridges according to CAN/CSA-S6-00*

3.10.2. *Concept of acceptable risk*

In the previous examples, the level of safety is defined by acceptable failure probabilities without explicitly introducing the consequences of the failure. There are, however, approaches which try to define this level of safety as a risk function, i.e. parameters representing the structure's value and importance.

An example of such a procedure is the proposal made in the British recommendations BD79 [BD 01] relating to the assessment of existing bridges; it does not directly propose acceptable reliability indexes for assessing an existing structure, but proposes categories of failure consequences.

Categories	Scenarios	C_0/C_e
High > 10 deaths	– Complete collapse of a very busy motorway bridge – Other motorway or a 4+ lane trunk road over a busy railway line – Bridges in a built-up area (e.g. over a station, near to shops, offices, schools, etc.)	1
Medium < 10 deaths	– Complete collapse of a motorway bridge <10 m span – Trunk road with < 4 lanes – Partial collapse of a bridge in the "high" scenario – Bridge on a minor road crossing a minor road	3
Low No death	– Element failure/local damage/partial collapse	10

Table 3.10. *Definition of consequence categories according to BD 79*

As a basis, this proposal uses the reliability index which is subjacent to the design rule and the cost of a highway bridge collapse. The risk can be calculated using this basis:

$$R_0 = P_f^0 \, C_0 \qquad\qquad [3.143]$$

For an existing structure, we try to maintain this risk level:

$$R_0 = R_e = P_f^e \, C_e \qquad\qquad [3.144]$$

where C_e is the cost of failure for this structure. This leads us to define, for this structure, an acceptable failure probability of:

$$P_f^e = P_f^0 \frac{C_0}{C_e}$$
[3.145]

A structure will, then, be classed into one of the three consequence categories (high, medium, low) corresponding to ratios 1, 1/3 and 1/10 for C_e / C_0 (Table 3.9). The BD79 proposal authorizes calculating the ratio C_e / C_0 more rigorously if necessary.

Another example of defining acceptable risks was proposed by Brühwiler and Bailey [BRU 02]. For all the identified risks, risk categories are determined, on the one hand linked to the magnitude of the induced damage, and on the other, the structure's economic importance. These risk categories enable us to define safety levels characterized into failure probabilities. Incidentally, the acceptable failure probabilities will be different according to the failure modes concerned.

The risk category R_{cd} related to damage is described in numbers of deaths induced by the structure's collapse (Table 3.11). As a function of the danger considered and the failure mode involved (with this couple defining a risk situation), this category also depends on the volume of traffic, the geometry and the structure's location. The economical importance can be estimated by the costs induced by failure.

Probable number of deaths	R_{cd}
<1	I
1	II
5	III
10	IV
50	V
100	VI
500	VII

Table 3.11. *Risk categories related to damage*

Safety measures	Consequences		
	Minor $\rho < 2$	Moderate $2 < \rho < 5$	Critical $5 < \rho < 10$
High	I	II	III
Normal	III	IV	V
Low	V	VI	VII

Table 3.12. *Risk categories related to structure's importance*

In Table 3.12, the risk categories R_{cu} related to the importance or the use of the considered structure are given according to the consequences of failure and the relative costs of safety measures implemented to reduced the danger. The consequences are expressed in the ratio between the failure cost C_f (cost of accident, reconstruction, indirect costs such as delays, etc.) and intervention costs for preventing failure C_i.

$R_c = \max(R_{cd}, R_{cu})$ corresponds to the risk category for the considered situation. Table 3.13 provides the relationship between this risk level and the acceptable failure probability.

R_c	P_f^0	β_0
I	10^{-3}	3.1
II	$5\,10^{-4}$	3.4
III	10^{-4}	3.7
IV	$5\,10^{-5}$	4.0
V	10^{-5}	4.2
VI	$5\,10^{-6}$	4.4
VII	10^{-6}	4.7

Table 3.13. *Annual target failure probability*

This approach provides target values for the structure or potentially an element, provided that the system's failure is dominated by the failure of this element. In a structure which is highly redundant, only Table 3.13 in connection with Table 3.12 is applied: the reader will note that this is equivalent to the values proposed by the *Joint Committee of Structural Safety* [JCS 00].

EXAMPLE 3.30.– Let us consider a highway viaduct over a railway line and a river; we will consider two cases of failure. The first deals with the partial failure of a beam, whilst the second involves span collapse (collapse mechanism). In the first case, the risk of accident is judged to be low, and only one lane will be blocked. In this case, R_{cd} and R_{cu} are respectively I and III (normal safety measures/low consequences). We therefore obtain that the risk level and the acceptable failure probability are III and 10^{-4}. In the second case, the consequences are more serious. The highway is closed off and the considered risk situation leads to nearly one hundred deaths following the collapse of a span. R_{cd} and R_{cu} are VI and III (important safety measures/important consequences). We therefore obtain that the risk level of the acceptable failure probability are VI and 510^{-6}.   □

3.10.3. *Remarks*

These regulation examples highlight the high level of disparity between acceptable probabilities. This is not surprising because, as we highlighted previously, they have been calibrated by using existing rules (see Appendix A in the commentaries on code CAN/CSA-S6 which succinctly presents this calibration procedure). More interesting is the modularity of acceptable reliability indexes introduced by certain standards for the same limit state (ultimate or serviceability). This modularity is particularly relevant for assessing existing structures. It is either explicit as in the CSA-S6 and NKB regulations, or JCSS, or it is implicit as in the BD79 proposal. In all cases, it relies on the consideration of failure consequences.

Chapter 4

Structural Assessment of Existing Structures

4.1. Introduction

Structure geriatrics may be defined as the discipline dealing with the set of procedures relating to monitoring old or existing structures, in other words, inspection, destructive and non-destructive testing, diagnosis, structural assessments, and maintenance. The assessment process for a structure is therefore crucially important in order to maintain it in terms of serviceability and safety conditions. These conditions are specified by the main contractor or the technical authorities responsible for the structure. These conditions may rely on:

– safety criteria which provide a level of safety for users;

– functionality criteria which ensure a continuous level of serviceability during particular events such as earthquakes;

– economic criteria or other criteria established by authorities.

Different cases must then be distinguished according to whether the structure is in a good condition, or not, but where public safety is not compromised or in danger. There are various options available: the structure can be maintained in that condition or it can be modified.

So that decisions regarding maintenance (repair, rehabilitation, strengthening or modification of the infrastructure's service conditions, like weight restriction for a bridge) are not excessive, the assessment procedure must not be unduly conservative [SHE 99]. We will limit the following sections to *structural assessment*, i.e. verification through calculating the structure's performance. There is another type of assessment, called *condition assessment*, which is provided by the inspections.

4.2. Assessment rules

Studying standards, recommendations and guidelines highlight the significant differences in the procedures and methods used for structural assessment. The reasons behind an assessment are:

– a need to carry exceptional operational loads;

– the structure has undergone a change, such as a degradation, mechanical damage, repair or use modification;

– an old structure built according to outdated rules or loads has not been assessed according to current rules.

Currently, the rules used for assessing civil engineering structures are mainly provided by design standards, with additional recommendations with regard to testing methods. In some countries, the design rules used may be current standards or those used when the structure was built. The current characteristics of load calculations can be used, although in some cases they may be modified specifically for assessment, and may include a reduction based on the knowledge of live loads. Additional specifications may also be given for exceptional live loads.

Design rules are mainly based on two alternative approaches: the principle of allowable stress, and the partial factor format. It is suitable to point out here that the rules stipulated in design standards are only valid in a certain context. For an assessment, some situations may occur which make the standards inapplicable due to particular structural conditions or non-standard construction details. In fact, partial factors are intended to cover a large set of uncertainties and may also prove not to be very representative of the real need for assessing the performance of a particular structure. They are supposed to take into account the evolution of materials and loads in a fixed way. For exceptional or damaged structures, assessing the reliability may be over or underestimated. Standards also include a large degree of generalization in terms of performance and loading. This may be useful for design studies because calculations are getting easier, and because the induced costs are marginal in the budget of a new structure. For repair or strengthening, the required level of safety or serviceability cannot be obtained by using these general design rules.

4.3. Limits when using design rules

Standards present performance levels which, generally, go beyond the margins which can reasonably be allowed for existing structures. This is explained by the level of knowledge about the structure's condition and external loads, when they can be observed or measured. Also, partial factors may be reduced whilst maintaining an identical degree of performance. Partial factors take into account variability in the structural behavior and loading pattern. The performance level required is also reduced with age.

The use of such principles gives a certain complexity to assessment calculations. The choice of an acceptable performance (safety, serviceability, durability) level is currently very difficult to fix down, due to the lack of rules on the way in which such a choice is made, particularly in terms of socio-economical consequences. In all cases, the methods implemented must take into account:

– *The construction mode*: for concrete structures, unacceptable detected cracks indicate potential deficiency in the structure's strength. Thus, ultimate and serviceability limit states must be considered. For steel structures, the ultimate limit states will be more suitable.

– *The type of structure*: uncertainties intervening before the construction phase, the risk of brittle rupture or the possibilities of load redistribution will be different according to structure type.

– *The part of the structure being considered*: the consequences of element failure may be more or less important according to the part of the structure involved and the limit state considered.

It is clear that establishing principles and procedures specific to the assessment of existing structures is desired because certain aspects of the assessment are based on an approach which differs substantially from design assessments, and requires knowledge which goes much further beyond application fields of design standards. However, it is not a given fact that structural assessment directly requires sophisticated calculation methods. The assessment must be carried out according to a growing level of sophistication, aiming at a greater level of precision at each higher level. In order to avoid pointless rehabilitation or replacement procedures (and therefore to reduce management costs), the engineer must be in a position to use all techniques, all methods and all information available efficiently. A simple analysis may be cost effective if it can demonstrate that the structure is satisfactory, but if it is not, its use may present large disadvantages, and therefore more sophisticated methods seem more relevant.

An approach considering the specificities of the structure which must be assessed is a realistic and rational procedure for evaluating the reliability and serviceability aptitude.

4.4. Main stages in structural assessment

Assessing a structure usually consists of assessing its condition, followed by a structural assessment. During a condition assessment, only the critical structural elements are inspected, in order to be ready to take safety measures if public safety is concerned (for example, weight restrictions, or in extreme cases, stopping use). This assessment is very basic and is limited to examining existing documents and visiting the structure for preliminary inspections. In many cases, this inspection may identify particular conditions which must be studied more thoroughly. The main stages for structural assessment are then the following [CAL 97]:

– preliminary *in situ* inspection to specify the condition, the environment, etc.;

– an analysis of existing documentation, including loading, maintenance and repair history, etc.;

– additional investigations;

– an analysis of the collected information so as to refine load and resistance models;

– a refined analysis of limit states and calculation variables;

– failure probability calculations, i.e. the probabilities of exceeding the limit states;

– decision making.

A *preliminary inspection* is a quick and low cost inspection carried out by an engineer or a skilled team. The aim of a preliminary inspection is to identify possible damage through visual observations, using simple methods. The information thus gathered is related to certain aspects, such as external characteristics, evident deformation, cracks, etc. The results of such an inspection indicate the condition of the structure in qualitative terms: for example, no damage, minor damage, serious damage, etc. The function of the preliminary inspection includes identifying the aspects linked to the structure's condition and its safety. This may lead to other investigations or special inspections, which are necessary for forming a specific diagnosis, for obtaining a quantitative assessment or for proposing a repair program.

Design and inspection documents contain important information which is necessary for a complete assessment. These documents must be checked regarding their precision, and in particular, if they have been updated after structural modifications.

Additional investigations may include various inspection methods (cores, non-destructive testing methods, etc.) thus updating dimensions and structural details, assessing material properties, etc. Additional investigations may be led by reviewing available documents: calculation notes, construction notes, inspection notes, modification details, standards and decisions, topography, underground condition, etc. The details and dimensions of a structure, such as characteristic values and mechanical properties, may be obtained by using design documents, if such documents exist and the information is reliable. In the case where some information may give cause for doubt, the structural dimensions and the assumed properties for the materials must undergo thorough tests and inspections. Planning such research is based on information which is already available. One of the major problems is the way in which this additional information is used. A probabilistic approach to assessment offers a useful framework for considering the information provided by additional investigations in order to refine loading and resistance models. The assessment must include the structure's condition by appropriately reducing resistance. For example, the cross-section of some metal structural elements should be reduced if there is significant corrosion. In the same way, concrete may suffer a loss of resistance due to physic-chemical attacks. Large crack growth may modify rigidity and therefore influence load distribution. The position of the strengthening may differ from that decided during the design phase. The structural condition must be determined by analysis or measurements. Static or dynamic tests can be used to characterize structural behavior and/or to envisage a load capacity when other approaches, such as detailed structural analysis or inspection, are not sufficient for demonstrating structural reliability. Additional investigations provide extra information which can be used to improve structural assessments.

Structural assessments must be carried out carefully because during the design stage, there is little information about the loads that the structure must carry during its service lifetime. Often, the assessment must take into account the fact that the structure was designed using old design rules. The structural analysis can be performed using basic principles of structural behavior or by more sophisticated methods such as finite element methods. Generally, the mathematical models are linear and elastic. The structural assessment must include the structure's real geometry, the real construction process and, if appropriate, the loss of resistance in the materials. Non-linear models can be used to provide other information, in order to explain special local phenomena. Other problems (crack growth, damage models, etc.) require special models.

At the end of the previous stages, the need for repair, rehabilitation or modification will be identified.

4.5. Structural safety assessment

All management activities associated with a structure are essentially intended to answer two questions. The first question is determining types of intervention, if necessary. The second is specifying whether management in itself is effective. To ensure the effectiveness of a management strategy, it is essential to have performance indicators.

4.5.1. *Basic concept*

The basic concept for assessment lies in the definition of *rating factors* $\mathbb{I}$ which represent a construction's performance. This rating factor may be deterministic or probabilistic and may or may not include specific data from the structure.

If this rating factor takes values which are higher than 1, then structural performance will be considered to be guaranteed. In the opposite case, the performance cannot be guaranteed without intervention. It is appropriate to point out here that the value of this rating factor may act as a "prioritization" tool for a structural element.

We may use four approaches, each one needing different levels of calculations and complexities. In practice, the simplest calculations will have to be carried out, as a priority, before going into a more complex approach. This pragmatic procedure, by using models and approaches with growing levels of sophistication, are especially found in British standards [BD 01] for instance.

4.5.2. *First approach*

The simplest approach is to perform a semi-probabilistic assessment based on the design rules. This method is particularly relevant when the structure has not been designed following current standards or has had modifications in its usage or structural integrity. Bailey [BAI 96] or Ryall [RYA 01] recommend using the partial factors from design rules whilst modulating, however, the partial factors related to live loads if there are restrictions.

Verifying a failure mode in relation to its limit state is expressed by the condition:

$$\gamma_G \, S(G_d) + \alpha_Q \, \gamma_Q \, S(Q_d) < \frac{R_d}{\gamma_R} \qquad\qquad [4.1]$$

where R is the resistance variable, $S(.)$ denoting the effects of dead loads G and live loads Q. $\gamma_G, \gamma_Q, \gamma_R, G_d, Q_d, R_d$ formally express the partial factors of load and resistance variables and the characteristic values of these same variables. Generally, they are partial factors and characteristic values from design rules, hence the introduction of coefficient α_Q, which balances load effects by a value lower than 1, in order to represent use (weight) restriction. The assessment rating factor $\mathbb{I}_{sp}$ (sp = semi-probabilistic) is therefore defined by:

$$\mathbb{I}_{sp,r} = \frac{\dfrac{R_d}{\gamma_R}}{\gamma_G \, S(G_d) + \alpha_Q \, \gamma_Q \, S(Q_d)}$$ [4.2]

If $\mathbb{I}_{sp} > 1.0$, then structural performance is considered as being verified. In the opposite case, two solutions may be envisaged: intervening on the structure (restrictions, strengthening, etc.) or refining the assessment.

EXAMPLE 4.1.– Let us consider a standard which consists of verifying the condition:

$$\gamma_Q \, Q_d + \gamma_g \, G_d \leq \frac{R_d}{\gamma_R}$$

where R_d, Q_d and G_d denote the characteristic values for the strength and the dead and live loads effects. The partial factors are worth $1/0.67$, 1.5 and 1.35 respectively.

Also, we presume that the ratio between live load effects and dead load effects is $\lambda = 2.5$. The characteristic values for R_d, Q_d and G_d are 16,800 kN.m, 5,000 kN.m and 2,000 kN.m respectively.

Therefore, we obtain a semi-probabilistic rating factor $\mathbb{I}_{sp}^i$ relating to this initial situation:

$$\mathbb{I}_{sp}^i = \frac{\dfrac{R_d}{\gamma_R}}{\gamma_Q \, Q_d + \gamma_G \, G_d} = \frac{16,800 \times 0.67}{1. \times 5,000 + 1.35 \times 2,000} \approx 1.1 \qquad \square$$

EXAMPLE 4.2.– We will take the conditions of example 4.1. We presume that the characteristic strength is, in fact, lower than the theoretical value of r%; the performance rating factor then takes a value $\mathbb{I}_{sp}^d$:

$$\mathbb{I}_{sp}^d = \frac{\dfrac{r\,R_d}{\gamma_R}}{\gamma_Q \, Q_d + \gamma_G \, G_d} = r\,\frac{\dfrac{R_d}{\gamma_R}}{\gamma_Q \, Q_d + \gamma_G \, G_d} = r\,\mathbb{I}_{sp}^i \approx 1.1\,r$$

The performance rating factor threshold $\mathbb{I}_{sp}^d = 1$ is therefore reached for $r = 0.91$. $\qquad \square$

4.5.3. *Second approach*

A second approach consists of using a probabilistic assessment. This method is relevant if specific information is already available on live load characteristics or resistance variables. In this case, the assessment rating factor $\mathbb{I}_p$ (p = probabilistic) is defined by:

$$\mathbb{I}_p = \frac{\beta_{assessment}}{\beta_0} \tag{4.3}$$

As for factor $\mathbb{I}_{sp}$, if $\mathbb{I}_p > 1.0$, then the structural performance is considered as being verified. In the opposite case, two solutions are possible: intervention on the structure (restrictions, strengthening, etc.) or refining the assessment. β_0 denotes the imposed target or acceptable reliability index for the assessment (it may differ from the one used in design rules).

EXAMPLE 4.3.– We will take example 4.1, presuming all the variables are normal. In the development of partial coefficients, it was assumed that the bias between characteristic values and mean values are worth respectively:

$$v_R = 0.95; v_G = 1.00; v_Q = 1.20$$

and that the coefficients of variation are: $\mathbb{COV}_R = 0.10; \mathbb{COV}_G = 0.10; \mathbb{COV}_Q = 0.3$.

It is then possible to calculate the target reliability index in relation to the conditions of the semi-probabilistic format:

$$\beta_0 = \frac{\dfrac{\gamma_R\left(\gamma_Q\lambda+\gamma_G\right)}{v_R} - \dfrac{1}{v_G} - \dfrac{\lambda}{v_Q}}{\sqrt{\left(0.1\dfrac{\gamma_R\left(\gamma_Q\lambda+\gamma_G\right)}{v_R}\right)^2 + \left(0.1\dfrac{1}{v_G}\right)^2 + \left(0.3\lambda\dfrac{1}{v_Q}\right)^2}} \approx 4.83$$

by taking into account the condition $\gamma_Q\, Q_d + \gamma_G\, G_d = \left(\gamma_Q\, \lambda + \gamma_G\,\right)G_d = \dfrac{R_d}{\gamma_R}$. This reliability index implies the semi-probabilistic rule. However, nothing is stopping us from introducing another acceptable reliability index. This is particularly recommended in certain standards (see Chapter 3):

The effective reliability index is given by:

$$\beta = \frac{\bar{R} - \bar{G} - \bar{Q}}{\sqrt{\left(0.1\,\bar{R}\right)^2 + \left(0.1\,\bar{G}\right)^2 + \left(0.3\,\bar{Q}\right)^2}}$$

$$= \frac{\dfrac{R_d}{v_R} - \dfrac{G_d}{v_G} - \dfrac{Q_d}{v_Q}}{\sqrt{\left(0.1\,\dfrac{R_d}{v_R}\right)^2 + \left(0.1\,\dfrac{G_d}{v_G}\right)^2 + \left(0.3\lambda\,\dfrac{G_d}{v_Q}\right)^2}}$$

$$= \frac{\dfrac{16,800}{0.95} - \left(1 + 2.5\dfrac{1}{1.20}\right) \times 2,000}{\sqrt{\left(\dfrac{1,680}{0.95}\right)^2 + \left(200\right)^2 + \left(\dfrac{0.3 \times 2.5}{1.20} \times 2,000\right)^2}} \approx 5.29$$

The initial probabilistic performance rating factor is then worth:

$$\mathbb{I}_p^i = 5.29/4.83 \approx 1.09$$

By taking $r = 0.91$, obtained in example 4.2, the reliability index is then worth $(\bar{R} = r\,R_d/v_R)$:

$$\beta = \frac{0.91 \times \dfrac{16,800}{0.95} - \left(1 + 2.5\dfrac{1}{1.20}\right) \times 2,000}{\sqrt{\left(0.91 \times \dfrac{1,680}{0.95}\right)^2 + \left(200\right)^2 + \left(\dfrac{0.3 \times 2.5}{1.20} \times 2,000\right)^2}} \approx 4.84$$

and therefore:

$$\mathbb{I}_p^d = 4.84/4.83 \approx 1.00$$

which is entirely coherent with the semi-probabilistic approach. This is not surprising because the calculations performed under semi-probabilistic and probabilistic methods are based on the same assumptions. ☐

EXAMPLE 4.4.– We will use example 4.3, assuming that additional tests show that the average resistance value is 9% lower than the theoretical value, and may reduce the variation coefficient from 0.1 to 0.05. The reliability index and the performance rating factor are then worth:

$$\beta = \frac{\dfrac{16,800}{0.95} \times 0.91 - \left(1 + 2.5\dfrac{1}{1.20}\right) \times 2,000}{\sqrt{\left(\dfrac{16,800}{0,95} \times 0.05 \times 0.91\right)^2 + \left(200\right)^2 + \left(\dfrac{0.3 \times 2.5}{1.20} \times 2,000\right)^2}} \approx 6.61$$

$\mathbb{I}_p^d = 6.61/4.83 \approx 1.36$

We must state many important aspects here:

– The acceptable reliability index is calculated with a 10% coefficient of variation for the resistance. Therefore, it is not recalculated with the new coefficient of variation. In fact, the target index is established using hypotheses on the uncertainties of the variables. It cannot be modified according to specific data, but must keep a general character.

– We notice that with the same level of reduction, but with a reduced coefficient of variation, the performance rating factor is higher than 1, and even higher than the initial performance rating factor. This example shows the superiority of the probabilistic rating factor in relation to the semi-probabilistic rating factor: it helps to introduce the improved knowledge of the coefficient of variation. In a semi-probabilistic format, this coefficient of variation is hidden in partial safety factors. Taking this into account would involve re-calibrating the partial factors. Example 4.6 gives an example of this application.

If the tests had led to an increase in the coefficient of variation (from 0.1 to 0.15 for example), the result would have been different:

$$\beta = \frac{\dfrac{16,800}{0.95} \times 0.91 - \left(1 + 2.5\dfrac{1}{1.20}\right) \times 2,000}{\sqrt{\left(\dfrac{16,800}{0.95} \times 0.15 \times 0.91\right)^2 + \left(200\right)^2 + \left(\dfrac{0.3 \times 2.5}{1.20} \times 2,000\right)^2}} \approx 3.64$$

$\mathbb{I}_p^d = 3.64/4.83 \approx 0.75 < 1$ □

4.5.4. *Third approach*

The objective is to reduce the uncertainties on the different variables so as to obtain a more relevant assessment. The assessment rating factor $\mathbb{I}_{sp,r}$ (sp = semi-probabilistic, r = recalibrated);

$$\mathbb{I}_{sp,r} = \frac{\dfrac{R_d}{\alpha_R\,\gamma_R}}{\dfrac{\gamma_G\,S(G_d)}{\alpha_G} + \dfrac{\gamma_Q\,S(Q_d)}{\alpha_Q}} \qquad\qquad [4.4]$$

Each loading effect (dead load, live load) is treated separately. The partial factors of the resistance variables are also modified by corrective factors. These are generally calibrated according to the nature and the values of the inspection results.

As for the previous rating factors, if $\mathbb{I}_{sp,r}$ takes a value lower than 1, a more detailed structural assessment, probabilistic but based on specific data, may improve the performance assessment.

EXAMPLE 4.5.– In example 4.4, we showed that a probabilistic assessment, in the case of reducing the coefficient of variation, would allow us to obtain a performance rating factor higher than 1, whereas we should obtain 1 exactly with a semi-probabilistic rating factor. This difference is explained by the fact that the partial factors are based on assumptions (particularly uncertainties) and should be modified consequentially when better knowledge is available on the parameters' variability. This involves calculating new partial factors. According to Chapter 3 (example 3.25), we obtain:

$$\frac{1}{\gamma_R^m} = \left(1 - \alpha_R^m\ \underbrace{\beta_0}_{4.85}\ \mathbb{COV}_R\right)\frac{1}{\nu_R}$$

$$= \left(1 - \frac{\bar{R}\ \mathbb{COV}_R}{\sqrt{\left(\bar{R}\ \mathbb{COV}_R\right)^2 + \left(\bar{G}\ \mathbb{COV}_G\right)^2 + \left(\bar{Q}\ \mathbb{COV}_Q\right)^2}}\ \beta_0\ \mathbb{COV}_R\right)\frac{1}{\nu_R}$$

$$= \left(1 - \frac{0.91\times\dfrac{16{,}800}{0.95}\times 0.05\times 4.83\times 0.05}{\sqrt{\left(\dfrac{16{,}800}{0.95}\times 0.05\times 0.91\right)^2 + \left(200\right)^2 + \left(\dfrac{0.3\times 2.5}{1.20}\times 2{,}000\right)^2}}\right)\frac{1}{0.95}$$

$$\approx 0.92$$

$$\gamma_G^m = \left(1 + \frac{200\times 4.83\times 0.1}{\sqrt{\left(\dfrac{16{,}800}{0.95}\times 0.05\times 0.91\right)^2 + \left(200\right)^2 + \left(\dfrac{0.3\times 2.5}{1.20}\times 2{,}000\right)^2}}\right)\frac{1}{1} \approx 1.06$$

$$\gamma_Q^m = \left(1 + \frac{\dfrac{0.3 \times 2.5}{1.20} \times 2,000 \times 4.83 \times 0.3}{\sqrt{\left(\dfrac{16,800}{0.95} \times 0.05 \times 0.91\right)^2 + \left(200\right)^2 + \left(\dfrac{0.3 \times 2.5}{1.20} \times 2,000\right)^2}}\right) \left|\frac{1}{1.2}\right. \approx 1.84$$

which gives:

$$\alpha_R = \frac{\gamma_R^m}{\gamma_R} = \frac{0.67}{0.92} \approx 0.73; \quad \alpha_G = \frac{\gamma_G}{\gamma_G^m} = \frac{1.35}{1.06} \approx 1.27; \quad \alpha_Q = \frac{\gamma_Q}{\gamma_Q^m} = \frac{1.50}{1.84} \approx 0.81$$

The updated semi-probabilistic rating factor is therefore:

$$\mathbb{I}_{sp,r} = \frac{\dfrac{16,800}{0.73 \times 1.5}}{\dfrac{1.35}{1.27} \times 2,000 + \dfrac{1.5}{0.81} \times 5,000} \approx 1.35$$

which is close to the probabilistic rating factor value in example 4.4.   □

Example 4.5 is, however, misleading; it may give the impression that partial factors can be unique. In section 3.7, we saw that there was an infinite set of partial factors (due to the lack of invariance in the semi-probabilistic format). The proposed solution in the previous example is just one amongst this infinite number of solutions. In fact, it consists of relocating the design point from $\beta = 6.61$ to $\beta_0 = 4.85$, and keeping the direction cosines, which is equivalent to a translation, and which varies the set of partial factors significantly. This change is not necessarily due to a change in the characteristic values and coefficients of variation, but is also induced by the fact that the initial partial factors do not come from a calibration which uses exactly the mathematical expressions from the previous example.

EXAMPLE 4.6.– We will take example 4.5 without modifying the resistance variable properties. By using the expression from Chapter 3 given for calibrating partial factors, we would obtain:

$$\frac{1}{\gamma_R^m} = \left(1 - \frac{\dfrac{16,800}{0.95} \times 0.1 \times 4.83 \times 0.1}{\sqrt{\left(\dfrac{16,800}{0.95} \times 0.1\right)^2 + \left(200\right)^2 + \left(\dfrac{0.3 \times 2.5}{1.20} \times 2,000\right)^2}}\right) \left|\frac{1}{0.95}\right.$$

$$\approx 0.64$$

$$\gamma_G^m = \left(1 + \frac{200 \times 4.83 \times 0.1}{\sqrt{\left(\frac{16,800}{0.95} \times 0.1\right)^2 + \left(200\right)^2 + \left(\frac{0.3 \times 2.5}{1.20} \times 2,000\right)^2}}\right) \frac{1}{1} \approx 1.04$$

$$\gamma_Q^m = \left(1 + \frac{\frac{0.3 \times 2.5}{1.20} \times 2,000 \times 4.83 \times 0.3}{\sqrt{\left(\frac{16,800}{0.95} \times 0.1\right)^2 + \left(200\right)^2 + \left(\frac{0.3 \times 2.5}{1.20} \times 2,000\right)^2}}\right) \frac{1}{1.2} \approx 1.53$$

which differs from initial partial factors, particularly for dead loads, whereas the other factors seem to be less heavily affected (to the nearest rounded errors).   □

An alternative when applying the previous expressions for recalibrating partial factors consists of retaining the target reliability index β_0 which we will write as β_{0c}, and looking for an updated set of partial factors $\{\gamma\}$, so that the difference between the target index $\beta_0(\{\gamma\})$ calculated with this set of partial factors is the lowest value possible from the target index β_{0c}. It is, then, a question of solving an optimization problem of the following type:

$$\begin{cases} J(\{\gamma\}) = \left(\beta_0(\{\gamma\}) - \beta_{0c}\right)^2 \\ \{\gamma\}_{recal} = \arg\min_{\{\gamma\}} J(\{\gamma\}) \end{cases}$$ [4.5]

A solution is certain, insofar as all the random variables considered generally follow continuous probability density functions. The function $\beta_0(\{\gamma\})$ is, then, continuous for the limit states studied and passes through the set of real positive numbers. Thus, we are guaranteed that a set of partial factors $J(\{\gamma\}) = 0$ exists. However, we must insist on the fact that there is absolutely no unicity in this solution, as the effect of increasing one of the coefficients can easily be compensated by decreasing another coefficient. To solve the problem [4.5], by a gradient method in particular, the objective function $J(\{\gamma\})$ must be differentiable and locally convex around the set of obtained solutions. Figure 4.1 shows the shape of the objective function $J(\{\gamma\})$, in the case of a linear limit state with normal random variables (two partial factors). This figure highlights a set of solutions.

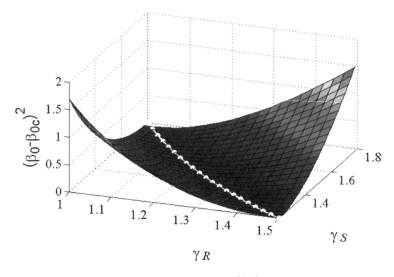

Figure 4.1. *Shape of objective function* $J(\{\gamma\})$ *and set of solutions*

The final objective of the third approach is to help the engineer to readjust these partial factors according to data obtained from the structure's inspection [CRE 09b]. But, even if standards do not mention this, the partial factors are interdependent via the design point. It is not, however, intuitive to readjust the value of a partial factor related to live loads from resistance-based data. Since a multitude of possible solutions exist for recalibrating the coefficients, it is wise to choose a solution which only modifies the partial factor(s) which is/are naturally related to the variable(s) concerned.

Let us remember that for assessing the performance level of a structure, we need to call upon a semi-probabilistic rating factor $\mathbb{I}_{sp}$ which depends on the values of partial factor $\{\gamma\}$. In light of the numerous sets of partial factors, it is necessary for the rating factors $\mathbb{I}_{sp}$ of each limit state to be invariant of the set of solutions $\{\gamma\}$. This invariance may be numerically verified for each limit state studied. It is difficult to demonstrate in a general case since it depends on the form of the rating factor and the chosen limit state.

We must tackle one more point before carrying out a coherent recalibration in agreement with the standards used by the engineer. A standard or a regulation often provides a single partial factor for all dead loads. Thus, loads due to self-weight and superstructures, which are *a priori* independent and display very different variables, are affected by the same coefficient γ_G. Even if we could question the legitimacy of this effect, the recalibration method must allow the engineer to procure results which are coherent with the standard's requirements.

EXAMPLE 4.7.– We will take example 4.6, looking only to update the partial coefficient related to resistance; the operational and dead load coefficients are retained. Similarly to example 4.3, we obtain:

$$\beta_0\left(\hat{\gamma}_R\right) = \frac{0.91\times\dfrac{\hat{\gamma}_R\left(\gamma_Q\lambda+\gamma_G\right)}{v_R}-\dfrac{1}{v_G}-\dfrac{\lambda}{v_Q}}{\sqrt{\left(0.05\times0.91\times\dfrac{\hat{\gamma}_R\left(\gamma_Q\lambda+\gamma_G\right)}{v_R}\right)^2+\left(0.1\dfrac{1}{v_G}\right)^2+\left(0.3\lambda\dfrac{1}{v_Q}\right)^2}}$$

$$\equiv \beta_{0c} = 4.83$$

We set out $\Gamma = 0.91\times\left(\gamma_Q\lambda+\gamma_G\right)$ and by squaring the previous expression, we obtain:

$$\left(\frac{\hat{\gamma}_R\Gamma}{v_R}-\frac{1}{v_G}-\frac{\lambda}{v_Q}\right)^2 = \beta_{0c}^2\left[\left(0.05\frac{\hat{\gamma}_R\Gamma}{v_R}\right)^2+\left(0.1\frac{1}{v_G}\right)^2+\left(0.3\lambda\frac{1}{v_Q}\right)^2\right]$$

which goes back to solving a second degree equation in $X = \dfrac{\hat{\gamma}_R\Gamma}{v_R}$:

$$X^2\left(0.05^2\beta_{0c}^2-1\right)+2X\left(\frac{1}{v_G}+\frac{\lambda}{v_Q}\right)-\left(\frac{1}{v_G}+\frac{\lambda}{v_Q}\right)^2$$

$$+\beta_{0c}^2\left[\left(0.1\frac{1}{v_G}\right)^2+\left(0.3\lambda\frac{1}{v_Q}\right)^2\right]=0$$

which gives: $X \approx 6.534$ or $\hat{\gamma}_R \approx 1.33$. In this example, the corrective factors are:

$$\alpha_R = \frac{\gamma_R^m}{\gamma_R}=\frac{1.33}{1.50}\approx 0.89;\ \alpha_G=\alpha_Q=1.0$$

The updated semi-probabilistic rating factor is now worth:

$$\mathbb{I}_{sp,r} = \frac{\dfrac{16,800}{0.89\times1.5}}{\dfrac{1.35}{1.0}\times2,000+\dfrac{1.5}{1.0}\times5,000}\approx 1.24$$

This value differs from the value of 1.35 calculated with the direction cosines in example 4.3. This difference highlights that there is no invariance in the semi-probabilistic approach; the difference between the probabilistic and the semi-probabilistic rating factor is justified because the partial factors are not updated by a translation, but by moving the design point from $\beta = 6.61$ to $\beta_0 = 4.83$. In the present example, the process does not maintain the direction cosines. The following figure illustrates this.

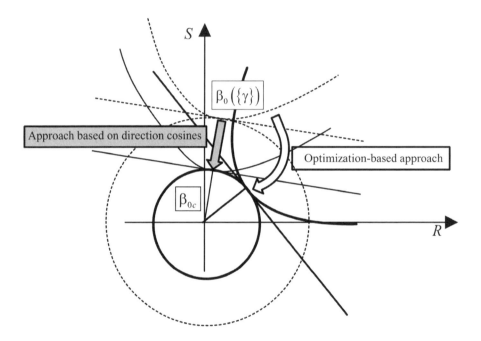

The reduced mean value does not play a major role in this example; only the modification of the coefficient with an influence on the limit state plays a significant part (provided that the bias it not modified).   □

4.5.5. *Fourth approach*

As for the third method, we will use the inspection and auscultation data to readjust the failure probability at the time of assessment:

$$\mathbb{I}_{p,r} = \frac{\beta_{\text{assessment, updated}}}{\beta_0} \qquad [4.6]$$

If this rating factor is lower than 1, more complete inspections or auscultations may be required: strengthening or live loads restrictions may also be necessary.

Calculating an *updated reliability index* uses a Bayesian analysis. When additional information is obtained during new investigations, it must be used to improve the *a priori* reliability assessment. The Bayesian analysis provides a useful framework for this consideration.

The Bayesian analysis uses conditional probability concepts (see Chapter 1). Let us consider, for example, that we assume (before any tests on a structure take place) a distribution for the strength R expressed by a probability density function $f_R^{priori}(r)$. The new information Θ introduces uncertainties that we may model by a density distribution $f_\Theta(\theta)$. By means of this new information, the strength distribution can be modified (updated): let $f_R^{posteriori}(r)$ be the new density function – or the *a posteriori density* – of R. Figure 4.2 illustrates the influence of new information on the *a priori* distribution of a variable.

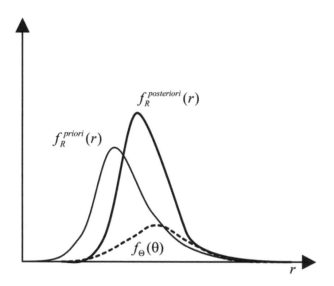

Figure 4.2. *A priori, a posteriori distributions*

The *Bayesian analysis* is a probabilistic framework in which the uncertainty on a variable and the existing knowledge on this variable are combined. By using this combination, knowing the variable's outcomes enables us to estimate the distribution parameters and thus to *conclude* on the occurrence of future events. Let us consider again the variable R described by its *a priori* density function

$f_R^{priori}(r/\Theta)$ conditioned by the "uncertain" parameters Θ. Imagine that we carry out n independent measurements $(r_1,\cdots,r_i,\cdots,r_n)$ which consequentially form an n-sample R_n. The Bayes theorem, applied to the case of a continuous distribution, allows us to calculate the *a posteriori* distribution of Θ:

$$f_\Theta(\theta/R_n) = \frac{\displaystyle\prod_{i=1}^{n} f_R^{priori}(r_i|\Theta)}{\displaystyle\int_{-\infty}^{+\infty}\prod_{i=1}^{n} f_R^{priori}(r_i|\Theta)\, f_\Theta(\theta)\, d\theta} \qquad [4.7]$$

We can therefore obtain the *a posteriori* density function of variable R:

$$f_R(r/R_n,\Theta) = \int_{-\infty}^{+\infty}\prod_{i=1}^{n} f_R(r_i|\Theta)\, f_\Theta(\theta|R_n)\, d\theta \qquad [4.8]$$

If the uncertainties on Θ are high, then the data will not contribute much, and therefore will hardly affect the *a priori* distribution. On the other hand, if the data presents very little dispersion, then the *a posteriori* distribution will differ from the *a priori* distribution.

The additional information may be singular in number, or may consist of many different observations. There are two types of observations provided by the inspection: quantitative or qualitative. Each of these results is an event associated with an *event margin H* (similar to the safety margin) and an occurrence probability. The qualitative results give data, for example, on the detection or non-detection of an event linked to a particular phenomenon. This information is expressed by:

$$H \leq 0 \qquad [4.9]$$

The quantitative inspection results correspond to measurements. The information is expressed by the event:

$$H = 0 \qquad [4.10]$$

The quantitative information $H=0$ will be used to re-assess the structural performance after a series of measurements. The quantitative information $H\leq 0$ will be used in conditional maintenance where only general results (such as detection or non-detection of a defect) are expected after an inspection.

We consider that the reliability of a failure component is described by its safety margin M and we presume that the qualitative and quantitative data is available and formulated in terms of event margins: $\left(H_{quant,i}\right)_{1 \leq i \leq n}$ and $\left(H_{qual,j}\right)_{1 \leq j \leq m}$. Knowing this information, the failure probability is then a conditional probability:

$$P_f^{updated} = \mathcal{P}\left(M < 0 / \left[\bigcap_i H_{quant,i} = 0\right]\left[\bigcap_j \bigcap H_{qual,j} < 0\right]\right) \qquad [4.11]$$

Calculating these probabilities is not easy when there are many events. When quantitative or qualitative inspection data is available separately, the calculations may be facilitated.

4.5.5.1. *Quantitative inspections*

Let us consider a set of quantitative data:

$$P_f^{updated} = \mathcal{P}\left(M < 0 / \left[\bigcap_i H_i = 0\right]\right) \qquad [4.12]$$

When a set of quantitative inspection data is available, it can be shown as [MAD 96]:

$$\beta_{updated} = \frac{\beta - {}^t\rho_{MH} \; \rho_{HH} \; \beta_H}{\sqrt{1 - {}^t\rho_{MH} \; \rho_{HH}^{-1} \; \rho_{MH}}} \qquad [4.13]$$

where β, β_H, ρ_{MH} and ρ_{HH} are respectively the reliability indexes of M before updating, the vector of indexes related to the event margins H, the correlation matrix between safety and event margins.

EXAMPLE 4.8.– The carbonation depth of concrete follows a square root of time law:

$$x(t) = k \sqrt{t}$$

Parameter k follows a lognormal law with a mean and standard deviation of $\bar{k}, \sigma_k$. The safety margin is the carbonation depth distance in relation to the concrete cover. It is defined by:

$$M = c - x(t) = c - k\sqrt{t}$$

The concrete cover is assumed to be lognormally distributed with mean $\bar{c}$ and standard deviation σ_c. The safety margin can therefore be rewritten as a logarithm:

$$\hat{M} - \ln c \ - \ln k - \ln \sqrt{t}$$

$$= \ln \bar{c} - \frac{\left(\sigma_c/\bar{c}\right)^2}{2} + \frac{\sigma_c}{\bar{c}} U_c - \ln \bar{k} + \frac{\left(\sigma_k/\bar{k}\right)^2}{2} - \frac{\sigma_k}{\bar{k}} U_k - \frac{1}{2}\ln t$$

(provided that the coefficients of variation are lower than 0.3).

The reliability index is therefore:

$$\beta = \frac{\ln \bar{c} - \frac{\left(\sigma_c/\bar{c}\right)^2}{2} - \ln \bar{k} + \frac{\left(\sigma_k/\bar{k}\right)^2}{2} - \frac{1}{2}\ln t}{\sqrt{\left(\sigma_c/\bar{c}\right)^2 + \left(\sigma_k/\bar{k}\right)^2}}$$

We presume that $\bar{k} = 2 \ 10^{-3}$ m/years$^{1/2}$, $\bar{c} = 2.5$ cm, $COV_k = 0.2$ and $COV_c = 0.1$. Therefore we obtain:

$$\beta(t) = \frac{\ln 0.025 - \frac{0.1^2}{2} - \ln\left(2 \ 10^{-3}\right) + \frac{0.2^2}{2} - \frac{1}{2}\ln t}{\sqrt{\left(0.1\right)^2 + \left(0.2\right)^2}}$$

$$= \frac{2.54 - 0.5 \ln t}{0.226} \approx 11.234 - 2.242 \ln t$$

At $t = 30$ years, a carbonation depth is assessed at $x_e = 2.0$ cm with an uncertainty measurement of 20% (lognormal distribution). The event margin is $H = x_e - x(t = 30 \text{ years}) = 0$. We therefore obtain:

$$\beta(t = 30 \text{ years}) = 11.234 - 2.242 \times \ln 30 \approx 3.61$$

$$\beta_H = \frac{-\ln \bar{k} + \dfrac{\left(\sigma_k/\bar{k}\right)^2}{2} - \dfrac{1}{2}\ln 30 + \ln \bar{x}_e}{\sqrt{\left(\sigma_k/\bar{k}\right)^2 + \left(\sigma_{x_e}/\bar{x}_e\right)^2}}$$

$$= \frac{-\ln\left(2\ 10^{-3}\right) + \dfrac{0.2^2}{2} - \dfrac{1}{2}\ln 30 + \ln\left(0.02\right)}{\sqrt{\left(0.2\right)^2 + \left(0.2\right)^2}}$$

$$\approx 2.19$$

The direction cosines for the common variables are:

$$\alpha_k^M = -\frac{0.2}{0.226} \approx -0.885;\ \alpha_k^H = \frac{0.2}{0.282} \approx -0.709 \text{ and thus } \rho_{MH} = 0.627$$

$$\beta_{updated}\,(t \geq 30 \text{ years}) = \frac{11.234 - 2.242\ln t - 0.627 \times 2.19}{\sqrt{1 - 0.627^2}}$$

$$= \frac{9.861 - 2.242\ln t}{0.779} \approx 12.659 - 2.878\ln t$$

We allow that the acceptable failure probability is from the Eurocode's category RC2 with the service limit state for a service lifespan of 100 years (this hypothesis is not completely justifiable according to Chapter 3, because the SLS targets are not necessarily applicable to limit states of durability. The index 2.90 is used as an application example here, but must be considered with care):

$$P_f^0 = 1 - \left(1 - \Phi\left(-2.90\right)\right)^{100} \approx 0.17,\ \text{or}\ \beta_0 \approx 0.95$$

The updated performance rating factor is:

$$\mathbb{I}_{p,r} = \frac{\beta_{assessment,\ updated}}{\beta_0} = 13.325 - 3.029\ln t$$

The following figure shows the evolution of the reliability index according to the initial prediction and updates. The initial model over-estimates the performance level, since the acceptable threshold for 100 years was reached. In truth, with the measurement at 30 years, we notice that the unitary performance threshold is reached before 60 years. This updating will be more noticeable when the measurement is higher (smaller coefficient of variation) or when the number of measurements is larger.

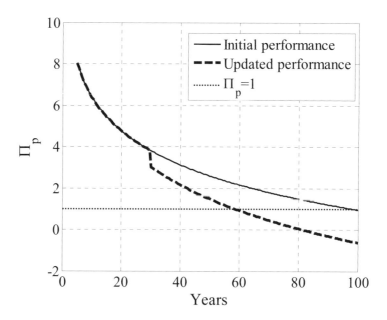

4.5.5.2. *Qualitative inspections*

When a set of qualitative inspection data is available, we also show:

$$P_f^{updated} = \frac{\Phi_{m+1}\left(-\left\{\begin{matrix} \beta \\ \beta_H \end{matrix}\right\}; \rho_{MH}\right)}{\Phi_m\left(-\beta_H; \rho_{HH}\right)}$$

[4.14]

where $\Phi_k(.)$ is the *multivariate normal distribution function* of dimension k. The additional information provided by the inspections and tests are often based on resistance or loading variables. Using more global results (for example, sagging under a moving load) with the aim of updating the reliability level is more difficult.

EXAMPLE 4.9.– Let us take again the hypotheses found in example 4.1. A loading test producing a force of $Q_e = 8,000$ kN.m (assumed deterministic) is carried out. This test is conclusive, i.e. we obtain a qualitative event margin $H = R - G - Q_e > 0$. In this case, the updated or readjusted failure probability is given by the bivariate normal distribution:

$$P_f^{updated} = \frac{\Phi_2\left(-\beta; -\beta_H; \rho_{MH}\right)}{\Phi\left(-\beta_H\right)}$$

The reliability index of the event margin is worth:

$$\beta_H = \frac{-\bar{R} + Q_e + \bar{G}}{\sqrt{\left(\bar{G}\;\mathrm{COV}_G\right)^2 + \left(\bar{R}\;\mathrm{COV}_R\right)^2}} = \frac{8,000 + 2,000 - \dfrac{16,800}{0.95}}{\sqrt{\left(\dfrac{1,680}{0.95}\right)^2 + \left(200\right)^2}} = -4.31$$

The direction cosines for the safety and event margin relating to the only common variables (R and G) are:

$$\alpha_G^M = -\frac{200}{\sqrt{\left(\dfrac{1,680}{0.95}\right)^2 + \left(200\right)^2 + \left(\dfrac{0.3 \times 5,000}{1.20}\right)^2}} = -0.09$$

$$\alpha_G^H = \frac{200}{\sqrt{\left(\dfrac{1,680}{0.95}\right)^2 + \left(200\right)^2}} = 0.11$$

$$\alpha_R^M = \frac{\left(\dfrac{1,280}{0.95}\right)}{\sqrt{\left(\dfrac{1,680}{0.95}\right)^2 + \left(200\right)^2 + \left(\dfrac{0.3 \times 5,000}{1.20}\right)^2}} = 0.81$$

$$\alpha_R^H = -\frac{\left(\dfrac{1,680}{0.95}\right)}{\sqrt{\left(\dfrac{1,680}{0.95}\right)^2 + \left(200\right)^2}} = -0.99$$

which gives $\rho_{HM} = -\left(0.09 \times 0.11 + 0.81 \times 0.99\right) = -0.81$.

$\Phi_2\left(-\beta; -\beta_H; \rho_{MH}\right)$ is calculated by recurrence using Hermite polynomials (see Chapter 3):

$$\Phi_2\left(-5.29; 4.31; -0.81\right) \approx 2.6\ 10^{-8}$$

which gives:

$$P_f^{updated} = \frac{2.6\ 10^{-8}}{0.9999} \approx 2.6\ 10^{-8}, \text{ or } \beta_{updated} \approx 5.44$$

The updated performance rating factor is therefore:

$$\mathbb{I}_{p,r} = \frac{5.44}{4.85} \approx 1.12$$ □

4.5.6. *Implementing rating factors*

As recalled in the previous sections, an assessment on an existing structure may be carried out for a number of reasons:

– the introduction of new loading regulations;

– modification of the structure's functions;

– degradations involving a loss of serviceability or load-carrying capacity.

The first stage in assessment methodology is a deterministic assessment using the verification criteria as defined in current design standards. This method is relevant when the assessed structure has not been designed following current regulations or has been subjected to changes in usage, or has had structural alterations. The applied performance rating factor is the simple semi-probabilistic factor for which reduced loads (if there is weight restriction for instance) will potentially be introduced. In cases where the performance is not verified (factor lower than 1), then a probabilistic reliability analysis can be used to indicate whether the performance is, in fact, adequate. This reliability analysis may also facilitate planning data collection [CRE 03b, CRE 04].

Assessments may be improved by using a reliability analysis using probabilistic models. Determining probabilistic models can be achieved using possible knowledge of the structure's properties (analyzing available documents) or by using databases. Calculating the probabilistic performance rating factor has the advantage of avoiding conservative simplifications which exist in deterministic models for strengths and loads. The assessment is stopped if the rating factor for all the failure components is verified.

If the performance is not verified by semi-probabilistic or simple probabilistic analyses by using load and resistance models, there are still two possible approaches; extra data can be gathered or an intervention must be put forward. However it is more appropriate to continue with the assessment since the gain in intervention costs will probably be larger than the cost of collecting data.

Additional data is gathered in order to update the semi-probabilistic or probabilistic models for strain and resistance. In these cases, we must determine the probabilistic characteristics of the variables. These will then be used directly in a reliability analysis or for calibrating reduction coefficients in order to be applied to partial factors. More sophisticated calculation methods may also be introduced and require the model coefficients to be modulated.

Once these specific characteristics are available, then updated semi-probabilistic models can be obtained to be used in a semi-probabilistic format. Using such an assessment at this stage can be explained by the predominant use of deterministic computing tools in engineering offices. If the performance is *not* verified using the updated semi-probabilistic performance rating factor, then the choice between an updated probabilistic analysis and an intervention must be analyzed.

The most rational approach lies in an updated probabilistic analysis. The updated semi-probabilistic performance rating factor requires the introduction of reduction coefficients to be calibrated which, if they can be associated with specificities on the structure, cover ranges of uncertainties, and therefore induce a loss of relevance in the assessment. The updated probabilistic rating factor then displays the best guarantees for considering the data available. When this performance rating factor is not acceptable (lower than 1), there are two strategies: reducing the uncertainties by using additional data or planning maintenance intervention.

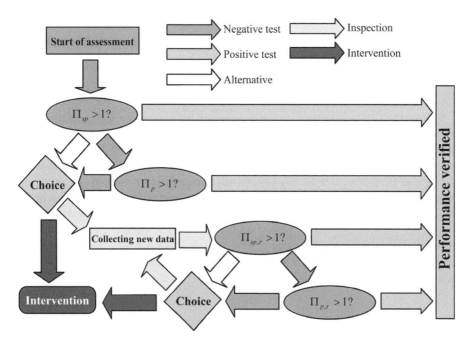

Figure 4.3. *Methodology for applying performance rating factors (adapted from [BAI 96])*

An intervention procedure may be decided on for any performance rating factor. But today it has been agreed that any extra information reduces the uncertainties linked to assessment processes. The available data must not only be usable for the models, but it must also show sufficient reliability levels. Figure 4.3 summarizes the different stages of this assessment process by adapting the methodology proposed by Bailey [BAI 96], using the four performance rating factors.

4.6. General remarks on the methods

Reducing partial factors for existing structures is already being practiced in certain technological sectors: aeronautics, pipelines, and oil rigs. We can make a certain number of remarks:

– The acceptable reliability indexes used for assessments cannot be as restrictive as those for design standards.

– Uncertainties on the resistances found in existing structures can be refined by inspection and monitoring; for healthy existing structures, these uncertainties are often less dispersed than those introduced implicitly into the standards for new structures. On the other hand, for degraded structures, dispersion may be much more important.

– Using system modeling on a structure, particularly if there are many failure mechanisms, enables us to increase its load-carrying capacity by taking into account its load redistribution when a section is failing. It is therefore essential to correlate the section's capacity with the structure's capacity as a system.

– Specific information available shows considerable benefits for the assessment. This concerns the condition of the structure but also the characteristics of the live load it must carry.

Canada [CAN 01] and the USA [AAS 10] have used probabilistic approaches to determine the partial factor applicable to the assessment of existing bridges based on LRFD (*load and resistance factor design*) methodologies, a method which is similar to the French and European approach of limit states. The analyses are primarily led for individual structural elements, although it is possible to perform a complete structural analysis while taking redundancy into account.

Chapter 5

Specificities of Existing Structures

5.1. Loads

5.1.1. *Introduction*

A *load* is an external or internal (chemical reaction) phenomenon acting on a structure which calls upon its resistance. It may be a set of concentrated or distributed forces (the cause of displacements or deformations), or an environmental phenomenon able to cause structural modifications over time.

In the design of a structure with regard to limit states, loads are often given by models taken from standards and regulations. These models are intended to cover the uncertainties on loads and their modeling through *partial factors* and *characteristic values*.

In some cases, it may be necessary to model these loads in a more accurate or relevant way: this is the case for designing exceptional structures, for which load regulations are not applicable, or for managing existing structures to avoid results which are too conservative. So, it is necessary to have probabilistic load models at hand which take into account the uncertainties specific to the phenomena in question. Mostly, these models are parametric, with their parameters being calibrated on experimental data. The problem with load modeling is similar in every respect to that for modeling resistances.

Figure 5.1 represents these loads. Generally, in a standard assessment phase of a bridge, the loads considered are often dead loads and live loads.

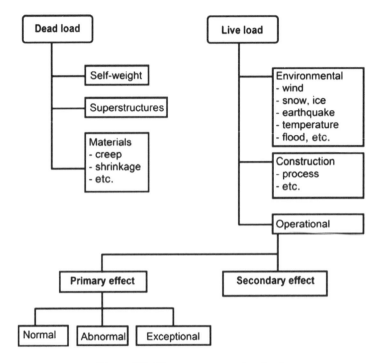

Figure 5.1. *Diagram representing loads*

The loads are classified according to many criteria:

– *duration*: permanent, variable and accidental (or exceptional);

– *structural response*: static, dynamic;

– *spatial variability*: concentrated or distributed;

– *nature*: environmental, operational, accidental.

We can achieve a simpler classification according to whether the loads come from environmental phenomena or whether they are created by humans. In each case, the amplitude of most loads varies in time and space. This implies that the loads are most often represented by stochastic processes or fields. For certain dynamic effects (like those induced by wind load), there may also be an interaction – or coupling – on the one hand between the structure, and between the loads on the other hand.

Many loads intervene in the dimensioning of structures: Figure 5.2 gives a few examples of this, particularly focusing on the nature of the time variation.

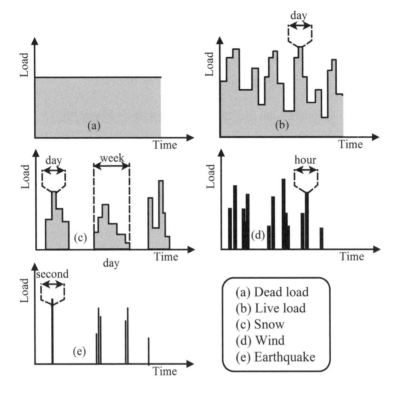

Figure 5.2. *Temporal occurrences of some loads*

Most of the time, load models are based on simple mathematical descriptions of their time, spatial, and directional properties relative to the structure. The level of detail for these models will be largely guided by the quality of data available, and by a realistic model of the effect of this load. The level of realism and precision of this effect is proportionate to the sensitivity of decisions made for structural assessment, and the cost of these decisions. As a matter of fact, a load will receive a model depending on the effect and the structure in question.

A complete load model is made of many elements which describe its amplitude, location, direction, time, etc. These components interact together but they may also interact with the structure's response (this is particularly the case for aeroelastic wind effects). In most cases, these models can be simplified and formalized using two variables $F_{ref,L}$ and E_L associated with the expression of the load independently from the considered structure, and the definition of a conversion factor respectively. Or, also, they may be associated with a transformation of the reference load to the load exerted on the structure.

$$F_L = \mathcal{F}\left(F_{ref,L}, E_L\right) \tag{5.1}$$

$\mathcal{F}$ is a suitable load enabling $F_{ref,L}$ to transform into F_L using the conversion coefficient E_L, a simple product in the vast majority of cases.

The time dependence of the load is carried by variable $F_{ref,L}$ making E_L independent of time. On the other hand, the spatial variability is apprehended by E_L. In accordance with the previous section, according to the level of sophistication required, many models can be used for the same load [5.1].

Within the framework of modeling these loads with a probabilistic approach, it will be necessary to use models of stochastic processes or fields (time or spatial variability).

5.1.2. Stochastic processes

5.1.2.1. Introduction

A process $Y(t)$ is a family of random variables, indexed by a continuous or discrete parameter[1] t. At $t = t_0$, $y(t_0)$ is an outcome of $Y(t_0)$. For a fixed time t, there are therefore many possible outcomes. By generalizing this fact for each time index, then many families of outcomes $y_i(t)$ can be generated. These correspond to many *histories* of the variable $Y(t)$. Figure 5.3 illustrates these histories for a signal.

We will consider p instants $(t_i)_{1 \leq i \leq p}$. The joint probability distribution of the random vector $(Y(t_i))_{1 \leq i \leq p}$ uses the distributions of variables $Y(t_i)$ as marginal distributions for each of its components. If the family $(t_i)_{1 \leq i \leq p}$ is shifted by a time shift τ, the random vector $(Y(t_i + \tau))_{1 \leq i \leq p}$ allows a new joint distribution. If for each family of p instants the joint distributions do not vary under any time shift τ, then the process $Y(t)$ is said to be *stationary*. If there is only invariance for the distributions and the joint distributions for $p = 2$, then the process is said to be second order stationary. In most applications, these conditions are often sufficient

_____

1 A spatial coordinate can be substituted to the time index in this section without loss of generality.

for stating the stationarity hypothesis for a process. There are many consequences of this property, in particular:

– that variables $Y(t)$ therefore all have identical distributions;

– that the correlation coefficient between two variables $Y(t)$ and $Y(t+\tau)$ is only dependent on the time shift τ.

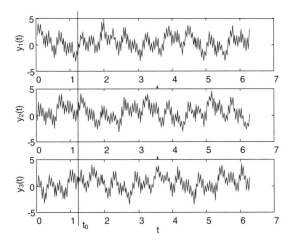

Figure 5.3. *Process histories*

If $Y(t)$ is a second order stationary process, it is said to be *ergodic* if it verifies the following conditions:

$$\mathbb{E}\big[Y(t)\big]_{stationary} = \mathbb{E}\big[Y(t')\big]_{ergodic} = \lim_{T\to+\infty} \frac{1}{T} \int_{-\frac{T}{2}}^{\frac{T}{2}} y(t)\,dt \qquad [5.2]$$

$$\mathbb{E}\big[Y(t)Y(t+\tau)\big]_{stationary} = \mathbb{E}\big[Y(t')Y(t'+\tau)\big]$$

$$\underset{ergodic}{=} \lim_{T\to+\infty} \frac{1}{T} \int_{-\frac{T}{2}}^{\frac{T}{2}} y(t)\,y(t+\tau)\,dt \qquad [5.3]$$

This property is important because it allows us to estimate the covariances and expected values only according to the knowledge of a single history. Ergodicity means that, for a stationary process, a history (a series of outcomes at n different instants) may be considered as an n-sample of a variable $Y(t)$ at a fixed time t.

If the process is stationary and ergodic, the mean and standard deviation are deduced by knowing a history $y(t)$ of the process $Y(t)$ using equations [5.2] and [5.3]:

$$\mathbb{E}\big[Y(t)\big] = \lim_{T \to +\infty} \frac{1}{T} \int_{-\frac{T}{2}}^{\frac{T}{2}} y(t)\, dt \tag{5.4}$$

$$\mathbb{V}\big[Y(t)\big] = \mathbb{E}\Big[\big(Y(t) - \mathbb{E}\big[Y(t)\big]\big)^2\Big]$$

$$= \lim_{T \to +\infty} \frac{1}{T} \int_{-\frac{T}{2}}^{\frac{T}{2}} y(t)\, y(t+\tau)\, dt \; - \mathbb{E}\big[Y(t)\big]^2 \tag{5.5}$$

The *cross-correlation function* between two processes $Y(t)$ and $Z(t)$ is defined in a similar way to the covariance between two random variables:

$$R_{YZ}(\tau) = \mathbb{E}\Big[\big(Y(t) - \mathbb{E}\big[Y(t)\big]\big)\big(Z(t+\tau) - \mathbb{E}\big[Z(t)\big]\big)\Big] \tag{5.6}$$

Due to the stationarity hypothesis, $R_{YZ}(\tau)$ does not depend on time t and is a symmetrical function because $R_{YZ}(\tau) = R_{YZ}(-\tau)$. If process $Y(t)$ is used instead of $Z(t)$, the cross-correlation function becomes the auto-correlation function. Finally, for $\tau = 0$, we can easily show:

$$R_{YY}(0) = \mathbb{V}\big[Y(t)\big] \tag{5.7}$$

We can also show that the *auto-correlation function*, also known as the *covariance function* verifies the double inequality:

$$-R_{YY}(0) \le R_{YY}(\tau) \le R_{YY}(0) \tag{5.8}$$

The auto-correlation function may be approached empirically by:

$$R_{YY}(\tau) \approx \frac{1}{N-k} \sum_{i=1}^{N-k} \big(Y_i - \mathbb{E}[Y]\big)\,\big(Y_{i+k} - \mathbb{E}[Y]\big) \tag{5.9}$$

with $Y_i = Y(i\,\Delta t)$ and $\tau = k\,\Delta t$. Δt is the time interval between two measurements of process Y, with N being the total number of measurement points. Equation [5.9] is only valid if the process is ergodic.

EXAMPLE 5.1.– Let us consider two random processes $x(t)$ and $y(t)$ formed by sinusoidal waves with constant amplitude and frequencies. A typical form of $x(t)$ is $x(t) = x_0 \sin(\omega t + \theta)$ where θ is a phase which is uniform distributed between 0 and 2π. The density function of θ is given by:

$$f_\theta(\lambda) = \begin{cases} \dfrac{1}{2\pi} & 0 \le \lambda \le 2\pi \\ 0 & \text{else} \end{cases}$$

Let us presume that process $y(t)$ is shifted from $x(t)$ by a constant phase $\phi : y(t) = y_0 \sin(\omega t + \theta - \phi)$. The inter-correlation function is then given by:

$$R_{xy}(\tau) = \int_0^{2\pi} x_0\, y_0 \sin(\omega t + \theta)\sin(\omega t + \theta - \phi)\frac{1}{2\pi}\, d\theta = \frac{1}{2}x_0\, y_0 \cos(\omega\tau - \phi)$$

$$R_{yx}(\tau) = \frac{1}{2}x_0\, y_0 \cos(\omega\tau + \phi)$$

□

EXAMPLE 5.2.– We consider a periodic signal $x(t)$ defined by the following figure:

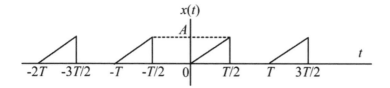

As the signal is periodic, the mean properties calculated over a long period of time are identical to those calculated over one single period. By limiting the analysis to the time interval $\left[-\dfrac{T}{2}, \dfrac{T}{2}\right]$, the signal $x(t)$ is written:

$$x(t) = \begin{cases} 0 & -\dfrac{T}{2} < t < 0 \\ \dfrac{2A}{T}t & 0 < t < \dfrac{T}{2} \end{cases}$$

The mean value of this signal is therefore easily calculated:

$$\mathbb{E}[x] = \frac{1}{T} \int_{-T/2}^{T/2} x(t)\,dt = \frac{1}{T} \int_{0}^{T/2} \frac{2A}{T} t\,dt = \frac{A}{4}$$

In order to calculate the autocorrelation function, we must distinguish the time intervals $0 < \tau < T/2$ and $T/2 < \tau < T$:

$$R_{xx}(\tau) = \frac{1}{T} \int_{-T/2}^{T/2} x(t)\,x(t+\tau)\,dt = \frac{1}{T} \int_{0}^{(T/2)-\tau} \frac{2A}{T} t \frac{2A}{T}(t+\tau)\,dt$$

$$0 < \tau < T/2$$

$$= \frac{A^2}{6}\left[1 - 3\frac{\tau}{T} + 4\left(\frac{\tau}{T}\right)^3\right]$$

$$R_{xx}(\tau) = \frac{1}{T} \int_{-T/2}^{T/2} x(t)\,x(t+\tau)\,dt = \frac{1}{T} \int_{T-\tau}^{T/2} \frac{2A}{T} t \frac{2A}{T}\big(t-(T-\tau)\big)\,dt$$

$$T/2 < \tau < T$$

$$= \frac{A^2}{6}\left[1 - 3\frac{(T-\tau)}{T} + 4\left(\frac{(T-\tau)}{T}\right)^3\right]$$

□

The *power spectral density* of a process $Y(t)$ is defined as the Fourier transform for the auto-correlation function of this variable:

$$S_Y(\omega) = \mathcal{F}\big(R_Y(\tau)\big) = \int_{-\infty}^{+\infty} R_Y(\tau)e^{-i\omega\tau}\,d\tau \qquad [5.10]$$

The auto-correlation function is therefore the inverse Fourier transform of the power spectral density:

$$R_Y(\tau) = \mathcal{F}^{-1}\big(S_Y(\omega)\big) = \int_{-\infty}^{+\infty} S_Y(\omega)e^{i\omega\tau}\,d\omega \qquad [5.11]$$

For $\tau = 0$, we obtain, according to [5.7]:

$$R_Y(0) = \sigma_Y^2 = \int_{-\infty}^{+\infty} S_Y(\omega)\,d\omega \qquad [5.12]$$

Similarly to the definition of power spectral density, the *power cross-spectral density* is defined as the Fourier transform of the cross-correlation function:

$$S_{YZ}(n) = \mathcal{F}\left(R_{YZ}(t)\right) = \int_{-\infty}^{+\infty} R_{YZ}(t)e^{-i2\pi nt}\,dt \qquad [5.13]$$

Contrary to the spectral density, the cross-spectral density function has a complex part. The concept of power spectral density is essential for studying many stochastic phenomena (wind, waves, etc.).

In the same way as calculating random variables, in order to characterize dispersion and symmetry, we may introduce the concept of *spectral moments*:

$$m_k = \int_{-\infty}^{+\infty} |\omega|^k \, S_Y(\omega)\,d\omega \qquad [5.14]$$

In particular, we obtain: $m_0 = \sigma_Y^2, m_1 = \sigma_{\dot{Y}}^2, m_2 = \sigma_{\ddot{Y}}^2$. However, these moments will exist (as for random variables) depending on the spectral density: thus, for white noise, m_0 does not exist because its variance is infinite. Figure 5.4 gives a few examples of processes, the auto-correlation function and the power spectral density.

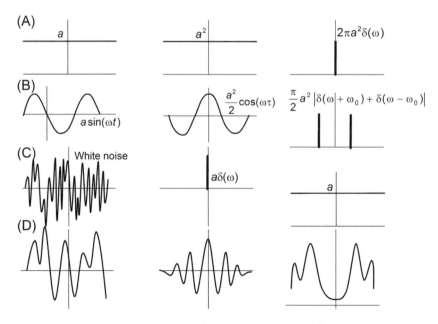

Figure 5.4. *Examples of processes and spectral densities*

5.1.2.2. *Time-dependent reliability*

In structural dynamics, it is traditional to focus on the reliability function $W_X(a,T)$ which is the probability that the structure's response $X(t)$ when subjected to random excitations remains lower than a level threshold a (deterministic) over a working or service lifetime T:

$$W_X(a,T) = P\left(X(t) \le a, 0 \le t \le T\right) \qquad [5.15]$$

The limit state function of this problem is, then, just $X(t) - a = 0$.

For stochastic processes, solving this problem consists of characterizing the statistics of extreme values. In particular, we should focus on the statistics of *threshold crossings*, and in the present case, statistics of the *first passage* or *excursion.* The probability density of this event is, then:

$$f_X(a,T) = \frac{\partial W_X(a,T)}{\partial T} \qquad [5.16]$$

A direct result of this problem is to study the distribution of *local extremes* reached over period T; its density function is then:

$$g_X(a,T) = \frac{\partial W_X(a,T)}{\partial a} \qquad [5.17]$$

The coefficient of variation of the variable defined by this density function is called the *peak factor.*

5.1.2.3. *Threshold crossings*

When studying the effects of a load, it is often useful to know the probability of an effect crossing a level over a fixed time period. Generally, higher levels relating to rare events are the most important. Crossing threshold a (for $a > 0$) at time t_1 means that the process $X(t) < x$ before t_1 and $X(t) > x$ after t_1.

The theory of threshold crossings is due to Rice [RIC 54]. The principle consists of defining a function for counting exceedences (crossings) $N_X(a,t_1,t_2)$ associated with a period of observation (Figure 5.5):

$$N_X(a,t_1,t_2) = \int_{t_1}^{t_2} |\dot{x}(t)| \, \delta\left(x(t) - a\right) dt \qquad [5.18]$$

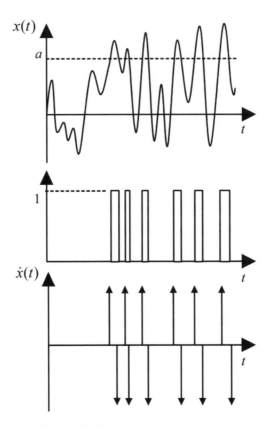

Figure 5.5. *Counting threshold crossings*

The mean of $N_X(a,t_1,t_2)$ is therefore given by:

$$\mathbb{E}\left[N_X(a,t_1,t_2)\right] = \int_{-\infty}^{+\infty} \int_{t_1}^{t_2} |\dot{x}(t)| \, p_{X\dot{X}}(a,\dot{x},t) \, d\dot{x} \, dt \qquad [5.19]$$

For a zero-mean stationary process, the joint probability density function is independent of time, and the number of level up-crossings (positive slope) is equal to the number of down-crossings (negative slope). The mean positive-slope (negative-slope) crossing rate is given by (number of positive-slope crossings per unit time):

$$v_{X,a}^+ = v_{X,a}^- = \frac{1}{2} \int_{-\infty}^{+\infty} |\dot{x}(t)| \, p_{X\dot{X}}(a,\dot{x}) \, d\dot{x} = \int_{0}^{+\infty} |\dot{x}(t)| \, p_{X\dot{X}}(a,\dot{x}) \, d\dot{x} \qquad [5.20]$$

For Gaussian processes, expression [5.20] can be explicitly calculated [LIN 76]:

$$p_{X\dot{X}}(a,\dot{x}) = \frac{1}{2\pi\,\sigma_X\,\sigma_{\dot{X}}}\exp\left[-\frac{1}{2}\left(\frac{x^2}{\sigma_X^2}+\frac{\dot{x}^2}{\sigma_{\dot{X}}^2}\right)\right] \qquad [5.21]$$

which gives:

$$v_{X,a}^{+} = \frac{1}{2\pi}\frac{\sigma_{\dot{X}}}{\sigma_X}\exp\left[-\frac{a^2}{2\,\sigma_X^2}\right] \qquad [5.22]$$

5.1.2.4. *Analysis of extremes*

A maximum of a continuous process (to at least the second derivative) corresponds to $\dot{x}(t)=0$ and $\ddot{x}(t)<0$. This implies that the distribution of local extremes must be obtained with the joint distribution of $X(t)$, $\dot{X}(t)$ and $\ddot{X}(t)$. Similarly to the threshold crossings, the number of extrema above the level a is given by:

$$M_X(a,t_1,t_2) = \int_{t_1}^{t_2} |\ddot{x}(t)|\,\delta(\dot{x}(t))\,\mathbb{H}(x(t)-a)\,dt \qquad [5.23]$$

where $\mathbb{H}(.)$ is the Heaviside function.

The mean of $M_X(a,t_1,t_2)$ is therefore given by:

$$\mathbb{E}[M_X(a,t_1,t_2)] = -\int_{t_1}^{t_2}\int_{a}^{+\infty}\int_{-\infty}^{0}\ddot{x}(t)\,p_{X\dot{X}\ddot{X}}(x,0,\ddot{x},t)\,dx\,d\ddot{x}\,dt \qquad [5.24]$$

The number of extremes, independently from their amplitude, is thus given by:

$$\mathbb{E}[M_X(t_1,t_2)] = -\int_{t_1}^{t_2}\int_{-\infty}^{+\infty}\int_{-\infty}^{0}\ddot{x}(t)\,p_{X\dot{X}\ddot{X}}(x,0,\ddot{x},t)\,dx\,d\ddot{x}\,dt \qquad [5.25]$$

The cumulative probability function of the extremes $F_E(a,t_1,t_2)$ is then given by:

$$F_E(a,t_1,t_2) = 1 - \frac{\mathbb{E}[M_X(a,t_1,t_2)]}{\mathbb{E}[M_X(t_1,t_2)]} \qquad [5.26]$$

which gives the following density function:

$$p_E(a,t_1,t_2) = -\frac{1}{\mathbb{E}[M_X(t_1,t_2)]} \int_{t_1}^{t_2} \int_{-\infty}^{0} \ddot{x}(t)\, p_{X\dot{X}\ddot{X}}(a,0,\ddot{x},t)\, d\ddot{x}\, dt \qquad [5.27]$$

In the case of a Gaussian stationary process, we can show quickly that:

$$p_{X\dot{X}\ddot{X}}(x,0,\ddot{x}) = \frac{1}{\sqrt{2\pi}\,\|[S]\|}\exp\left[-\frac{1}{2\|[S]\|}\left(m_2\, m_4\, x^2 + 2m_2^2\, x\,\ddot{x} + m_0\, m_2\, \ddot{x}^2\right)\right] \quad [5.28]$$

with the following for covariance matrices:

$$[S] = \begin{bmatrix} m_0 & 0 & -m_2 \\ 0 & m_2 & 0 \\ -m_2 & 0 & m_4 \end{bmatrix} \qquad [5.29]$$

The previous expressions are integrated analytically, which enables us to obtain the mean maxima frequency (time is eliminated due to the stationary nature of the process):

$$v_{ext} = \frac{1}{2\pi}\sqrt{\frac{m_4}{m_2}} \qquad [5.30]$$

It is also not difficult to assess the extrema density function. By setting out $\alpha = v_a^+/\mathbb{E}[M_X] = m_2/\sqrt{m_0\, m_4}$, we obtain:

$$p_E(a) = \frac{1}{\sqrt{2\pi}}\frac{\sqrt{1-\alpha^2}}{\sqrt{m_0}}\exp\left(-\frac{a^2}{2m_0\left(1-\alpha^2\right)}\right)$$

$$+\frac{\alpha\, a}{2\sqrt{m_0}}\left[1+erf\left(-\frac{a}{\sqrt{m_0\left(\frac{2}{\alpha^2}-2\right)}}\right)\right]\exp\left(-\frac{a^2}{2m_0}\right) \qquad [5.31]$$

If the stochastic process is a wide band process, the extrema are symmetrically distributed on both sides of the mean and the distribution tends towards a Gaussian distribution ($\alpha = 0$):

$$p_E(a) = \frac{1}{\sqrt{2\pi}} \frac{1}{\sqrt{m_0}} \exp\left(-\frac{a^2}{2m_0}\right)$$ [5.32]

On the other hand, if it is a narrow band process, there are no negative maxima, and the probability density tends towards the Rayleigh distribution:

$$p_E(a) = \frac{a}{2\sqrt{m_0}}[1+1]\exp\left(-\frac{a^2}{2m_0}\right) = \frac{a}{\sqrt{m_0}}\exp\left(-\frac{a^2}{2m_0}\right)$$ [5.33]

5.1.2.5. *Poisson's reliability model*

By allowing crossing independence,[2] when a process X is Gaussian and stationary with a zero mean, the number of crossings over the positive threshold a is a process Q which is Poisson distributed with $\lambda = v_{X,a}^+$ for crossing rate (if we observe thresholds $\pm a$, $\lambda = 2v_{X,a}^+ = 2v_{X,a}^-$). The probability of n crossings over a period T is therefore (see Chapter 1):

$$\mathcal{P}(a,T) = \exp\left(-v_{X,a}^+ T\right)\frac{\left(v_{X,a}^+ T\right)^n}{n!}$$ [5.34]

The probability of non-crossing, corresponding to the desired reliability, is therefore given for $n = 0$, which gives the following with expression [5.22]:

$$W(a,T) = \exp\left(-v_{X,a}^+ T\right) = \exp\left[-v_{0,X}^+ T \exp\left[-\frac{a^2}{2\sigma_X^2}\right]\right]$$ [5.35]

with $v_{0,X}^+ = \frac{1}{2\pi}\frac{\sigma_{\dot{X}}}{\sigma_X}$. The term $2v_{0,X}^+ T$ represents the mean number of half cycles.

The failure probability over period T is:

$$\mathcal{P}_f(a,T) = 1 - \exp\left[-v_{0,X}^+ T \exp\left[-\frac{a^2}{2\sigma_X^2}\right]\right]$$ [5.36]

---

2 The difficult aspect of the previous analysis is the independence of the level crossings. This independence is particularly unacceptable for narrow band processes, as in structural dynamics. Generally, the incursion into the failure domain of the envelope process precedes the first passage for each bundle of crossings. It is, therefore, more reasonable to work with envelope processes than with the initial process.

and the density function of the first passage is:

$$p_T(a) = v_{0,X}^+ \exp\left[-\frac{a^2}{2\,\sigma_X^2}\right]\exp\left[-v_{0,X}^+ \exp\left[-\frac{a^2}{2\,\sigma_X^2}\right]T\right] \qquad [5.37]$$

The mean and time variance of the first passage are then:

$$\mathbb{E}[T_a] = \frac{1}{v_{0,X}^+}\exp\left[\frac{a^2}{2\,\sigma_X^2}\right]$$

$$\qquad\qquad\qquad\qquad\qquad\qquad\qquad\qquad\qquad\qquad [5.38]$$

$$\mathbb{V}[T_a] = \left(\frac{1}{v_{0,X}^+}\right)^2\exp\left[\frac{a^2}{\sigma_X^2}\right]$$

Equation [5.36] assumes that the failure probability a $t = 0$ is zero. We can modify it by taking into account the initial probability (using a conditional probability calculation) [MAD 86]:

$$P_f(a,T) = 1 - P\big(X(0) < a\big)\exp\left[-v_{0,X}^+ T\exp\left[-\frac{a^2}{2\,\sigma_X^2}\right]\right] \qquad [5.39]$$

The previous expression can be generalized for any stochastic process:

$$P_f(a,T) = 1 - P\big(X(0) < a\big)\exp\left[-\int_0^T v_{X,a}^+(t)\,dt\right] \qquad [5.40]$$

with:

$$v_{X,a}^+ = \int_0^{+\infty} |\dot{x}(t)|\, p_{X\dot{X}}(a,\dot{x})\, d\dot{x} \qquad [5.41]$$

EXAMPLE 5.3.– We will consider a cracked steel component. The theory of fracture mechanics helps to assess the strength of a cracked component by comparing the stress intensity factor at time t with critical stress intensity factor K_{Ic}:

$$K_I = S\,Y(a)\sqrt{\pi a} \leq K_{Ic}$$

where a is the size of the defect and S is the far field stress. There is no failure over period T if the stress (over this period) does not exceed:

$$S > \frac{K_{Ic}}{Y(a)\sqrt{\pi a}}$$

The failure probability is approached by expression [5.39] if the stress follows a stationary Gaussian process:

$$P_f(T) = 1 - F_{S(0)}\left(\frac{K_{Ic}}{Y(a)\sqrt{\pi a}}\right)\exp\left[-v_{0,S}^+ T \exp\left[-\frac{\left[\frac{K_{Ic}}{Y(a)\sqrt{\pi a}} - \mathbb{E}(S)\right]^2}{2\,\sigma_S^2}\right]\right]$$

□

5.1.2.6. Reliability model for studying extrema

Another kind of approach for studying reliability is to focus on a discrete time process $Y(t_k)$ made up of maxima. If we call $h(t_k)$ the conditional probability of having the first passage of level a at time t_k, then the reliability is written as:

$$W(a,T) = \prod_{k=1}^{N}\left(1 - h(t_k)\right) \qquad\qquad [5.42]$$

where N is the number of maxima. The conditional probability $h(t_k)$ is defined by:

$$h(t_k) = \mathcal{P}\left(Y(t_k) \geq a \middle| \bigcap_{i=1}^{k} Y(t_k) < a\right) \qquad\qquad [5.43]$$

Assuming that the extrema are independent, we obtain (narrow band process, therefore the maxima are Rayleigh distributed):

$$h(t_k) = \mathcal{P}\left(Y(t_k) \geq a\right) = \int_a^{\infty} p_E(x)\,dx$$

$$= \int_a^{\infty} \frac{x}{\sqrt{m_0}}\exp\left(-\frac{x^2}{2m_0}\right)dx = \exp\left(-\frac{a^2}{2m_0}\right) \qquad\qquad [5.44]$$

We will note here that the independence hypothesis for the *extrema* results in the fact that the probability of each extremum to be the first is therefore identical for each of them. The reliability is therefore given by:

$$W(a,T) = \left(1 - \exp\left(-\frac{a^2}{2m_0}\right)\right)^N \qquad [5.45]$$

With the term $2v_{0,X}^+ T$ representing the mean number of half cycles, we obtain that the mean number of maxima is $2v_{0,X}^+ T$:

$$W(a,T) = \left(1 - \exp\left(-\frac{a^2}{2m_0}\right)\right)^{2v_{0,X}^+ T} = \left(1 - \frac{v_{a,X}^+}{v_{0,X}^+}\right)^{2v_{0,X}^+ T} \qquad [5.46]$$

5.1.2.7. *Return period*

This is a concept often used by engineers in design problems. Let us consider the event A describing the fact that a process X either reaches or exceeds a given level a_0. The return period $R(a_0)$ is the *mean time* between two consecutive occurrences of event A. Let T be the structure's *working* or *service life* and α the probability of event A being exceeded over the period T. The return period is given by the following ratios:

$$R(a_0) = -\frac{T}{\ln(1-\alpha)} \cong \frac{T}{\alpha} \qquad [5.47]$$

if probability α is low. An exceedence probability of 1% over a year is equivalent to a return period of 100 years. But the probability of exceeding this value over 100 years is, on the other hand, $1 - F_X(a_0)^{100} = 1 - 0.99^{100} \approx 0.634$.

If the process is stationary and Gaussian, then equation [5.36] gives for the level a_0:

$$P_f(a_0,T) = 1 - \exp\left[-v_{0,X}^+ T \exp\left[-\frac{a_0^2}{2\,\sigma_X^2}\right]\right] = \alpha \qquad [5.48]$$

i.e.

$$-\frac{\ln(1-\alpha)}{T} = \frac{1}{R(a_0)} = v_{0,X}^+ \exp\left[-\frac{a_0^2}{2\,\sigma_X^2}\right] = \frac{1}{2\pi}\frac{\sigma_{\dot{X}}}{\sigma_X}\exp\left[-\frac{a_0^2}{2\,\sigma_X^2}\right] \qquad [5.49]$$

We then obtain:

$$1 = \frac{1}{2\pi}\frac{\sigma_{\dot{X}}}{\sigma_X}\exp\left[-\frac{a_0^2}{2\sigma_X^2}\right]R(a_0) = v_{X,a_0}^+ R(a_0)$$

which gives a good explanation of the fact that, over the period $R(a_0)$, the process only exceeds the threshold a_0 once [CRE 01].

EXAMPLE 5.4.– Consider a stationary standard normal process (zero mean and unit standard deviation). In this case, the mean number of crossings for level x, per time unit, is:

$$v(x) = v_0 \exp\left[-\frac{x^2}{2}\right]$$

with $v_0 = \frac{\sigma_{\dot{X}}}{2\pi}$. If $R_t(x)$ is the return period for value x, the mean number of crossings $v(x)$ over period $R_t(x)$ is given by the following definition:

$$v(x).R_t(x) = 1$$

According to equation [5.48], we will write:

$$\alpha = 1 - \exp\left(-v_0 T \exp\left[-\frac{x^2}{2}\right]\right)$$

The previous equation clearly expresses that the value x of the peak factor has a probability α of being exceeded over the period T. The derivative of this expression therefore provides the density function of the maxima of X over T:

$$f(x) = \frac{d\left[\exp\left(-v_0 T\exp\left(-\frac{x^2}{2}\right)\right)\right]}{dx}$$

$$= -\frac{d\left[v_0 T\exp\left(-\frac{x^2}{2}\right)\right]}{dx}\exp\left(-v_0 T\exp\left(-\frac{x^2}{2}\right)\right)$$

$$= x v_0 T x\exp\left(-\frac{x^2}{2}\right)\exp\left(-v_0 T\exp\left(-\frac{x^2}{2}\right)\right) = x\xi\exp(-\xi)$$

with $\xi = v_0 T \exp\left(-\dfrac{x^2}{2}\right)$. The mean value of the maxima of X over T, as well as the standard deviation, are given by the following expressions:

$$\mathbb{E}[X_{max}] = \int_0^{+\infty} x^2 \, \xi \exp(-\xi) \, dx$$

$$\mathbb{V}[X_{max}] = \int_0^{+\infty} (x - \overline{x}_{max})^2 \, x \, \xi \exp(-\xi) \, dx$$

But:

$$x = \sqrt{2 \ln(v_0 T) - 2 \ln(\xi)}$$

which can be expanded into a series in the following way:

$$x = \sqrt{2 \ln(v_0 T)} - \frac{\ln \xi}{\sqrt{2 \ln(v_0 T)}} + \cdots$$

$$x^2 = 2 \ln(v_0 T) - 2 \ln \xi + \cdots$$

The mean value for the maxima of X is simplified as:

$$\mathbb{E}[X_{max}] \approx \sqrt{2 \ln(v_0 T)} - \frac{1}{\sqrt{2 \ln(v_0 T)}} \int_0^{+\infty} \ln \xi . \exp(-\xi) \, d\xi$$

$$\approx \sqrt{2 \ln(v_0 T)} + \frac{\gamma}{\sqrt{2 \ln(v_0 T)}}$$

because $\int_0^{+\infty} \ln \xi \exp(-\xi) \, d\xi = -\gamma$ ($\gamma \approx 0.5772$ is the Euler constant). The variance of the maxima of X is obtained in an identical way:

$$\mathbb{V}[X_{max}] \approx -\frac{\gamma^2}{2 \ln(v_0 T)} + \frac{1}{2 \ln(v_0 T)}\left(\frac{\pi^2}{6} + \gamma^2\right) \approx \frac{1}{2 \ln(v_0 T)} \frac{\pi^2}{6}$$

because $\int_0^{+\infty} (\ln \xi)^2 \exp(-\xi) \, d\xi = \frac{\pi^2}{6} + \gamma^2.$ $\qquad\qquad\square$

5.1.2.8. *Introducing time into reliability studies*

When there is a degradation process, the uncertain nature is modeled by random variables combined with deterministic functions which are time dependent. The processes are, then, not ergodic. As these degradation processes are generally monotonic according to time (a crack increases in size, corrosion expands, etc.), it is therefore possible to only focus on the instant when performance is assessed. These problems are called time-independent reliability problems. The vast majority of the problems fulfill this characteristic. The second type of problem involves reliability related to level crossing: a problem tackled in the previous sections. This reliability is said to be time dependent.

Consider the set of random variables $\{X(t)\}$ and a performance function or limit state $g(\{X(t)\})$. A reliability analysis at time T amounts to calculating:

$$P_{f,inst}(t) = \mathcal{P}\big(g(\{X(t)\}) \leq 0\big) \tag{5.50}$$

This instantaneous failure probability differs from the accumulated failure probability $P_{f,cum}$ over $[0,t]$:

$$P_{f,cum}(0,t) = \mathcal{P}\big(\exists \tau \in [0,t] / g(\{X(\tau)\}) \leq 0\big) \tag{5.51}$$

When g decreases over $[0,t]$, $P_{f,inst}$ and $P_{f,cum}$ coincide:

$$P_{f,cum}(0,\tau) = P_{f,inst}(\tau) \quad \forall \tau \leq t \tag{5.52}$$

When this property is not applicable, the current approach for assessing $P_{f,cum}$ consists of assessing the crossing rates which can be expressed by:

$$\mathrm{v}^+(t) = \lim_{\Delta\tau \to 0, \Delta\tau > 0} \frac{\mathcal{P}\big(\{g(\{X(t)\}) > 0\} \cap \{g(\{X(t+\Delta\tau)\}) \leq 0\}\big)}{\Delta\tau} \tag{5.53}$$

The accumulated failure probability can then be bounded by:

$$\max_{0 \leq \tau \leq t} P_{f,inst}(\tau) \leq P_{f,cum}(0,t) \leq P_{f,inst}(0) + \int_0^t \mathrm{v}_X^+(\tau)\,d\tau \tag{5.54}$$

The PHI2 method [AND 02] deals with the upper limit of the previous interval; it consists of calculating the crossing rate [5.53] by expressing it as a parallel system, which involves knowing the following correlation:

$$\rho(t, \Delta\tau) = -\sum_k \alpha_k(t)\alpha_k(t + \Delta\tau)$$

[5.55]

where $\alpha_k(t), \alpha_k(t + \Delta\tau)$ are the components of the direction cosines at the design point. With this correlation being determined, the crossing rate is deduced using the bi-variate distribution:

$$\mathcal{P}\left(\{g(\{X(t)\}) > 0\} \cap \{g(\{X(t + \Delta\tau)\}) \le 0\}\right) = \Phi_2\left(\beta(t), -\beta(t + \Delta\tau), \rho(t, \Delta\tau)\right)$$

[5.56]

i.e.

$$v_{PHI2}^+(t) \approx \frac{\Phi_2\left(\beta(t), -\beta(t + \Delta\tau), \rho(t, \Delta\tau)\right)}{\Delta\tau}$$

[5.57]

This equation makes the crossing rate calculation sensitive to the time interval $\Delta\tau$. Sudret [SUD 05] proposes an alternative based on the time derivatives of the direction cosines and the reliability index.

When this crossing analysis is not necessary, we should note that the evolution of the failure probability can be interpreted as the distribution function of the component's lifespan:

$$P_f(t) \equiv \mathcal{P}(T \le t) = F_T(t)$$

[5.58]

Expression [5.58] enables us to give a more concrete meaning to failure probability, connecting it to the lifespan of the studied component.

EXAMPLE 5.5.– We will use example 5.3 and presume that the size of defect a varies over time due to another phenomenon (fatigue, corrosion, etc.). The failure probability is written according to equation [5.40]:

$$P_f(T) = 1 - F_{S(0)}\left(\frac{K_{Ic}}{Y(a_0)\sqrt{\pi a_0}}\right) \exp\left[-\int_0^T v_{X,a(t)}^+\left(\frac{K_{Ic}}{Y(a(t))\sqrt{\pi a(t)}}\right) dt\right]$$

where $a(0)$ is the size of the defect at $t = 0$. This failure probability is indeed bounded by:

$$P_f(T) \leq P_{f,inst}(0) + \int_0^T v_X^+(\tau)\, d\tau$$

$$= F_{S(0)}\left(\frac{K_{Ic}}{Y(a_0)\sqrt{\pi a_0}}\right) + \int_0^T v_{X,a(t)}^+\left(\frac{K_{Ic}}{Y(a(t))\sqrt{\pi a(t)}}\right) dt \qquad \square$$

EXAMPLE 5.6.– Let us consider a welded joint for which the Wöhler curve is given by:

$$\ln(N) = \ln(A) - B\ln(S) = \ln(f(S))$$

Deterministic calculation

The fatigue damage over a reference period τ (a few weeks or a few days) is given by the Palmgren-Miner's law:

$$d_p = \sum_{i=1}^n d(S_i) = \sum_{i=1}^n \frac{n_i}{f(S_i)}$$

The total damage D over a period T is the sum of elementary damages $(s = T/\tau$ being the number of reference periods τ over the duration $T)$:

$$D(T) = \sum_{j=1}^s d_{p_j} = \frac{T}{\tau}\left(\sum_{j=1}^s d(S_i)\right) = \frac{T}{\tau} d_p$$

Based on the Miner's law, the limit state function is given by:

$$M = 1 - D(T) = 1 - \frac{T}{\tau} d_p$$

Traffic measurements over a week reference period and fatigue data provide the mean value $d_p = 1.48\ 10^{-4}$. The joint lifetime is therefore given by the condition

$$1 - D(T) = 0, \quad \text{or} \quad T_f = \frac{\tau}{d_p} = \frac{1}{1.48\,10^{-4}} \approx 6{,}756 \text{ weeks} \cong 135 \text{ years. This calculation}$$

does not allow us to identify the dominant uncertainties and the reliability level: a probabilistic model is consequently more appropriate.

Probabilistic calculation

Let us randomize the Wöhler's curve:

$$\ln(N) = \ln(A) - B\ln(S) + \varepsilon = \ln(f(S)) + \varepsilon$$

where ε is a random variable with zero mean and standard deviation σ_ε. Each couple $\ln(S)$, $\ln(N)$ is a random vector. A statistical estimation of $\ln(A)$, B and σ_ε is obtained from linear regressions based on a sample of independent outcomes $(\ln(S_i), \ln(N_i))$.

The fatigue damage is therefore rewritten:

$$D_p = \sum_{i=1}^{n} d(S_i) = \left(\sum_{i=1}^{n} \frac{n_i}{f(S_i)} \right) \exp(-\varepsilon)$$

$$= \left(\frac{n_1}{f(S_1)} + ... + \frac{n_n}{f(S_n)} \right) \exp(-\varepsilon) = d_p \exp(-\varepsilon)$$

assuming that $d_p = \dfrac{n_1}{f(S_1)} + ... + \dfrac{n_n}{f(S_n)}$ is normally distributed. The total damage D over the period T is given:

$$D(T) = \left(\sum_{i=1}^{n} d_{p_j} \right) \exp(-\varepsilon) = \frac{T}{\tau} \left(\sum_{i=1}^{n} d(S_i) \right) \exp(-\varepsilon) = \frac{T}{\tau} d_p \exp(-\varepsilon)$$

The safety margin is:

$$M(T) = 1 - \frac{T}{\tau} d_p \, e^{-\varepsilon}$$

From the traffic measurements and fatigue data, the following data is obtained:

$$m_\varepsilon = 0.0; \sigma_\varepsilon = 0.31; m_{d_p} = 1.48 \; 10^{-4}, \sigma_{d_p} = 3.97 \; 10^{-5}$$

The variables d_p and ε are respectively lognormally and normally distributed. An alternative safety margin can be expressed in logarithm terms:

$$\hat{M}(T) = \ln(1) - \ln\left(\frac{T}{\tau}d_p\ e^{-\varepsilon}\right) = -\ln\left(\frac{T}{\tau}\right) - \ln(d_p) + \varepsilon$$

with:

$$\sigma_{\ln(d_p)} = \mathbb{COV}[d_p] = 0.27$$

$$m_{\ln(d_p)} = \ln(m_{d_p}) - \frac{1}{2}\sigma^2_{\ln(d_p)} = -8.85$$

The reliability index is therefore given by:

$$\beta(T) = \frac{-\ln(T) - m_{\ln(d_p)} + m_\varepsilon}{\sqrt{\sigma^2_{\ln(d_p)} + \sigma^2_{\ln(\varepsilon)}}} = \frac{-\ln(T) + 8.85}{\sqrt{0.27^2 + 0.31^2}}$$

$$= \frac{-\ln(T) + 8.85}{0.41}$$

The probability of failure is nothing less than the cumulative probability function of joint lifetime:

$$F_T(t) = \Phi(-\beta(t)) = P_f(t)$$

The following figure illustrates the evolution of this probability function. It is interesting to note that, for $T_f = 135$ years, the probability of failure is roughly 50%.

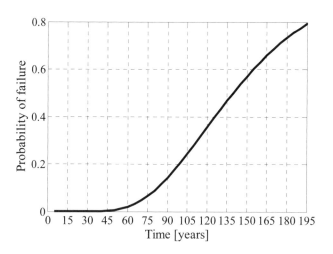

$\square$

5.1.3. *Spatial variability*

A load cannot only be time-dependent, but also space-dependent. Generally, the stochastic field $q(x,y,t)$ is replaced by an equivalent uniformly distributed load:

$$q_{\text{EUDL}}(t) = \frac{\int q(x,y,t)\, l(x,y)\, dx\, dy}{\int l(x,y)\, dx\, dy} \qquad [5.59]$$

where $l(x,y)$ is an influence function for a particular loading effect (the moment at mid-span, for example).

Let us break this stochastic field down into two terms (for simplicity reasons, we presume time-independence): the first term represents the variability between structural elements or structures, and the second denotes variability between the points of the same structure:

$$q(x,y) = q_0 + \tilde{q}(x,y) \qquad [5.60]$$

For a spatially homogeneous field, the statistical properties are independent of the loading point. It is, then, possible to show that the mean and variance of q_{EUDL} are given by:

$$\mathbb{E}\left[q_{\text{EUDL}}\right] = \mathbb{E}\left[q_0\right]$$

$$\mathbb{V}\left[q_{\text{EUDL}}\right] = \mathbb{V}\left[q_0\right] + \mathbb{V}\left[\tilde{q}\right] \frac{\iint \iint l(x_1,y_1)\, l(x_2,y_2)\, \rho(d)\, dx_1\, dx_2\, dy_1\, dy_2}{\left[\iint l(x_1,y_1)\, dx_1\, dx_2\right]^2} \qquad [5.61]$$

$\rho(d)$ is a correlation function which describes the correlation of field $\tilde{q}(x,y)$ between two distinct points. It is usual to model it by a decreasing exponential function of the difference between the two points:

$$\rho(d) = \exp\left(-\frac{d^2}{d_c^2}\right) = \exp\left(-\frac{(x_1-x_2)^2 + (y_1-y_2)^2}{d_c^2}\right) \qquad [5.62]$$

d_c is the correlation length. The correlation tends towards 0 for distances between points higher than the correlation length.

5.1.4. *Load combinations*

All the loads can be combined. The problem of combining loads consists of finding an equivalent load which represents two or many stochastic load processes, acting either in combination or individually in an additive way. We must, then, distinguish between the same types of loads and different loads. The combination of snow and wind is a typical example of different loads, whereas permanent loads are often of the same type, even if they are different.

If loads are the same, we often prefer to consider them as the components of a single load. The different components are usually described by similar probabilistic models. Dependence between components is generally expressed by cross-spectral power densities or cross-correlation functions.

If the loads are different, then the interaction may prove to be complex. Thus, the simultaneous load of wind and traffic effects must express the wind load according to traffic, and in return, relate the traffic intensity with wind speed: traffic on a structure may in fact change the wind effect on the structure. However, strong winds will also reduce the rate of vehicles. The combined action of snow and wind may also lead the wind to reduce snow accumulation, as well as producing heavier loads.

When studying the interaction of many loads, it is useful to have a load model on the one hand, and conditional probability models for this load according to the occurrence of other loads on the other hand. In most cases, this amounts to defining one of the loads as a dominant load, and to describe the arrival and intensities of other loads in a way that is conditional to the occurrence and amplitude of the first load.

The need for solving this type of problem has been introduced into regulation calibration and standard. Typically, the codes use simplified rules for combining loads. The most well-known are the Turkstra rule and the Ferry Borges-Castanheta model, respectively.

The first rule, developed by Turkstra in 1972, has often been used in practice. The principle is based on the choice of a primary load. When this load reaches its maximum, the values of the other loads are recorded. The procedure is reapplied for each load which becomes a primary load. If $\left(X_1, \cdots, X_j, \cdots, X_n \right)$ are independent loads, the Turkstra rule gives n scenarios:

$$\left(X_1(t_1) = \max_t X_1, \cdots, X_j(t_1), \cdots, X_n(t_1) \right)$$

$$\vdots$$

$$\left(X_1(t_j), \cdots, X_j(t_j) = \max_t X_j, \cdots, X_n(t_j) \right) \qquad [5.63]$$

$$\vdots$$

$$\left(X_1(t_n), \cdots, X_j(t_n), \cdots, X_n(t_n) = \max_t X_n \right)$$

The scenario producing the least satisfactory result is then retained. The Turkstra rule gives results which, from a theoretical point of view, are often unsafe. In fact, it is entirely possible for the most unfavorable situation to occur at an instant when none of the elementary loads reach their maximum. However, in practice, this rule gives back a scenario which is rather close to reality.

Often, in order to replace the Turkstra rule, load combinations are based on the *Ferry Borges-Castanheta* model. This model represents each stochastic process by a series of rectangular pulses (Figure 5.6). The values of these pulses are the loading intensities over a period τ. Thus, $\left(X_1, \cdots, X_j, \cdots, X_n \right)$ are independent loads translated into pulse processes over periods $\left(\tau_1, \cdots, \tau_j, \cdots, \tau_n \right)$

Periods τ_j are chosen so that the pulses may be considered as independent occurrences of the respective loads. The pulse processes thus formed are then put on a hierarchical basis according to pulse duration ($\tau_1 \geq \cdots \geq \tau_j \geq \cdots \geq \tau_n$, for example). In order to facilitate the calculations, a period of τ_j is always a multiple of the previous period ($\tau_j = k\ \tau_{j-1}$, k as a whole constant for all the series of loads).

The Ferry Borges-Castanheta model then considers the maximum of X_n together with X_{n-1} over a period τ_{n-1} as a new variable. The maximum of the new variable $X_{n-1} + \max_{\tau_{n-1}} X_n$ is then sought over a period τ_{n-2}, and so on until reaching the structure's working life T. The resultant load is then given by the following expression:

$$Y = \max_T \left(X_1 + \left(\cdots + \max_{\tau_{n-2}} \left(X_{n-1} + \max_{\tau_{n-1}} X_n \right) \right) \right) \qquad [5.64]$$

Thus, for three loads, equation [5.65] is written as:

$$Y = \max_T \left(X_1 + \max_{\tau_1} \left(X_2 + \max_{\tau_2} \left(X_3 \right) \right) \right) \qquad [5.65]$$

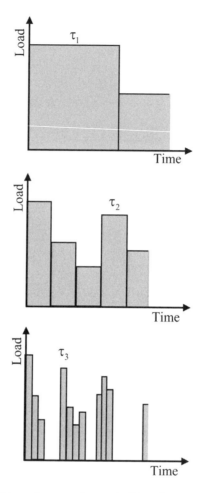

Figure 5.6. *Load hierarchy according to the Ferry Borges-Castanheta model*

5.1.5. *Permanent loads*

The main characteristics of permanent loads are:

– The probability of occurrence at any time is close to 1;

– The time-variability is negligible;

– The uncertainties regarding amplitude are low in comparison to other loads.

Permanent loads are not supposed to vary over time. In truth, they may vary greatly at certain instants when, for example, a bridge is reloaded by a new road surface. Another effect is related to delayed effects (creep, shrinkage, relaxation) which are time-dependent and closely related to the material: they are usually considered as loads.

Quantifying permanent loads and their variability is a function of the variability of the weight density, but also the dimensions of the element. The following data takes these two variabilities into account; as the dimensional aspect is also taken into account in element resistance, this chapter later discuss the uncertainties related to geometrical data.

5.1.5.1. *Dead loads*

Uncertainties linked to these loads are due to natural variation, modeling and estimation errors. Natural variations and estimation errors are the primary consequences of material density, construction and manufacturing. The uncertainty of modeling is relatively low insofar as certain parts of the structure are ignored. Uncertainties related to natural variations cannot be reduced by data collection, but estimations and modeling hazards may be reduced. Finally, it would be appropriate to distinguish variation within the same element, variation between structural elements within the same structure, and variation between different structures of the same type.

It is often standard practice to model material density (geometrical aspects will be taken up again further on in this chapter) by normal distributions. Table 5.1 gives a few useful values for different materials, but these often result from scattered data and a certain amount of care must be taken when using these values. They can, however, be used as a basis for *a priori* reliability analysis, with the engineer being responsible for updating them through a Bayesian analysis.

Material		Mean value [kN/m^3]	Variation coefficient
Steel		77	<1%
Concrete			
	Ordinary	24	4%
	High strength	24-26	3%
		(according to composition)	
	Light	(according to composition)	4–8%
Ballast		16-20	10–15%
Wood			
	Spruce	4.4	10%
	Pine	5.1	10%
	Oak	6.5	10%

Table 5.1. *Some examples of statistical characteristics of material density*

Table 5.2 also gives a few uncertainty values related to permanent loads taken from various references and summarized by Bailey [BAI 96] for dead loads and road bridge superstructures.

Load	Variability							
	Natural		Estimate		Model		Total	
	Bias	COV	Bias	COV	Bias	COV	Bias	COV
Metal	1.00	0.02	1.00	0.01	1.01	0.02	1.01	0.03
Concrete	1.05	0.06	0.99	0.05	1.03	0.05	1.07	0.10
Ready to use concrete	1.02	0.02	0.99	0.03	1.02	0.02	1.03	0.04
Roadways	1.08	0.14	1.09	0.14	1.02	0.05	1.20	0.25
Lighting, sidewalks, etc.	1.01	0.02	1.00	0.05	1.03	0.03	1.04	0.06

Table 5.2. *Variability associated with permanent loads*

Permanent loads related to dead loads and non structural elements are generally modeled by normal distributions. Uncertainty regarding permanent loads is given in terms of *bias* and *coefficients of variation*. The bias represents the relationship between the mean value and the nominal or design value. A bias higher than 1 means that the design value reduces the real mean load. Variability in structural element dead loads depends on the material type and the manufacturing method. For large elements, it is advisable to condition the dead load variability by geometric size variability. Superstructure loads are often badly estimated and under assessed. This explains why biases higher than 1, and coefficients of variation are needed for these variables.

The spatial correlation for a structural element is often represented by an exponential function of the following type:

$$\rho(\Delta r) = \rho_0 + (1 - \rho_0) \exp\left(-\left[\frac{\Delta r}{d}\right]^2\right) \qquad [5.66]$$

where Δr, ρ_0 and d represent the distance between two points of an element, the distant asymptotic correlation and the correlation length, respectively.

5.1.5.2. *Prestressed loads*

In concrete structures, prestressed effects are generally considered as a force being applied to the structure. A prestressed force $P(x,t)$ at a distance of x from the anchorage at time t may be expressed by the following [JCS 01]:

$$P(x,t) = P_0 - \Delta P(x,t) \qquad [5.67]$$

where $P_0, \Delta P(x,t)$ are the initial prestressed force and the prestressed losses, respectively. In practice, it is normal to assess prestressed loss over two different instants: instantaneous loss immediately after the tension has been set up, and delayed loss over an infinite time.

Due to the many factors affecting prestressed loss and the existing relationships between some of them, there are no analytical models enabling us to predict loss with a high degree of accuracy. Standards and empirical models, nonetheless, allow us to assess them. Uncertainties linked to prestressing depend on the uncertainties of the models used. According to [JCS 01] there is only a small amount of data available to assess the uncertainties of the models and associated parameters. However, [JCS 01] proposes that we retain the value given by the standard or regulation as a mean value for the losses, and to take those found in Table 5.3 as coefficients of variation. These coefficients of variation are very high and we should highlight here that they can be relatively excessive.

Mirza [MIR 80] suggests that we retain different coefficients of variation: 1.5–2% for instantaneous loss, and 15–20% for delayed losses. Cremona and Byrne [CRE 09] propose 10% coefficients of variation for both loss types. Normal distributions are generally recommended in order to characterize these losses.

Parameter	Variation coefficient	
	$t = t_0$	$t = \infty$
Prestressed losses $\Delta P(x,t)$	30%	30%
Prestressed force $P(x,t)$	4–6%	6–9%

Table 5.3. *Coefficients of variation for prestressed losses and prestressing force*

5.1.6. *Live loads*

A *live load* is, naturally, a load which varies over time. The variation frequency depends on the load type (snow, temperature, traffic). In each case, the representative value to be taken into account in the design or assessment is the value which displays a probability (*a priori* accepted) of being reached or exceeded over a reference period. Generally, the probability and reference period are replaced by a single parameter, the *return period*:

$$R = -\frac{T}{\ln(1-\alpha)} \qquad [5.68]$$

with T, as the reference period and α as the exceedence probability. As an example, Eurocode specifies a reference period of 50 years and an exceedence probability of

0.02 for environmental live loads (snow, wind, temperature, etc.). For road traffic, these parameters are fixed at 100 years and 0.1 respectively, meaning a return period of 1,000 years.

For a structure influenced by many live loads, we should not assume that the total characteristic load is the sum of the individual characteristic load values. This is due to the fact that the simultaneous occurrence of these values has a very low probability. For example, it is not possible for the characteristic snow load to occur simultaneously with the characteristic temperature! The adopted approach in standards is to consider load combinations by successively combining the characteristic values for one single variable with the mean values of others.

5.1.6.1. *Floor loads*

Floor loads in buildings come from equipment, stored goods and people. We should distinguish them according to the building's intended use: hospitals, hotels, office blocks, schools, shops, residential areas. In the design phase, it is vital to take into account any potential changes in their usage over their working life. These loads vary with time and space and therefore require probabilistic modeling. As an initial approximation, spatial variations are generally presumed to be homogeneous. Time variations may be broken down into two components; a quasi-permanent load (equipment, furniture, etc.) and an exceptional load (exceptional events, specific temporary use, etc.) which are sometimes difficult to assess. The duration of such an exceptional load is generally very short.

For the quasi-permanent load, spatial variation is described by a Gamma distribution, whereas exceptional loads will be exponentially distributed. In the first case, the load's variance is modeled by:

$$\mathbb{V}[Q] = \sigma_V^2 + \sigma_U^2 \, \kappa \min\left(\frac{A_0}{A}; 1\right)$$ [5.69]

where A is the load-carrying surface, and A_0 is a reference surface. Factor κ depends on the shape of the influence line (Figure 5.7). The time variation associated with this quasi-permanent load is Poisson distributed. The probability function for the maximum load is given by:

$$F_{q_{max}}(x) = \exp\left[-\lambda T \left(1 - F_q(x)\right)\right]$$ [5.70]

where $F_q(x)$ is the distribution function of the quasi-permanent load. λT is the mean number of occupancy changes.

Table 5.4 and Figure 5.7 give a few characteristics of floor loads [CIB 89, JCS 95].

Use	A_0 [m²]	$\mathbb{E}[q]$ [kN/m²]	σ_U [kN/m²]	σ_V [kN/m²]	$1/\lambda$ [year]
Offices	20	0.5	0.3	0.6	5
Residential	20	0.3	0.15	0.3	7
Hotels	20	0.3	0.05	0.1	10
Libraries	20	1.7	0.5	1.0	>10
Schools	100	0.6	0.15	0.4	>10
Shops (floors)	100	0.9	0.6	1.6	1-5
Shops (storage)	100	3.5	2.5	6.9	0.1-1.0
Light industry	100	1.0	1.0	2.8	5-10
Heavy industry	100	3.0	1.5	4.1	5-10

Table 5.4. *Characteristic parameters of floor loads*

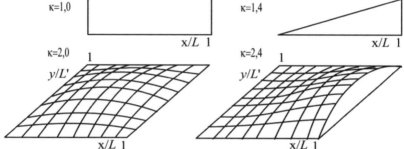

Figure 5.7. *Parameter* κ *according to the influence line/surface*

We can distinguish many kinds of car parks:

– residential areas;

– office areas;

– business areas;

– areas for special events (stadiums, theatres, etc.);

– transit zones (airports, train stations, etc.).

Light vehicles are generally modeled by punctual forces following a normal law with a mean of 15 kN and a variation coefficient of 15 to 30% according to the car park's use and the type of vehicles using it.

A parking space generally covers a surface of 2.4 to 5 m². The distribution function of the maximum effect on a car park floor [CIB 89] is given by:

$$F_{Q_{max}}(x) = \exp\left[-\lambda t_a t_h T \Phi\left(-\frac{x - \left[\sum_j I_j \mathbb{E}[q_j]\right]}{\sqrt{\sum_j I_j^2 \mathbb{V}[q_j]}}\right)\right]$$ [5.71]

where L_j is the influence coefficient relating to load q_j on the car park. Table 5.5 gives the parameter characteristics of expression [5.72] for different types of car parks.

Area	Number of peak days per year t_a [day]	Number of peak hours per day t_h [hour]	Average staying time T [hour]	Number of vehicles per day λ [day^{-1}]
Business	312	8	2.4	3.2
Transit	30	14-18	10-14	1.3
Events	50-150	2.5	2.5	1.0
Offices	260	8-12	8-12	1.0
Residential	360	17	8	2.1

Table 5.5. *Characteristics of the time-variability of car park loads (light vehicles)*

5.1.6.2. *Traffic load*

For road bridges, the predominant live load is the traffic. It induces static effects which must be modified in order to take into account dynamic effects (Table 5.6). The static effects presume that the vehicles are moving on a smooth surface, and the dynamic effects come from the effects of interaction between vehicles (by their suspension system) and the bridge (using its dynamic characteristics and the roadway's roughness). So as to lead a relevant study on safety and serviceability of a bridge, we must have continuous recordings of traffic weight (weight-in-motion data), or traffic simulations based on statistical data of the variables representing vehicles and flows. These recordings or simulations are limited to describing the static effects of traffic. They are then corrected by dynamic impact factors to take into account dynamic effects.

Static effects	Dynamic effects
Traffic composition	Vehicle speed
Distance between vehicles	Vehicle suspension
Weight of axle groups	Vehicle mass
Vehicle geometry	Number of vehicles on roadway
Weight variation according to lane	Roadway roughness
	Bridge modal characteristics

Table 5.6. *Static and dynamic effects*

The objective of a rational approach to assessing a structure is to apply traffic loads which represent the real loads acting on the bridge. Since traffic varies from one site to another, the uncertainties associated with this variable are relatively high. The values given in design standards may be very pessimistic in many cases, because they are based on heavy traffic. Consequentially, on other sites, there is a potential reduction in traffic loads which must be taken into account when assessing existing structures.

Uncertainties related to traffic are due to natural variability, estimation and models. Natural variability appears from one site to the next, and on the same site over time. Variability between sites is removed when specific data (weight-in-motion data) is available for the site of the structure being studied. In the same way, these specific data enable us to reduce estimation errors. Within the framework of a linear model for structural behavior, the effect of traffic on a highway lane is the sum of the effects of different axles which form an *axles' flow*. This axles' flow comes from a random flow of vehicles with random characteristics: traffic composition, distance between vehicles, axle group weight, and vehicular geometry.

The different vehicular categories are represented by the number of axles, the total load weight of each vehicle and the distribution of this weight on each axle. The basic categories are vehicles, vans, buses, trucks, with the latter being those causing the heaviest effects. Traffic composition on a structure may be expressed in percentages for each vehicle category. The traffic flow varies, which involves a variation in the distances between vehicles. Generally, the slower the traffic flow is, the more the distance between vehicles is reduced and vice versa. The distance between vehicles may have three configurations: fluid, congested with light vehicles and congested without light vehicles. Distance is often modeled by Beta distributions (see Chapter 1). A vehicle's axle group presumably acts as a punctual load. Experience has shown that they seem to be modeled correctly by a Beta type bimodal distribution.

The relevant data (bending moment, etc.) for assessing a structure (whether it is extrapolated from weight data or simulations) is presented as an extreme value distribution (Chapter 1). The maximum annual effect of normal traffic S_T is calculated as follows:

$$S_T = Q_T \, X_T \, \Phi \tag{5.72}$$

where Q_T, X_T and Φ respectively denote the reference traffic effects (i.e. from standards or regulations), a random error multiplier and a dynamic impact factor. The reference traffic effect is determined by the loading on the structure's influence lines according to Table 5.7 (taken from [EN 03]) and Figure 5.8.

	Load concentrated by axle (kN)	Distributed load (kN/m²)
Lane 1	300	9
Lane 2	200	2.5
Lane 3	100	2.5
Other lanes	0	2.5
Residual surface	0	2.5

Table 5.7. *Basic intensities of traffic loads according to [EN 03]*

For verifying the load-carrying capacity of British bridges, Cooper [COO 97] proposed a probabilistic modeling of [5.73]. This modeling assumes that the maximum effects of traffic loads on a bridge element are distributed according to a Gumbel extreme value distribution for variable X_T:

$$f_{X_T}(x) = \exp\left(-\exp\left[-\alpha_{X_T}\left(x - U_{X_T}\right)\right]\right) \tag{5.73}$$

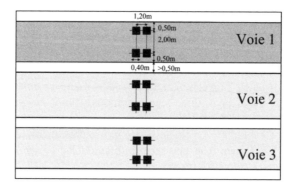

Figure 5.8. *Model of characteristic traffic load*

The mean value of X_T is given in Table 5.8 for $T = 1$ year and depends on loaded lengths and daily flows. For a duration T (in years), parameters are deduced from Table 5.8 according to:

$$U_{X_T} = U - \frac{\ln\left(-\ln\left[1 - 1/T\right]\right)}{\alpha} \underset{T \to \infty}{\approx} U + \frac{\ln(T)}{\alpha} \qquad [5.74]$$

	All lanes loaded							
Loaded length [m]	Moment				Shear force/End reactions			
	50,000 v/day (a)	10,000 v/day (b)	2,000 v/day (c)	α	50,000 v/day (a)	10,000 v/day (b)	2,000 v/day (c)	α
5	0.311	0.275	0.250	64	0.307	0.271	0.246	64
10	0.341	0.302	0.275	58	0.334	0.294	0.265	56
16	0.388	0.344	0.313	52	0.389	0.349	0.322	58
20	0.386	0.341	0.310	52	0.386	0.344	0.315	55
40	0.376	0.332	0.301	52	0.353	0.315	0.289	61
	Single lane loaded							
5	0.423	0.383	0.354	56	0.443	0.400	0.370	53
10	0.387	0.350	0.324	62	0.407	0.368	0.340	58
16	0.444	0.399	0.368	52	0.490	0.435	0.396	42
20	0.411	0.381	0.360	77	0.421	0.381	0.354	58
40	0.373	0.351	0.335	102	0.382	0.347	0.323	67

(a) highway; (b) main road; (c) secondary road – for one direction

Table 5.8. *Mean value of variable X_T*

Loaded length [m]	Weight restriction			
	38 t	25 t	17 t	7.5 t
5	0.92	0.92	0.66	0.38
10	0.93	0.93	0.82	0.48
16	0.97	0.97	0.77	0.41
30	1.00	1.00	0.82	0.41
40	1.00	1.00	0.82	0.41

Table 5.9. *Mean value of the reduction coefficient for variable X_T*

Loaded length [m]	Good road surface		Mediocre and poor road surface	
	All lanes loaded			
	Mean	Coefficient of variation	Mean	Coefficient of variation
10	1.08	0.065	1.13	0.088
25	1.08	0.065	1.13	0.088
30	1.12	0.071	1.17	0.095
35	1.16	0.077	1.22	0.102
40	1.20	0.083	1.27	0.110
	Single lane loaded			
10	1.08	0.093	1.13	0.133
25	1.08	0.093	1.13	0.133
30	1.12	0.104	1.17	0.141
35	1.16	0.114	1.22	0.148
40	1.20	0.125	1.27	0.157

Table 5.10. *Dynamic impact factors*

When a load restriction is imposed on the structure, a reduction coefficient λ is applied to the mean value of X_T, following Table 5.9. A standard deviation of 0.02 is proposed in [BD 79]. The dynamic impact factors are normally distributed, as a function of the span length (and not the loaded length), the number of lanes and the condition of the roadway (Table 5.10).

5.1.7. *Environmental loads*

Environmental loads are the consequence of physical or chemical phenomena, which are independent of human activity. Environmental loads are mainly random and require appropriate modeling. Moreover, there is a high level of time-dependency as soon as there are dynamic variable loads: wind, waves, earthquakes. We must, therefore, use the theory of random processes in order to describe them. However, in most problems encountered by engineers, we only need simple descriptions: power spectral density, level crossing, maxima and minima. Data which is useful for representing variables must be collected over appropriate periods of time, with care taken to ensure that the phenomenon is stationary. We must, therefore, use statistically representative databases. In the case of extreme values, which are essential for designing structures, we encounter the problem of model extrapolation. For sufficiently long return periods, stationarity can be allowed, but we must be aware that extrapolation will only take place using current observations, and not by using particular unobserved phenomena in current situations. This is the case for tropical cyclones. Modeling these loads, then, requires us to introduce multimodal distributions where the last mode represents these rare and exceptional phenomena. Finally, many natural loads are correlated and must be treated as such: this is the case for wind and waves, rains and floods (positive correlation) or temperature and snow (negative correlation).

The main climatic loads are wind, precipitation (rain, snow), ice and temperature. We may also include waves in these loads as they are a result of a wind load on the sea's surface. Although meteorological models generate deterministic descriptions, these loads are random by nature and variable over time. It is traditional to describe them by stochastic processes. They are also complex, from a time point of view, because they contain many scales of fluctuation:

– long duration: hourly, daily, seasonal, annual variations;

– short duration: turbulence.

It is therefore necessary to have as much short term data (high frequency sampling) as long term data (low frequency sampling). This distinction involves using different models. This is why the loads considered in the long term will be modeled by extreme value distributions (random variables), whereas short term loads will be described by power spectral variables (stochastic processes).

Load type	Model
Short load periods (stochastic processes)	
Winds	Power spectral densities (Karman, Kaimal, Davenport spectra, etc.)
Swell	Power spectral densities (Pierson spectra)
Long load periods (*extrema*)	
Wind	Gumbel distribution
Swell	Rayleigh distribution
Hydrological or meteorological	Fréchet distribution Gumbel distribution

Table 5.11. *Probabilistic models for some envrionmental loads*

Some environmental loads are, however, rarely modeled by stochastic processes insofar as they are the extreme values which enter primarily into structural design: snow, temperature and thermal gradients, rain. Table 5.11 indicates a few statistical models which can be used in studies.

5.1.7.1. *Snow*

Snow load is an important load for certain countries. Each year, many roofs cave in under its load. Thus, light constructions in areas with high snow fall are particularly critical in relation to this load. Snow loads depend on many parameters: climatic conditions (temperature, rain, sun), roof shape, heating, etc.

The load induced by snow on roofs is generally given by an expression of the following type:

$$Q_t = Q_s \, r \, k^{\frac{h}{h_r}} \qquad\qquad\qquad [5.75]$$

where Q_s, r, h and h_r are the ground load, a conversion factor between the ground load and the roof load, the building's altitude and a reference altitude (generally 300 m) respectively. Load Q_t acts vertically; Q_s is time-dependent, but not space dependent within a given climatic area for the same altitude.

To characterize the roof load, firstly, the ground load Q_t is estimated. The snow's thickness and density are the random basic variables. In addition, the dispersion on the density is very important between powder snow and water saturated snow (slush). It is, then, preferable to use a water content parameter rather than density. The probabilistic model for describing ground load must, on the one hand, specify the cumulative probability function of the load's duration (or occurrence), and on the other hand the cumulative probability function for the maximum load $Q_{s,max}$. In this last case, it is traditional to use Gamma distributions.

The second stage is considering the snow's load on the structures. For many reasons, the transfer from the ground load to the roof load differs considerably. In fact, it is necessary to take into account the exposure and the many phenomena coupled with heating and insolation. Finally, we must add the problem of accumulating different types of snow. The conversion factor r may be broken down into the sum of many terms:

$$r = \eta_a \, C_e \, C_t + C_r \qquad\qquad\qquad [5.76]$$

where η_a, C_e, C_t, C_r are shape coefficients, exposure coefficients, thermal and redistribution coefficients related to wind. Coefficients η_a and C_e are reduction factors taking into account a building's exposure to wind and the roof's slope α (see Table 5.12). A 15% coefficient of variation for $\eta_a C_e$ is prescribed by [JCS 01].

$\alpha = 0°$	$\eta_a C_e = 0.1 + 0.6\exp\left(-0.1U(H)\right)$
$\alpha = 25°$	$\eta_a C_e = 0.7 + 0.3\exp\left(-0.1U(H)\right)$
$\alpha = 60°$	$\eta_a C_e = 0.0$

Table 5.12. *Coefficient values $\eta_a C_e$ according to the roof's angle α and the average wind speed $U(H)$ over a period of one week (H is the roof level)*

Coefficient C_t represents the reduced snow load through heat transmission via the roof. For unheated buildings or highly insulated roofs, this coefficient is worth 1; a value of 0.8 is generally retained in practice. A deterministic value is generally used for this variable. Finally, the coefficient C_r takes into account the redistribution of the snow on the roof caused by the wind. For buildings with simple roofs, this effect may be ignored. For double sloped roofs, the effect is taken to be constant and equal to $\pm C_{r0}$ for each slope, according to Figure 5.9. The coefficient C_{r0} is Beta distributed with a mean given by the variation curve according to the roof's angle (Figure 5.9). The coefficient of variation is taken to be equal to 1.0. For other types of roofs, the values given in EN1991-1-4 and ISO 4355 may be used[3].

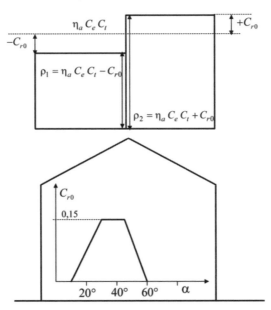

Figure 5.9. *Redistribution of snow on a symmetrical double sloped roof*

3 These values relate to mean values which are increased by a standard deviation.

5.1.7.2. *Wind loads*

Wind load is also a phenomenon which may induce strong effects on structures. For moderate speeds, the wind may be considered as an accompanying load. For high speeds (gusts), it is preferable to consider them as primary loads. These peak values are generally modeled by Gumbel distributions. The ground roughness, buildings or open spaces may favor high wind speeds and cause serious complications for modeling. Once the distribution of the mean wind speed has been calibrated, the wind effect distribution must be calculated. The wind effects generally vary according to the square of the speed and a pressure coefficient. These factors are given in standards and it is recommended to consider them as mean values with a 15% coefficient of variation.

The forces generated by the wind on a structure essentially depend on the relative air/structure motion. Two types of structures must then be distinguished:

– Structures known as *rigid,* for which the relative air/structure motion is mainly due to the wind itself. In this case, the forces of inertia are insignificant due to the wind-induced forces. Most civil engineering structures can be found in this category; it is generally sufficient to consider wind effects as static loads which are equivalent and proportional to the mean speed of the reference wind. The general principles for modeling *equivalent static load effects* are well adapted to these structures. Therefore, there is no need to use any particular modeling methods to take these effects into account; these effects being static loads are therefore similar to others.

– *Flexible structures*, for which relative air/structure motion is the combination of air movement and the structure's own movement caused by the wind. The forces of inertia should no longer be neglected and dynamic effects must be taken into account. There are some very complex phenomena due to the fact that the oscillating structure disturbs the wind flow, and as a consequence, disturbs the forces generated on the structure itself. These phenomena may produce *aeroelastic instabilities.*

This section only deals with rigid structures and the reader may refer to [CRE 03a], [SIM 98] for additional elements on flexible structures.

5.1.7.2.1. Wind characterization

The wind, defined as the air movement relative to the Earth's surface, is caused by temperature gradients. The causes for these gradients are mainly the differences in solar radiation between various latitudes, marine currents, and the fact that dry land warms up and cools down quicker than the oceans. Along with these "thermal" effects, the Earth's rotation produces inertia, known as the Coriolis effect. Friction reduces the flow speed around the Earth's surface. This effect results in the formation of a turbulence zone known as the *atmospheric boundary layer*; inside this layer, space/time variations occur from the wind speed. The thickness of this layer varies by a few hundred meters to many kilometers according to the wind speed, the ground roughness and the site's latitude.

Wind speed recordings in this atmospheric boundary layer show that these speeds may be represented by superposing a *mean wind speed* and a *fluctuating wind speed* (turbulence). Wind speeds, which are of interest for designing civil engineering structures, are speeds which are high enough for us to disregard convection phenomena. With these same wind speeds being lower than the speed of sound, we may reasonably allow the hypothesis for the incompressibility of air [CRE 03a], [SIM 98].

The origin of these fluctuations can be explained by considering the wind (or rather a gust of wind) as a set of vibrating air vortices, of different sizes and oscillation frequencies which move forward (Figure 5.10). These fluctuations induce motions for flexible structures. Thus, in a Cartesian reference, with the X axis parallel to the mean wind direction, the wind speed at a point P is given by:

$$U(P,t) = \overline{U}(P) + u(P,t)$$
$$V(P,t) = v(P,t) \tag{5.77}$$
$$W(P,t) = w(P,t)$$

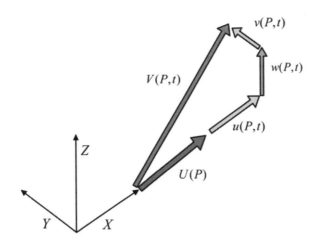

Figure 5.10. *Wind speed components*

The parameters which characterize the random nature of the turbulent fluctuations of wind speed, and which are obtained from experimental data, are the following:

– mean wind speed U;

– standard deviation of fluctuations $\sigma_u, \sigma_v, \sigma_w$;

– turbulence intensities;

– turbulence length scales;

– spectral density for each component;

– cross-spectral density.

At a given point in space, longitudinal, vertical and transverse wind fluctuations are stochastic processes. The kinetic energy distribution of the speed fluctuations according to the frequency presents two peaks separated by a very low energy zone. Low frequency energy is centered over a period of four days and one year (time scale of geostrophic movements) and high frequency energy is centered over a period of 1 mn (turbulence scale). Therefore, by averaging the wind speed over a period included in the energy hole (from 10 minutes to 2 hours), the wind speed may be considered as locally stationary.

For this stationarity hypothesis used for wind fluctuations, we should point out that the ergodic nature of these stochastic processes is also assumed. This additional hypothesis is rarely clarified correctly. It has, however, important consequences, since it enables us to determine the statistical and spectral properties of the fluctuation using one or a few recordings.

The average speed U is, by definition, independent of time but is a function of the point in space. There are many profiles characterizing this vertical spatial dependency. One of the frequently used models is the logarithmic profile according to height:

$$U(z) = U_{ref}\, k_T(z_0)\, \ln\left(\frac{z}{z_0}\right) \qquad\qquad [5.78]$$

The *reference mean speed* U_{ref} relates to conventional measuring conditions (average over 10 minutes, height of 10 m, flat area, low level of roughness, such as "flat lands", "airports"), independent of the real conditions of the construction site. The reference mean speed characterizes the "force" of the storm which is considered when designing a structure. It is treated as a random variable whose high values are represented by their annual exceedence probability. The most commonly used formulation, particularly for characterizing the mean wind speed, is the Gumbel distribution. With this model for extreme wind values, we are then focusing on the probability of the maximum speed value not crossing a (high) threshold, over a given time period (a year, the design working life, etc.). For problems in relation to structural safety, only the first up-crossing passage is critical even if this crossing is a short term one. To deal with problems of fatigue in materials or user comfort, we must also focus on the crossing duration over a period of time. For example, we will look at the frequency but also the foreseeable average traffic interruption on a bridge due to wind, or even the cumulated duration of strong storms (not only the strongest), which are likely to induce important stress variations on the materials, and therefore induce fatigue phenomena. These two statistics are entirely different and we cannot simply go from one to the other. In the second case, we consider the statistical distribution of mean speeds (over ten minutes) observed generally with a periodicity of three hours. The cumulative probability function is estimated by the cumulated frequency of no level crossing observed over the longest period possible, potentially chosen according to one or many criteria, depending on the problem studied. For example, the wind effects on a structure (stresses, overturning moments applied to vehicles, etc.) vary greatly with wind direction. We will choose speeds observed by sectors of 30 or 45 degrees, and we will estimate the frequency of each of these sectors and the distribution function for each one of them. Practice shows that this probability function of the mean wind speed is often correctly described by the Weibull distribution.

The *roughness length* z_0 widely varies according to site, in the order of 1 mm for wind coming from the sea, to several meters within an urban area (Table 5.13). It depends on the height of ground obstructions and how many there are; paradoxically, the sea's roughness is very low, even in strong wind, although the water's surface is very rough (water/air interaction).

Roughness categories		z_0 (m)	$k_T(z_0)$	$U(10)/U_{ref}$
1.	Sea, lakes and water levels subjected to the wind over a distance of at least 5 km	0.005	0.16	1.22
2.	Flat land, with or without some isolated obstructions (trees, buildings, etc.) airports	0.05	0.189	1.00
3.	Countryside with hedges, orchards, small woods, thickets, dispersed settlements	0.20	0.21	0.82
4.	Urbanized, industrial or forest areas	0.75	0.23	0.60
5.	Urban areas where at least 15% of the surface is taken up by buildings with an average height higher than 15 m.	2.00	0.25	0.40

Table 5.13. *Roughness categories according to Eurocode 1 [EN 05]*

The mean fluctuation speeds are zero: $\bar{u} = 0, \bar{v} = 0$, and $\bar{w} = 0$. Because of this, the longitudinal, lateral and vertical *fluctuation standard deviations* are easily written as:

$$\sigma_u = \sqrt{\overline{u^2}}, \quad \sigma_v = \sqrt{\overline{v^2}} \text{ and } \sigma_w = \sqrt{\overline{w^2}} \tag{5.79}$$

Turbulence intensity is defined as the ratio between the standard deviation of the fluctuation speed, and the mean wind speed. By considering the three fluctuating components, the three *turbulence intensities* are defined by:

$$I_u(z) = \frac{\sigma_u}{U(z)}, \quad I_v(z) = \frac{\sigma_v}{U(z)} \text{ and } I_w(z) = \frac{\sigma_w}{U(z)} \tag{5.80}$$

Turbulence intensities are non-dimensional parameters which characterize the turbulence level. Their values generally range between 3 and 20%. *Turbulence length scales* allow us to measure the average size of the fluctuations. There are 9 in total, three for each direction. The longitudinal length scale $u(x, y, z, t)$ is written as:

$$L_u^X = \frac{1}{\sigma_{u(P_1)} \, \sigma_{u(P_2)}} \int_0^\infty R_{u(P_1)u(P_2)}(\tau) \, d\tau \tag{5.81}$$

where $R_{u(P_1)u(P_2)}(\tau)$ is the *cross-correlation function* of the fluctuation $u(x, y, z, \tau)$. P_1 and P_2 denote two distinct points in space on the horizontal axis. The vertical and lateral length scales of component u may be obtained by taking the cross-correlation coefficient between two points on the vertical and lateral dimension. It is normal practice to take:

$$\begin{aligned} L_u^Z &= 0.5 \, L_u^X \\ L_u^Y &= 0.3 \, L_u^X \end{aligned} \tag{5.82}$$

The longitudinal length scale of vertical and lateral components increases with height, but does not vary with roughness. It is current practice to take:

$$\begin{aligned} L_v^X &= 0.77 \, L_w^X \\ L_w^X &= 0.4 \, z \end{aligned} \tag{5.83}$$

The turbulence length scale accounts for the size of the wind vortices. For example, L_u^X, L_u^Y and L_u^Z are respectively the longitudinal, transverse and vertical dimensions of the vortices associated with the fluctuations u of the speed in direction X. These sizes range from few hundred meters (longitudinal scales) to several meters (lateral scales).

Fluctuating wind components in a given point are characterized by *power spectral densities* (assuming the stationarity hypothesis) which represents the kinetic energy contribution for each vortex range which makes up the turbulence; each range has a different oscillation frequency. In other words, the spectral density characterizes the vibratory energy of the turbulence and its distribution in the frequency bandwidth. There are a few models which can characterize the set of wind components (Table 5.14). We will point out here that the majority of the wind energy is contained in a frequency bandwidth which does not exceed 1-1.5 Hz.

Model	Frequency range	Expression
Von Karman	$0.007 \leq n \leq 1.0$ Hz	$$S_u(n) = \frac{4L_u^X / \overline{U}}{\left(1 + 70.7(nL_u^X / \overline{U})^2\right)^{5/6}} \sigma_u^2$$ $$S_\xi(n) = \frac{4L_x^X}{U(z)} \frac{1 + 188.4\left(\frac{2nL_\xi^X}{U(z)}\right)^2}{\left(1 + 70.7\left(\frac{2nL_\xi^X}{U(z)}\right)^2\right)^{11/6}} \sigma_\xi^2$$ $(\xi = v, w)$
Kaimal	Medium and high frequencies	$$S_u(n) = \frac{17\dfrac{z}{U(z)}}{\left(1 + 33.00\dfrac{nz}{U(z)}\right)^{5/3}} \sigma_u^2$$ $$S_v(n) = \frac{4.2\dfrac{z}{U(z)}}{\left(1 + 9.50\left(\dfrac{nz}{U(z)}\right)\right)^{5/3}} \sigma_v^2$$ $$S_w(n) = \frac{1.4\dfrac{z}{U(z)}}{\left(1 + 5.30\left(\dfrac{nz}{U(z)}\right)\right)^{5/3}} \sigma_w^2$$

Table 5.14. *A few empirical expressions of the spectral densities for wind fluctuations*

Similarly to the power spectral density definition, the *cross-spectral power density* is defined as the Fourier transform of the cross-correlation function. But, contrary to the first definition, the cross-spectral density function has complex values. Unfortunately, we do not yet have a model for the imaginary parts. This explains why this part is neglected in the calculations. The following expression gives an approximation of this function:

$$\left\| S_{u(P_1)u(P_2)}(n) \right\| = \gamma_{u(P_1)u(P_2)}(n) \sqrt{S_{u(P_1)}(Pn) S_{u(P_2)}(P_2 n)} \qquad [5.84]$$

where $\gamma_{u(P_1)u(P_2)}(n)$ is the *coherence function* at points P_1 and P_2, which are often approached by:

$$\gamma_{u(P_1)u(P_2)}(n) \cong \exp\left(-\frac{2n\sqrt{\left(C_u^Z (z_1 - z_2) \right)^2 + \left(C_u^Y (y_1 - y_2) \right)^2 + \left(C_u^X (x_1 - x_2) \right)^2}}{\bar{U}(z_1) + \bar{U}(z_2)} \right) \qquad [5.85]$$

where:

- x_1, y_1, z_1 are the coordinates from point P_1;
- x_2, y_2, z_2 are the coordinates from point P_2;
- C_u^X are the *coherence coefficients*, determined from experiments.

Table 5.15 gives an example of a wind model from *in situ* measurements.

Wind component	u	v	w
$\dfrac{\sigma}{\bar{U}}$	0.090	0.076	0.050
L^X	200.00	65.00	35.00
L^Y	85.00	95.00	35.00
L^Z	40.00	30.00	20.00
C^Y	11.00	4.50	12.00
C^Z	11.00	4.50	12.00

Table 5.15. *Turbulence characteristic on the site of the Normandy bridge* $(U_{10} = 35.3 \ m/s; \ z_0 = 0.05 \ m)$

There are very few works on the coherence function for lateral and vertical fluctuations. However, Simiu and Scanlan [SIM 98] propose taking the same coherence function as the longitudinal component by reducing the coherence coefficients by 30%. We should also specify that where there is a lack of information relating to cross-spectral densities between wind components, these components being consequently disregarded.

The combination of wind and snow can be dangerous. The geometry of snow may give larger wind load areas and therefore lead to greater effects than previously envisaged.

5.1.7.2.2. Wind effects

For most structures, the resonant part of the wind effects is insignificant and average effects are more important. Dynamic pressure at an altitude z is given by (first order approximation):

$$Q(P,t) = \frac{1}{2}\rho \|V(P,t)\|^2$$
$$= \frac{1}{2}\rho\left(\left[U(P)+u(P,t)\right]^2 + \left[v(P,t)\right]^2 + \left[w(P,t)\right]^2\right) \qquad [5.86]$$
$$\approx \frac{1}{2}\rho\left(U(P)^2 + 2U(P)u(P,t)\right)$$

where ρ is the air density (1.25 kg/m^3). The transfer from this dynamic pressure to the pressure applied onto a surface brings the *pressure coefficient* into play. The wind-induced pressure is then assumed to be perpendicular to the surface. Through convention, this pressure coefficient takes positive values for pressures and negative values for depressions (Figure 5.11).

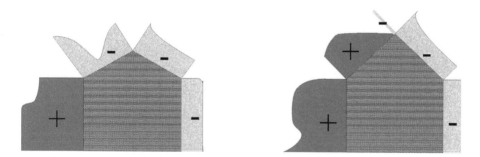

Figure 5.11. *Examples of wind pressure on a rigid building*

The pressure coefficients are not uniformly distributed over the total surface area of a structure and must be assessed separately. Moreover, they depend on the building's shape and the wind's angle of attack (i.e. relative position of the body in the airflow) and other characteristics (roughness, Reynolds number, etc.). In some cases, the external pressure coefficients must be combined with internal pressure coefficients.

The pressure at point P on the surface is then given by the relationship:

$$P_r(P,t) = C_p Q(P,t) = \frac{1}{2}\rho C_p \left(U(P)^2 + 2U(P)u(P,t)\right) \tag{5.87}$$

The standard deviation of the fluctuations can be obtained using the expression [5.12]:

$$\sigma_{P_r}^2 = \int_0^\infty S_Q(n)\, dn = \int_0^\infty \left(C_p \rho U(P)\right)^2 S_u(n)\, dn$$

$$= \left(C_p \rho U(P)\right)^2 \sigma_u^2 = 4\left(\frac{1}{2}C_p \rho U(P)^2\right)^2 \frac{\sigma_u^2}{U(P)^2} \tag{5.88}$$

$$= 4\left(\frac{1}{2}C_p \rho U(P)^2\right)^2 I_u^2$$

The *peak factor* is defined by the ratio:

$$k_p = \frac{P_r - \mathbb{E}\left[P_r\right]}{\sigma\left[P_r\right]} \tag{5.89}$$

As the fluctuating wind component is stationary and Gaussian, the result from example 5.1 is applied [CRE 03a]. The peak factor is Gumbel distributed with mean and variance given by:

$$\mathbb{E}\left[k_{p,\max}\right] \approx \sqrt{2\ln\left(v_0 T\right)} + \frac{\gamma}{\sqrt{2\ln\left(v_0 T\right)}} \tag{5.90}$$

$$\mathbb{V}\left[k_{p,\max}\right] \approx \frac{1}{2\ln\left(v_0 T\right)}\frac{\pi^2}{6} \tag{5.91}$$

The maximum pressure at point P is therefore written as:

$$P_{r,\max} = \frac{1}{2}C_p \rho U(P)^2 + k_p \sigma_{P_r} = \frac{1}{2}C_p \rho U(P)^2 \left(1 + k_p 2 I_u\right) \tag{5.92}$$

Its statistical characteristics are deduced from those for the peak factor.

By allowing a perfect correlation of the pressure and the same mean wind speed at any point U, then integrating the pressure on a surface A gives the following spectral density of the resulting pressure:

$$S_P(n) = A^2 (C_p \rho U)^2 S_u(n)$$ [5.93]

The perfect correlation hypothesis does not take into account the size of the vortices which constitute the turbulence; the correlation, in fact, decreases according to the distance between two points on the surface according to equation [5.86] but also according to frequency content (high frequencies). The spectral density for the total pressure on surface A is then modified in the following way:

$$S_P(n) = A^2 \left(C_p \rho U\right)^2 S_u(n)\, \Theta\!\left(\frac{n\, l_1}{U}, \frac{n\, l_2}{U}\right)$$ [5.94]

$\Theta\!\left(\dfrac{n\, l_1}{U}, \dfrac{n\, l_2}{U}\right)$ is called *aerodynamic admittance*: this function, which integrates

the spatial correlation, decreases according to frequency. This gives the following expression for the standard deviation of total pressure P:

$$\sigma_P^2 = \left(C_p \rho U\right)^2 \int_0^\infty \Theta\!\left(\frac{n\, l_1}{U}, \frac{n\, l_2}{U}\right) S_u(n)\, dn$$ [5.95]

The extreme effect is given similarly to equation [5.92] by:

$$P_{max} = \frac{1}{2} A C_p \rho U^2 + k_F \sigma_F = \frac{1}{2} C_p \rho U^2 \left(1 + k_F\, 2\, I_u\, \sqrt{k_b}\right)$$

$$k_b = \int_0^\infty \Theta\!\left(\frac{n\, l_1}{U}, \frac{n\, l_2}{U}\right) \frac{S_u(n)}{\sigma_u^2}\, dn$$ [5.96]

5.1.7.3. *Gravitational load*

Gravity plays an essential, predominant role in gravitational loads: for example, currents (tides, floods), avalanches, falls and landslides, rock falls, etc. Avalanches, landslides or falls are occasional and exceptional loads: they are often wrongly considered as being accidental due to their consequences. Table 5.16 gives a few of these models encountered in studies.

Load type	Model
Periodic phenomena	
Tide	Deterministic (tables)
Flood	Fixed, expert judgment (rare events)
Occasional phenomena	
Avalanches, falls, landslides	Occurrence probabilities of rare scenarios

Table 5.16. *Probabilistic models of some gravitational loads*

Loads due to ground pressure should be dealt with here. The effects largely depend on the material and the filling height, and the internal friction coefficient. The average values are provided by guidelines, and a 7% coefficient of variation for unitary weight and a 10% value for the friction coefficient must be envisaged.

5.1.7.4. *Seismic load*

A *seismic load* is a variable dynamic load according to risk areas. Statistical studies using seismographic networks enable us to establish laws for the occurrence of earthquakes according to the intensity, magnitude and the level of risks. An earthquake is a disturbance which starts its life in the Earth's crust and which causes a movement in the ground, able to reach strong amplitudes in the three directions from the place where it is felt. It is a random, transitory, non-stationary phenomenon, still relatively unknown, whose length varies from a few seconds to many tens of seconds, but whose number of "strong" cycles which can damage structures is generally low. The excitation caused by an earthquake is described by a response spectrum which enables us, for time-domain analyses, to reconstitute accelerograms, i.e. acceleration variation curves according to time. The *accelerograms* are curves which fluctuate quickly and irregularly around the zero value. The corresponding spectra are mostly coded in standards and regulations and on maps which indicate the risks and intensity (for a return period of 400 years in general). The position of the epicenter, the geology, and the ground and structure type, are the most important parameters when analyzing a structure's seismic behavior.

In structural design, we must distinguish *service earthquakes* and *design earthquakes*. When a structure is confronted with the first kind of earthquake, both the structure and its installations must be protected and left undamaged. For the second kind, we allow possible local damage (the extent needs to be fixed) without the collapse of the entire structure. Earthquakes are always primary loads. The uncertainties which affect it come from two origins:

– an input signal (called seismic excitation, related to the source or non-homogeneous term of the equations of motion);

– the structure (dimensions, constitutive material laws, i.e. the non-linear term of the homogeneous part of the equations of motion). In particular, according to the degree of sophistication sought in ground modeling (no ground/structure interaction, full modeling of the ground/structure interaction, modeling by dynamic impedance), an ever growing number of variables will be required (Figure 5.13).

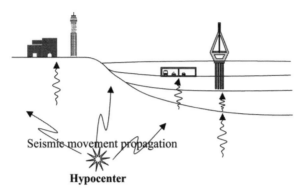

Figure 5.12. *Seismic wave propagation*

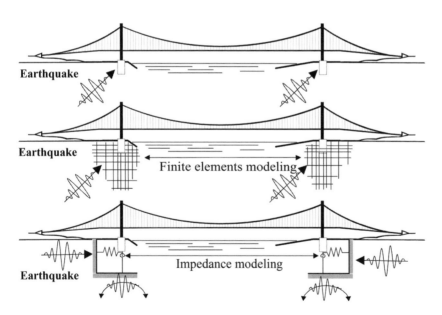

Figure 5.13. *Modeling approaches for seismic calculations*

The vertical and horizontal components of a seismic movement may be represented in many ways and are related to the site being observed. Accelerograms have become the basis for all seismic analyses since recordings have been possible[4]. The power spectra also make up a frequency representation of the earthquake's energy content. Finally, the response spectra represent the maximum response of an oscillator with 1 degree of freedom subjected to seismic excitation.

5.1.7.4.1. Probability assessment of seismic hazards

The *probability assessment of seismic hazards* aims to predict large scale seismic movement occurring in a given site. The procedure consists of:

– identifying independent sources of seismic activity and determining a model for seismic magnitude by the contribution of each source;

– adjusting an attenuation law, a function of the ground's characteristics for acceleration peaks on the site;

– calculating strong movements in a site with a non-exceedence probability over a given time period.

Magnitude measures the energy released during an earthquake. The higher the magnitude, the more energy is released by the earthquake. The classic expression of an earthquake's magnitude is owed to Gutenberg and Richter (1944). It involves a relationship linking together the earthquake's magnitude to the number of times this magnitude was exceeded in a given place over a given time period:

$$\log N(M) = a - b\,M \qquad\qquad [5.97]$$

where $N(M)$ is the number of earthquakes per time and volume unit with a magnitude higher than M. Parameters a and b depend on the area in question. Parameter a depends on the observation period and the seismic activity for the area, whereas parameter b only varies a little (generally taken to be equal to 1.0). It follows from this that the magnitude for a given region is characterized by a truncated exponential distribution, i.e. that the distribution function of M for $M_1 \le M \le M_2$ is given by:

$$F_M(m) = \frac{1 - \exp\left[-\beta\left(M - M_1\right)\right]}{1 - \exp\left[-\beta\left(M_2 - M_1\right)\right]} \qquad\qquad [5.98]$$

---

4 The first seismic recordings were made in the 1920s during earthquakes in Tokyo (1923) and Santa Barbara (1925).

with $\beta = b \ln 10 \approx 2.30$ (si $b = 1.0$). M_1, M_2 are the minimum and maximum magnitudes for a given site (for structural calculations, these upper and lower bounds generally correspond to 4 and 8 on the Richter scale).

There is a high correlation between that which is produced at a given site and that produced at the source of the earthquake. The intensity of the movement decreases with the distance Δ between these two points. This relationship is described by *attenuation laws*. These laws are based on wave propagation theory in elastic solids. A general formulation for the attenuation law of acceleration peaks A (expressed in fractions of g) is given by:

$$A = b_1 e^{(b_2 - M)} \left(\Delta + k \right)^{-b_3} \varepsilon_A \qquad\qquad [5.99]$$

Δ, k are expressed in km, parameters b_1, b_2, b_3 and k are constants which depend on the site in question. ε_A is an error (log-normal distribution) which takes into account the variability of the attenuation law. The expression [5.99] introduces larger variability and is not universally accepted. Table 5.17, adapted from [JCS 01] gives a few values of these parameters for given sites. For most of these sites, the lack of information on residual errors means that values of 1.0 and 2.0 are taken for the mean and coefficient of variation of ε_A.

Site	Reference	b_1	b_1	b_1	k	$\mathbb{COV}(\varepsilon_A)$
California Central America	Esteva and Villaverde (1973)	5.7	0.8	2.0	40	0.64
Japan	Katayama and Saeki (1978)	0.02	0.7	0.8	0	-
Taiwan	Mau and Kao (1978)	0.38	0.876	1.836	20	-
Central Europe	Ahorner and Rosenhauer (1975)	1.28	0.8	2.0	13	-
Greece	Makropoulos (1978)	2.2	0.7	1.8	20	0.50
Italy	Sabetta and Pugliese* (1987)	0.014	0.363	0.5	25	0.20

$* A = b_1 e^{(b_2 - M)} \left(\Delta^2 + k \right)^{-b_3} \varepsilon_A$

Table 5.17. *Parameters of attenuation laws for seismic calculations*

In the most simplified models, the source of an earthquake may be considered as a point which represents the nucleation position (also called the *focus* or *hypocenter*). This focus is the point where the earthquake is generated. The focal distance Δ is, then, the distance between the hypocenter/focus and the site. It is a function of the horizontal distance R to the site's focus and the epicentral depth (Figure 5.14):

$$\Delta = \sqrt{R^2 + H^2} \qquad\qquad [5.100]$$

The distribution of R may be determined under the hypothesis that the various sources are uniformly distributed over the different areas of homogeneous seismic activity. For the focal depth, a lognormal distribution may be considered.

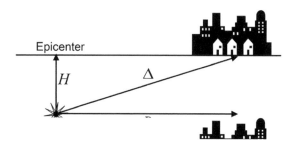

Figure 5.14. *Hypocenter and epicenter*

It is normal to assume that earthquakes occur according to a Poisson process with v as an occurrence rate, although physical considerations suggest a certain dependence between consecutive events. The exceedence probability $P_f(a,T)$ over a reference period T is given by:

$$P_f(a,T) = 1 - \exp\left(-vT P'(a,T)\right) \; P_f(a,T) = 1 - \exp\left(-vT P'(a,T)\right) \quad [5.101]$$

where v is the number of earthquakes with magnitudes higher than M_1 per time unit and $P'(a,T)$ is the probability of exceeding an amplitude a for an earthquake with a magnitude higher than M_1 over a given area. The uncertainty over the average rate is very important, particularly if the observations only cover a short period of time or its seismic activity is low. A Bayesian updating is recommended. The following simplified model owed to Benjamin [BEN 68], can be used; it is deduced from the Bayes theorem:

$$-vT = \ln\left[\frac{1}{\left(1 + \dfrac{T}{\tau}\right)^{n+1}}\right] \qquad\qquad [5.102]$$

where n is the number of earthquakes observed over a period τ.

In the case where different sources must be considered simultaneously, the exceedence probability will be written as:

$$P_f(a,T) = 1 - \prod_{i=1}^{m} \left(1 - P_{f,i}(a,T)\right)$$

[5.103]

where m is the number of sources.

5.1.7.4.2. Acceleration and associated response

A typical accelerogram $\gamma(t)$ (Figure 5.15) is a curve with rapid and irregular fluctuations around the value 0; the total duration T varies from a few seconds to a few tens of seconds. But, we can distinguish, inside this time period, a *strong motion part* during which the acceleration amplitude is notably higher than for the rest of the time. For rocky grounds and magnitudes of $M = 5.5 \div 7.5$, the increasing part, the duration of the strong part (strong motion) and the decreasing part can be taken as being equal to 1-3 s, 5-15 s, and 4-10 s. The seismometers simultaneously record the motion components according to three orthogonal directions (two horizontal, and one vertical), thus giving three accelerations. Besides the duration T, an accelerogram provides:

– the value A of the maximum acceleration modulus (expressed in fractions of g);

– an estimation of the dominant frequency f_d of the signal, by counting in the strong motion part the number of crossings the acceleration goes through 0 per time unit; this frequency is generally in the order of a few Hz, and a time step in the order of 1/100 of a second is accurate enough for sampling.

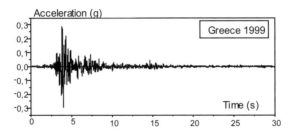

Figure 5.15. *Example of seismic acceleration*

Let us consider the differential equation:

$$\ddot{X}(t) + 2\zeta\omega\dot{X}(t) + \omega^2 X(t) = -\gamma(t)$$

[5.104]

with the following for initial conditions $\dot{X}(0) = 0; X(0) = 0$. The solution for this equation is written $X_\gamma(\omega, \zeta, t)$ and represents the response of an oscillator, moved by the acceleration $\gamma(t)$. The *response spectrum* $X_{\gamma,max}(\omega, \zeta) = \max_t \left(X_\gamma(\omega, \zeta, t) \right)$ is the graph for maximum values reached over time. ω, ζ respectively are the circulatory frequency and the damping ratio. By studying the statistics of a set of accelerograms, we can define a *design spectrum* with a regular shape $X_d(\omega, \zeta)$ (Figure 5.16). This distinction between design spectrum and response spectrum is important; in practice, seismic motion is given from a design spectrum. From this, it is possible to generate synthetic accelerograms whose response spectrum will reproduce the design spectrum with a good level of approximation.

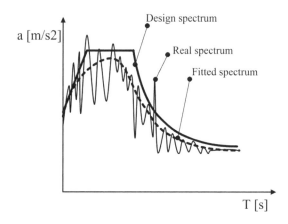

Figure 5.16. *Real, smooth and dimension spectra*

The design spectrum is the result of a statistical analysis, with an "engineering judgment" with concerns for usage comfort. The basic representation of seismic load at a given point from the ground's surface is expressed by an elastic response spectrum for the acceleration. The hazard is characterized by the parameter "maximum reference acceleration" a_{gR} for a class A soil (Table 5.18). This acceleration is modulated by an importance coefficient γ of the structure (Table 5.19).

From this response spectrum, we can deduce the *pseudo-velocity* and the *pseudo-acceleration*:

$$PSV = \omega X_{\gamma,max}(\omega, \zeta)$$
$$PSA = \omega^2 X_{\gamma,max}(\omega, \zeta)$$

[5.105]

Ground and site	Type 1 earthquake				Type 2 earthquake			
	S	T_B [s]	T_C [s]	T_D [s]	S	T_B [s]	T_C [s]	T_D [s]
A: Rock or other geological formation of this type with a 5 m maximum surface layer of a less resistant material	1.0	0.15	0.4	2.0	1.0	0.05	0.25	1.2
B: Stiff deposits of sand, gravel, over-consolidated clay, at least many tens of meters in thickness, characterized by a gradual increase in mechanical properties with depth	1.2	0.15	0.5	2.0	1.35	0.05	0.25	1.2
C: Deep deposits of average density sand, averagely stiff gravel or clay of a few tens to many hundreds of meters in thickness	1.15	0.20	0.6	2.0	1.5	0.10	0.25	1.2
D: Deposits of soil without cohesion with low to average density (with or without coherent soft layers) or including soft to hard coherent soils	1.35	0.20	0.8	2.0	1.8	0.10	0.30	1.2
E: Soil profile including a surface alluvia layer with grade C or D for v_S values and a thickness between approx. 5 and 20 m, on a stiffer material with $v_S > 800$ m/s	1.4	0.15	0.5	2.0	1.6	0.05	0.25	1.2
S1: Deposits made of, or containing, a layer of at least 10 m in thickness of soft clay/silt with a high plasticity index (> 40) with a high water content.	Particular studies							
S2: Deposits of liquefiable soils, sensitive clays or any other soil profile not included in classes A to E or S1	Particular studies							

Table 5.18. *Parameters of Eurocode 8 according to soil and site conditions*

Importance category	Buildings	Γ
I	Buildings of low importance for personal safety: for example, agricultural buildings, etc.	0.8
II	Current buildings not belonging to the other categories	1.0
III	Buildings whose resistance to earthquakes is important, taking into account the consequences of a collapse: for example, schools, meeting rooms, cultural institutions, etc.	1.2
IV	Buildings whose integrity in the case of an earthquake is vital for public protection: hospitals, fire stations, power plants, etc.	1.4

Table 5.19. *Importance categories according to Eurocode 8*

The pseudo-velocity is the speed value which gives a kinetic energy value of the oscillator mass which is equal to the maximum value of the elastic energy stored in the spring. Pseudo-acceleration is the modulus value of the absolute acceleration at the time when the displacement modulus reaches its maximum. We therefore have:

$$\left| X_{\gamma}(\omega,\zeta,t) \right| = X_{\gamma,\max}(\omega,\zeta)$$
$$\dot{X}_{\gamma}(\omega,\zeta,t) = 0 \tag{5.106}$$

which gives:

$$\left| \ddot{X}_{\gamma}(\omega,\zeta,t) + \gamma(t) \right| = \omega^2 X_{\gamma,\max}(\omega,\zeta) = PSA \tag{5.107}$$

The Eurocode 8 spectrum [EN 04a] distinguishes "boundary" periods T_B, T_C, T_D and a basic schematization for the reference response spectrum for the acceleration. The design spectrum can be corrected by coefficient η and is mathematically unique for Europe:

$$
\begin{aligned}
0 \le T \le T_B \quad & X_d(T) = a_g\, S \left[1 + \frac{T}{T_B}(2.5\eta - 1) \right] \\
T_B \le T \le T_C \quad & X_d(T) = a_g\, S\, 2.5\eta \\
T_C \le T \le T_D \quad & X_d(T) = a_g\, S\, 2.5\eta \left[\frac{T_C}{T} \right] \\
T_D \le T \le 4[s] \quad & X_d(T) = a_g\, S\, 2.5\eta \left[\frac{T_C T_D}{T^2} \right]
\end{aligned}
\tag{5.108}
$$

with:

$$a_g = \Gamma a_{gR}; \eta = \sqrt{\frac{10}{5+\zeta}} \geq 0.55 \qquad\qquad [5.109]$$

S is a parameter related to the soil defined in Table 5.18.

A given peak acceleration a_{gR} at a given place may be generated by different types of earthquake: a strong earthquake whose epicenter is far away, or a weaker earthquake whose epicenter is close. The real earthquake affecting an area is a geological function, close and far.

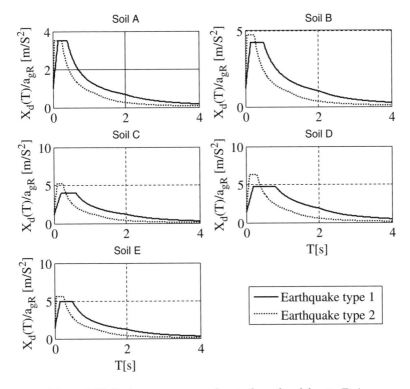

Figure 5.17. *Design spectra according to the soil and the site* $\Gamma = 1$

But design spectra corresponding to the two types of earthquakes mentioned above are different, because the waves propagated close or far away produce different effects. In Eurocode 8, this possibility is considered, and type 1 and 2 spectra are defined. Type 1 relates to faraway earthquakes with a magnitude high enough (higher than 5.5) to generate significant accelerations on the construction site, whose contribution is predominant. Type 2 should be considered if earthquakes with a magnitude lower than 5.5 constitute the prominent risk factor. Figure 5.17 gives an example of the Eurocode 8 design spectra. In practice, these spectra are used, but there is also a whole variety of spectra models which can be used in order to reproduce an earthquake's true dynamics, such as the Kanai and Tajimi spectrum [TAJ 60]:

$$S(\omega) = \frac{S_0 \left[1 + 4\varsigma_s^2 \left(\dfrac{\omega}{\omega_s} \right)^2 \right]}{\left[1 - \left(\dfrac{\omega}{\omega_s} \right)^2 \right]^2 + 4\varsigma_s^2 \left(\dfrac{\omega}{\omega_s} \right)^2}$$
[5.110]

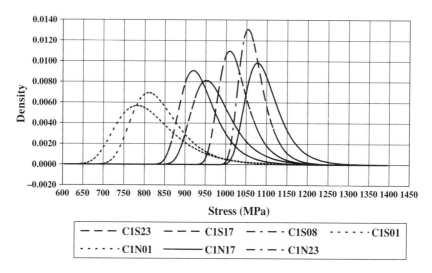

Figure 5.18. *Distribution of extreme values for stay cable tension on the Rion-Antirion bridge*

Peak acceleration may be modeled by a Type 1 or Type 2 distribution of extreme values. The complexity of these phenomena (in terms of load, but also in terms of load effect) makes it relatively difficult to model this phenomenon probabilistically, when there is a lack of sufficient information. However, we should also point out that using Gumbel models for expressing load effects, fitted on accelerograms simulated using real recordings or normalized spectra. Figure 5.18 gives an example of extreme value distributions for the stresses in a bridge's stay cables located in a seismic zone. Finally, let us point out that the accelerations are also correlated spatially, according to an exponential coherence function:

$$\rho(r) = \exp\left[-\alpha\, r^2\right] \tag{5.111}$$

where r varies from 10^{-6} m (for firm grounds) to 10^{-7} m (for soft grounds). For $r < 100$ m, we may consider a perfect correlation [JCS 01].

EXAMPLE 5.7.– A tall building is built in a seismic zone. Let us presume that the ductility demand in terms of displacement D_i at the critical point on floor i, given as the ratio from the required displacement to that which corresponds to the elastic limit, is expressed as a function of the applied load S using a deterministic performance function $g(.)$: $D_i = g(S)$, with load S being random.

In the case of a seismic load, we often use one single parameter: the peak acceleration a_g. However, this acceleration, with its frequency content, cannot singularly determine an acceleration time history. Several signal histories can be related to the same peak acceleration. Consequently, for a given peak acceleration, the response remains random and can only be expressed by introducing conditional probabilities. Let us consider a peak acceleration $A_g = a_g$ and let us assume that the conditional density function of the ductility demand is lognormal [ZHO 84]:

$$f_{D_i/A_g}(d_i) = \frac{1}{d_i\, \sigma_{\ln D_i}\, \sqrt{2\pi}} \exp\left(-\frac{\left(\ln d_i - m_{\ln D_i}\right)}{2\, \sigma_{\ln D_i}^2}\right)$$

with:

$$m_{\ln D_i} = 27 - 26 \left[\frac{32 - \dfrac{a_g}{a_{gy,i}}}{32}\right]^6 \qquad 1 \le \frac{a_g}{a_{gy,i}} \le 33$$

$$\frac{\sigma_{\ln D_i}}{m_{\ln D_i}} = \begin{cases} 0.2 & \dfrac{a_g}{a_{gy,i}} \le 1 \\[2ex] 0.1 + 0.4\left(\dfrac{a_g}{a_{gy,i}} - 1\right) & 1 < \dfrac{a_g}{a_{gy,i}} \le 2 \\[2ex] 0.6 & 2 < \dfrac{a_g}{a_{gy,i}} \end{cases}$$

$a_{gy,i}, a_g$ respectively denote the acceleration which induces yielding in the critical point on the i-th floor, and the peak acceleration. The ultimate ductility capacity is modeled by a lognormal distribution:

$$f_{C_i}(c_i) = \frac{1}{c_i \, \sigma_{\ln C_i} \sqrt{2\pi}} \exp\left(-\frac{\left(\ln c_i - m_{\ln C_i}\right)}{2 \, \sigma^2_{\ln C_i}}\right)$$

For any given floor, failure occurs when the ductility demand exceeds the ultimate capacity:

$$P_{f,i}(a_g) = \mathcal{P}\left(D_i > C_i\right) = \mathcal{P}\left(\frac{D_i}{C_i} > 1\right)$$

By setting out $Z_i = D_i/C_i$ and $\ln Z_i = \ln D_i - \ln C_i$, as the variables are lognormal, according to Chapter 3 we obtain:

$$P_{f,i}(a_g) = \mathcal{P}\left(\frac{D_i}{C_i} > 1\right) \equiv \mathcal{P}\left(\ln Z_i > 1\right) = \Phi\left(-\frac{m_{\ln C_i} - m_{\ln D_i}}{\sqrt{\sigma^2_{\ln C_i} + \sigma^2_{\ln D_i}}}\right)$$

With the building being considered as a series system of critical points, the building's failure probability can be bounded by the simple bounds of a series system (see Chapter 3):

$$\max_i P_{f,i}(a_g) \le P_f(a_g) \le 1 - \prod_i \left(1 - P_{f,i}(a_g)\right)$$

We will estimate the building's failure probability for a given peak acceleration by using the upper bound. The total failure probability is, then, easily obtained by applying the total probability theorem:

$$P_f = \int_{a_g} P_f(a) \, f_{A_g}(a) \, da$$

By assuming that $a_y = 0.15\,g$, $m_{\ln C_i} = 10.0$ and $\sigma_{\ln C_i} = 4.0$, we obtain:

PSA	$P(a)$	a/a_{y_i}	$m_{\ln D_i}$	$\sigma_{\ln D_i}/m_{\ln D_i}$	β_i	$P_{f,i}(a)$	$P_f(a)$
0.1 g	0.35	0.67	< 1	0.2	> 5	~ 0	~ 0
0.2 g	0.37	1.33	2.6	0.6	2.7	0.0035	0.034
0.3 g	0.15	2.00	5.5	0.6	1.7	0.0446	0.366
0.4 g	0.06	2.67	8.1	0.6	0.64	0.262	0.952
0.5 g	0.04	3.33	10.5	0.6	0.05	0.480	0.998
0.6 g	0.03	4.00	12.6	0.6	-0.40	0.660	~ 1

The total probability is therefore: $P_f = \sum_a P(a)\,P_f(a) \approx 19.4\,\%$.   □

5.1.8. *Exceptional loads*

5.1.8.1. *Impacts*

The basic model used for characterizing an impact load is (Figure 5.19):

– the presence of bodies (or projectiles) potentially able to collide with the structure; these projectiles have a reference trajectory (traffic lane, seaway, air traffic lane, etc.) but there may be a deviation from this reference trajectory according to a normal distribution;

– human or mechanical failure which may lead to a trajectory deviation; these occurrences are described by homogeneous Poisson processes;

– the trajectory after dysfunction which depends on the body's and environmental properties;

– mechanical impact between the body and the structure, with the kinetic energy being partly transferred across elastoplastic deformations or ruptures located in structural elements as well as in the projectile.

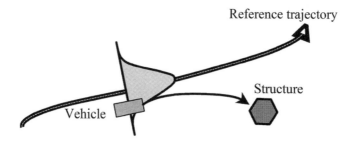

Figure 5.19. *Probabilistic collision model*

The probability of a body, moving in direction X, undergoing human or mechanical failure in a square with dimensions $dx \times dy$ and causing the structure's failure is given by:

$$p_{f,q}(x,y) = \mathcal{P}_{f,q}(x,y)\,f_l(y)\,dy\,\lambda(x)\,dx \qquad [5.112]$$

where $f_l(y), P_{f,q}(x,y), \lambda(x,t)$ respectively denote the density function of the body's position in the direction which is perpendicular to the displacement direction, the structure's failure probability related to the impact, and the probability of the body deviating from its path to position x.

The failure probability over a period T and for a number n of moving bodies per time unit is written:

$$\mathcal{P}_f(T) = 1 - \exp\left(-nT \iint \mathcal{P}_{f,q}(x,y)\,f_l(y)\lambda(x)\,dx\,dy\right) \qquad [5.113]$$

which, for low probabilities (and presuming a constant deviation probability along the trajectory – which is likely to be for a straight line trajectory, but which is less so for curved trajectories) is simplified as:

$$\mathcal{P}_f(T) = nT\,\lambda \iint \mathcal{P}_{f,q}(x,y)\,f_l(y)\,dx\,dy \qquad [5.114]$$

In principle, an impact is the interaction between a body and a structure. Therefore, it is not possible to formulate a load which is separated from resistance (this problem can be found in seismic loads). However, an upper limit of the impact load may be provided using the "rigid structure" hypothesis. If a projectile is modeled by an oscillator with one degree of freedom, with a mass m and a stiffness k, then the maximum interaction force is:

$$F_c = v_c\,\sqrt{k\,m} \qquad [5.115]$$

v_c is the projectile's impact speed. On this basis, the distribution function of the collision load F_c is:

$$\mathcal{F}_{F_c}(F_c < u) = 1 - n\lambda T \iint P\left(v_c(x,y)\sqrt{k\,m} > u\right) f_l(y)\,dx\,dy \qquad [5.116]$$

EXAMPLE 5.8.– Let us consider a bridge pile in the immediate vicinity of a straight road. The impact occurs if a vehicle leaves the road with a high speed at a given point on the road. The following figure describes the case study.

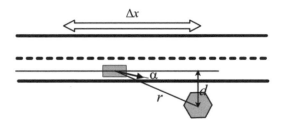

By using equation [5.117] and disregarding the following variable Y, we obtain:

$$\mathcal{F}_{F_c}\left(F_c < u\right) = 1 - n\lambda T\, \mathcal{P}\left(v_c \sqrt{k\,m} > u\right) \Delta x$$

Δx is the part of the road where collision may be envisaged. If r is the distance from the vehicle in relation to the bridge pile, this distance being a distance d from the road, then the collision speed can be written as:

$$v_c = \sqrt{v_0^2 + 2\,a\,r}$$

where a represents deceleration before impact. $\lambda\Delta x$ is the probability of a vehicle leaving the road over interval Δx. The following table gives a few numerical values and probabilistic models which can be used.

Variable	Definition		Type	Mean	Standard deviation
λ	Accident rate		Deterministic	10^{-10} m^{-1}	-
α	Collision angle		Rayleigh	$10°$	$10°$
v	Vehicle speed				
		Motorway	Lognormal	80 km/h	
		Urban	Lognormal	40 km/h	
		Car park	Lognormal	10 km/h	
a	Deceleration		Lognormal	4 m/s^2	1 m/s^2
m	Vehicle mass				
		HGV	Normal	20,000 kg	12,000 kg
		Light vehicle	Normal	1,500 kg	400 kg
k	Stiffness		Lognormal	300 kN/m	60 kN/m

Impact is a horizontal force and only the component which is perpendicular to the structure is to be considered. For lightweight vehicles, the collision force influences the structure at a height of 0.50 m above ground. For trucks, this height is raised to 1.25 m. The application surface area of the impact area is 0.25 m in height, and 1.50 m in width.   □

5.1.8.2. *Fires*

The probability of a fire starting in a given area A is modeled as a Poisson process with constant occurrence rate [JCS 01]:

$$P\left(\text{ignition in } [t, t+dt]\right) = v_{fire}\, dt \qquad\qquad [5.117]$$

The occurrence rate v_{fire} is the summation of local values over the area:

$$v_{fire} = \iint_A \lambda(x, y)\, dx\, dy \qquad\qquad [5.118]$$

where $\lambda(x, y)$ corresponds to the probability of fire ignition per year per m^2 for a given occupancy. Table 5.20 provides values for homogeneous local ignition rates.

Type of building	λ [m^{-2} year^{-1}]
dwelling/school	0.5 to 4 10^{-6}
shop/office	1 to 3 10^{-6}
industrial building	2 to 10 10^{-6}

Table 5.20. *Example values of annual fire probabilities λ per unit floor area*

After ignition a fire can develop through various ways. It may extinguish itself after a certain period of time because no combustible material is longer present. The fire may be detected very early and be extinguished manually (firemen) or automatically (sprinkler). In a minority of cases a fire may develop over an extended area from the initial one. From the structural point of view these extended fires (flashovers) may lead to local or global failure. The occurrence rate of flashover is given by:

$$v_{flashover} = P\left(\text{flashover/ignition}\right) v_{fire} \qquad\qquad [5.119]$$

The probability of a flashover once a fire has taken place, can obviously be influenced by the presence of sprinklers and fire brigades. Numerical values for the analysis are presented in Table 5.21 [JCS 01].

Protection method	$\mathcal{P}$(flashover/ignition)
Public fire brigade	10^{-1}
Sprinkler	10^{-2}
High standard fire brigade on site, combined with alarm system (industries only)	10^{-3} to 10^{-2}
Both sprinkler and high standard residential fire brigade	10^{-4}

Table 5.21. *Probability of flashover for given ignition depending on the type of protection measures*

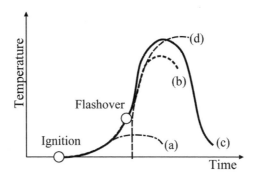

(a): temperature-time curve with functioning sprinkler system
(b): limited influence of a fire brigade arriving after a flashover
(c): temperature-time relation for a fully developed fire
(d): ISO-standard temperature curve (ISO 834)

Figure 5.20. *Temperature-time curves*

The combustible material can be considered as a random field, which in general is non-homogeneous and non-stationary. The intensity of this field at a point in space and time is defined as:

$$c = \frac{\sum_i m_i M_{k,i} H_i}{A} \qquad [5.120]$$

$m_i, M_{k,i}, H_i$ are respectively the combustion factor (ranging from 0 to 1), the combustible mass, and the specific combustible energy for material i. Probabilistic models for c are presented in Table 5.22.

Type of area	$\mathbb{E}[c]$ [MJm^{-2}]	$\mathbb{COV}[c]$
1. Dwellings	500	0.20
2. Offices	600	0.30
3. Schools	350	0.20
4. Hospitals	450	0.30
5. Hotels	300	0.25

Table 5.22. *Recommended values for the fire load intensity c*

For known characteristics of both the combustible material and the area, the post flashover period of the temperature time curve can be calculated from energy and mass balance equations. Many variables can be introduced as random in the model: amount and spatial distributions of combustible material, effective energy value, rate of combustion, ventilation characteristics, air use and gas production parameters, thermal conductivity properties and model uncertainties. The fire development depends on events like collapse of windows or containments, which may change the ventilation conditions or the available amount of combustible material respectively. The following simplified assumptions can be nevertheless used [JCS 01]:

– the combustible material is wood;

– the wood is spread uniformly over the floor area;

– the area is of a standard building material (brick, concrete);

– the fire is controlled by ventilation and not by the amount of fuel load (this is conservative);

– the initial temperature is 20°C.

In this case the temperature time curve depends only on the floor fire load density c and the opening factor f:

$$f = \frac{\sum_i A_i}{A_t} \sqrt{\frac{\sum_i A_i h_i}{\sum_i A_i}}$$ [5.121]

A_v and A_t are the area of the vertical opening i in area A, and the total internal surface area of the fire area (walls, floor and ceiling, including the openings). For a fire area which also contains horizontal openings, the opening factor can be calculated from a similar expression. In calculating the opening factor, it is assumed that ordinary window glass is immediately destroyed when fire breaks out. In many cases it will be possible to indicate a physical maximum f_{max}. The actual value of f in a fire is modeled by a random variable such as $f = f_{max}(1-X)$, X being a truncated lognormal variable (Table 5.23) cut off at $X = 1$.

In many engineering applications, use is made of the equivalent standard temperature-time relationship according to ISO 834:

$$T = T_0 + T_a \log(\alpha t + 1) \quad 0 < t < t_{eq}$$

$$t_{eq} = \frac{\Theta_{model} \, c \, A}{A_t \sqrt{f}}$$

[5.122]

T and T_0, are respectively the temperature in the area and the temperature at the start of the fire. T_a and α are parameters and Θ_{model} is a model uncertainty coefficient.

Variable	Distribution	Mean	Standard deviation
X	truncated lognormal	0.2	0.2
Θ_{model}	lognormal	4.0 s m$^{2.25}$/MJ	1.0
T_0	deterministic	20°C	-
T_a	deterministic	345 K	-
α	deterministic	0.13 s^{-1}	-

Table 5.23. *Recommended parameters for the ISO 384 model*

5.2. Resistance

Modeling resistance variables uses different sources of information:

– *Experimental data*: based on experimental data, statistical methods may be used to fit density functions and thus to estimate characteristic values. One of the problems here is that the quantity of information necessary to adjust such functions is limited. Moreover, it is concentrated on the central part of the considered resistance variable distribution, whereas the most interesting zone is the low tail of this distribution in order estimate a fractile or characteristic value.

– *Physical reasoning*: in certain cases, based on the physical phenomenon studied and past knowledge, it is possible to use an appropriate distribution which is suitable for describing the variable. In particular, it concerns the normal or lognormal distributions. Such a hypothesis enables us to solve the problem related to the absence of information on the tail distribution.

– *Subjective reasoning*: in many cases, the lack of data and references in order to apply physical reasoning make it difficult to model a variable. In such cases, only subjective reasoning can solve this problem. The distribution and its parameters will therefore be based on data taken from a literature survey. The choice of such an approach therefore involves the need to establish standards or guidelines where distribution types have to be proposed for certain resistance variables.

Describing material properties generally consists of a mathematical model (constitutive behavior law, creep model, etc.), adjusted by variables or random fields (elasticity modulus, creep coefficient, etc.). The ratios between the different variables (for example, compressive strength/tensile strength) may be part of the material's model.

In general, the designer is interested in the material's response to static and/or dynamic loads. But, the response to biological or physicochemical loads is equally as important for a manager because these phenomena can affect the material's mechanical properties.

Models and parameters are often taken from (standardized) tests which represent environmental and loading conditions in the best way possible. Samples must represent the manufacturing sites, must cover a long time period, and must include the effects of standardized quality testing measures. We must also take care about the differences between the conditions in which the tests are carried out and the structural environment.

For standard construction materials, information on various mechanical properties is generally available using experiments and tests. For new and innovative materials, the models and parameters must have a well defined experiment plan.

5.2.1. *Material properties and uncertainties*

Material properties are characterized using data taken from samples with well defined sizes, sampling rules specified by standards, and specific procedures for making use of the results.

For any given material, mechanical behavior is often described by a uniaxial constitutive law (Figure 5.21). The elasticity modulus and ultimate strength are often the basic data used to characterize the material (compressive and tensile strength). Other important parameters for this uniaxial constitutive law may be provided by tests: elasticity limit, ultimate strain or yielding strain, etc.

Alongside this uniaxial law, other properties, and therefore other parameters, may be important: delayed effects, thermal and moisture effects, chemical influence effects, effects of localized defects, etc.

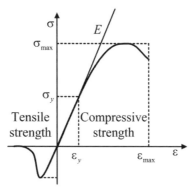

Figure 5.21. *Example of a uniaxial "stress/strain" curve*

In general, all properties are statistically correlated and randomly vary in time and space. Independently this, it is important to apprehend the uncertainties introduced by the deviation between the properties obtained on the samples, and the real properties of the structure:

– systematic deviations between the structural properties which are observed and predicted, which suggests a bias in the prediction;

– random deviations between the observed and predicted properties, which suggests incomplete models;

– uncertainties in the relationship between the material inside the structure and the material inside the samples;

– deviation of implementing the material in the structure in relation to the material in the samples;

– uncertainties related to degradations which affect the material.

In the case of spatial variations, it is useful to distinguish three hierarchical variation levels: global (macro), local (meso) and micro. Global variability, for example, relates to site production technology and strategy. Local variability may come from two issues: variability within as structure, or variability between production batches. Finally, local variability focuses on rapid variations and losses in homogeneity, as in concrete where the random distribution of the size and spacing between granulates and pores is considered.

The modeling process means that an object is taken as an arrangement of a large number of small elements. The statistical properties of these elements and their interactions can only be assessed qualitatively; this is often sufficient to draw global performance ratings. Using central limit theorems (Chapter 1) widely facilitates this analysis. By the asymptotic results, we are able to reduce the description of these objects' properties by a reduced number of characteristics. The central limit theory, the concept of extreme values, the convergence theorems, etc., play important roles.

Another useful approach is to introduce a reference volume, which is generally chosen according to practical principles. In general, it relates to the volume of test samples. Unfortunately, this volume does not often correspond to the volume for characterizing the behavior of elements on a micro-scale, or the volume for characterizing behavior *in situ.*

The reason for introducing this multi-scale modeling is a necessity, not only in terms of probabilistic calculations, but also for sampling, estimating and quality controls. The performance of reference volume samples, considered as micro-scale systems of elements, may be interpreted by various models (the weakest link model, the Daniel model, etc.).

By applying these models to systems with a large number of elements, they lead to specific distributions for the properties of the local scale system. The weakest link model leads to a Weibull distribution (Chapter 1) whereas the two other models give normal distributions. For high coefficients of variation, the normal distribution can be replaced by a lognormal one, in order to avoid negative values.

On the global (meso) scale, the structural element will be seen as a set of reference volume systems. The properties for each of these reference volumes can be correlated between them, with correlation coefficients which are dependent on distance. A correlation law which is often used takes the following form:

$$\rho\left(\Delta r_{ij}\right) = \rho_0 + \left(1 - \rho_0\right)\exp\left[-\left(\frac{\Delta r_{ij}}{d_c}\right)^2\right] \qquad [5.123]$$

5.2.2. *Properties of reinforcing and prestressing steel*

Figure 5.22 gives the typical diagram of a stress-strain law for a steel in reinforced concrete. Choosing a distribution for the yield strength is difficult, and studies highlight that no fit to a normal, lognormal or extreme values distribution seems satisfactory for describing tails.

Mirza and McGregor [MIR 79b] suggested using a Beta distribution for the yield strength:

$$f_{f_y}(\sigma) = A\left(\frac{\sigma - a}{c}\right)^B \left(\frac{b - \sigma}{c}\right)^C \qquad [5.124]$$

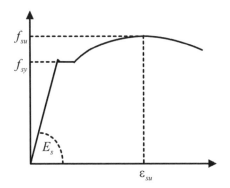

Figure 5.22. *Typical stress-strain diagram for a reinforcing steel*

where the constants (*A, B, C, a, b, c*) are obtained through fittings using available data. They are given in Table 5.24 for steels of 300 MPa and 400 MPa, with $a \leq \sigma \leq b$ as a validity range.

Category	Bias	VC	A	B	C	a	b	c	Unit
300	1.03	35	4.106	2.21	3.82	228	428	200	MPa
400	1.15	38	7.587	2.02	6.95	372	703	331	MPa

Table 5.24. *Coefficients of the probability density function f_{f_y}*

[JCC 01] proposes using – for the yield strength f_y – a normal variable function of diameter *d* of the steel bar, broken down into three variables which characterize the variation between various production sites, between batches and within the same batch:

$$f_y(d) = f_{y,1}(d) + f_{y,2} + f_{y,3}$$ [5.125]

where $f_{y,1} \sim N(\mu_1(d), \sigma_1)$, $f_{y,2} \sim N(0, \sigma_2)$ and $f_{y,3} \sim N(0, \sigma_3)$. The parameters for these distributions can be chosen in Table 5.25.

$\mu_1(d)$ [MPa]	μ_1 [MPa]	σ_1 [MPa]	σ_2 [MPa]	σ_3 [MPa]
$\dfrac{\mu_1}{\left(0.87 + 0.13\exp\left[-0.08d\right]\right)}$	$S_{nom} + 2\sqrt{\sigma_1^2 + \sigma_2^2 + \sigma_3^2}$	19	22	8

Table 5.25. *Characteristic of the JCSS probabilistic model [JCS 01]*

Table 5.26 gives some additional elements for reliability calculations (the distributions are taken to be normal).

Variable	Mean	Standard deviation	COV	Correlation			
Steel section [mm²]	Nominal section	-	2%	1.0	0.5	0.35	0.00
Yield strength [MPa]	$S_{nom} + 2\sigma$	σ	-		1.0	0.85	-0.50
Ultimate strength [MPa]	-	40	-			1.0	-0.55
Ultimate strain	-	-	9%				1.00

Table 5.26. *Correlation between variables for reinforcing steels [JCS 01]*

Within a reinforced beam, a correlation of 0.9 can be used for the yield strength because the reinforcements generally come from the same batch. For different diameters or between sections with different reinforcements, the correlation is smaller (in the order of 0.4). Along the same reinforcement, the correlation is taken at 1.0 over a distance of 10 m.

Reference	Distribution	Characteristic or nominal value [MPa]	Bias	Coefficient of variation [%]
Denmark [SOB 93]	Lognormal	370 - 420	-	4 – 5 * 10 **
Sweden [SOB 93]	Lognormal	-	-	8 **
USA, Canada, Europe [MIR 79b]	Beta	280 410	1.20 1.20	10.7 ** 9.3 **
France [HEN 98]	Lognormal	500	1.20	8.1 **
Europe [PIP 95]	Lognormal	400 500	1.24 1.17	4.7 ** 5.2 **
Europe [STR 03]	Lognormal	500	1.156	2.3 **
USA [NOW 03]	Normal	420	1.145	5.0 **
Portugal [WIS 07]	Lognormal	550 575	1.28 1.20	5.9 ** 4.9 **
Czech Republic [MEL 04]	Normal	235 355	1.20 1.11	7.5 ** 6.4 **

*: same batches
**: different batches

Table 5.27. *Statistical parameters of steel's yielding strength*

Table 5.27 summarizes different models taken from literature for the yield strength, whereas Table 5.28 provides the statistical characteristics for other variables. These various properties may obviously be affected by degradation (corrosion, etc.) or by the stress history (fatigue). Wisniewski [WIS 07] highlights a high correlation between the yield strength and ultimate strength, but a low correlation between the other properties.

Variable	Distribution	Min. value	Max. value	Mean value	Coefficient of variation [%]
Ultimate strength [MPa]	Normal Lognormal	614[a]	856[a]	690[a]	7.8[a]
		552[b]	646[b]	598[b]	3.3[b]
		613[c]	752[c]	680[c]	4.2[c]
		320[d]	580[d]	422[d]	4.6[d]
		461[e]	665[e]	566[e]	4.4[e]
		573[f]	822[f]	703[f]	5.9[f]
		622[f]	795[f]	691[f]	4.9[f]
Yield strength [MPa]	Normal Lognormal	511[a]	770[a]	602[a]	8.1[a]
		431[b]	544[b]	496[b]	4.7[b]
		519[c]	656[c]	585[c]	5.2[c]
		204[d]	399[d]	284[d]	7.5[d]
		325[e]	483[e]	393[e]	6.4[e]
		507[f]	723[f]	603[f]	6.0[f]
		507[f]	667[f]	578[f]	5.5[f]
Ultimate strain [%]	Normal Lognormal	12.6[a]	30.3[a]	23.3[a]	12.7[a]
		7.5[b]	16.0[b]	11.8[b]	14.3[b]
		6.0[c]	13.0[c]	9.4[c]	14.9[c]
		_[d]	_[d]	_[d]	_[d]
		_[e]	_[e]	_[e]	_[e]
		6.2[f]	26[f]	13.5[f]	24.5[f]
		9.6[f]	19.9[f]	13.1[f]	19.2[f]
Young's modulus [GPa]	Normal Lognormal	_[a]	_[a]	_[a]	_[a]
		217[b]	502[b]	300[b]	22.0[b]
		272[c]	533[c]	351[c]	15.0[c]
		_[d]	_[d]	_[d]	_[d]
		_[e]	_[e]	_[e]	_[e]
		179[f]	237[f]	205[f]	4.9[f]
		195[f]	208[f]	201[f]	1.0[f]

[a]S500 [SOB 93]; [b]S400 [PIP 95]; [c]S500 [PIP 95]; [d]S235 [MEL 04]; [e]S355 [MEL 04]; [f]S500 [WIS 07]

Table 5.28. *Statistical parameters for steel properties*

The constitutive law for a prestressing steel proves to be different from that of reinforcing steel (Figure 5.23).

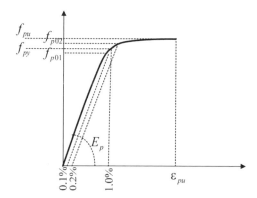

Figure 5.23. *Typical stress-strain diagram for a prestressing steel*

As prestressing steels have no distinct point for the yield strength, this value is determined on the basis of a strain/deformation criterion. This yield strength may also be defined as the stress $f_{p,0.1}$ which corresponds to 0.1% residual deformation, the stress $f_{p,0.2}$ corresponding to 0.2% residual deformation, or the stress corresponding to 1% total deformation[5]. The reference [JCS 01] proposes modeling the mechanical properties of prestressing steel using normal distributions (Table 5.29). Due to high correlations between the ultimate strength and the other stresses, [JCS 01] proposes expressing the stresses f_{py}, $f_{p,0.1}$ and $f_{p,0.2}$ as a function of f_{pu} (Table 5.30).

Property	Mean	Standard deviation	Variation coefficient
Ultimate strength f_{pu}	$1.04\,f_{pk}$ [MPa]	-	2.5%
	$f_{pk} + 66$ [MPa]	40 MPa	-
Young's modulus E_p	$195 - 200$ GPa	-	2.0%
Ultimate strain ε_{pu}	0.05	0.0035	-

Table 5.29. *Statistical parameters of the ultimate strength for prestressing steel*

---

5 This value is given for prestressing wires and strands. For prestressing bars, this value can be taken at 0.75%.

Element	$f_{p,0.1}$	$f_{p,0.2}$	f_{py}
Wire	$0.86\ f_{pu}$	$0.90\ f_{pu}$	$0.85\ f_{pu}$
Strand	$0.85\ f_{pu}$	$0.90\ f_{pu}$	$0.85\ f_{pu}$
Bar	-	-	$0.85\ f_{pu}$

Table 5.30. *Relationships between ultimate and yield strengths*

Nowak and Szerszen [NOV 03] recommend using a normal distribution for describing ultimate strength. Sobrino [SOB 93] suggests a lognormal distribution whereas Strauss [STR 03] concludes that normal, lognormal or Gamma distributions can be used indifferently. The lognormal and normal distributions, however, seem the most suitable for characterizing ultimate and yield strengths. Table 5.31 summarizes the various data in literature. Other parameters have also been studied, particularly by Sobrino [SOB 93] and Wisniewski [WIS 07]; they are presented in Table 5.32.

Reference	Variable	Characteristic value [MPa]	Bias	Coefficient of variation [%]
USA [DEV 92]	f_{py}	1,670	1.06	1.3
	f_{pu}	1,860	1.02	1.1
Spain [SOB 93]	f_{py}	1,670	1.04 – 1.06	1.7 – 2.5
	f_{pu}	1,860	1.04 – 1.06	1.8 – 2.0
Europe [STR 03]	f_{py}	1,570	1.07 – 1.14	0.6 – 2.3
	f_{pu}	1,770	1.03 – 1.08	0.5 – 2.2
USA [NOV 03]	f_{py}	1,720	1.07 – 1.14	1.0 – 3.0
	f_{pu}	1,860	1.04 – 1.06	1.0 – 3.0
Portugal [WIS 07]	f_{py}	1,520	1.05	1.7
	f_{pu}	1,770	1.02	1.2

Table 5.31. *Statistical parameters of ultimate and yield strengths*

Property	Mean value	Coefficient of variation [%]
Yield strength 0.1% [MPa]	1,766[a] 1,556[c] 1,718[d]	3.2[a] 1.7[c] 2.3[d]
Yield strength 0.2% [MPa]	1,823[a] 1,766[b]	2.9[a] 2.2[b]
Yield strength 1.0% [MPa]	1,778[a] 1,734[b] 1,592[c] 1,741[d]	2.5[a] 1.7[b] 1.7[c] 2.9[d]
Ultimate strength [MPa]	1,973[a] 1,942[b] 1,800[c] 1,934[d]	2.0[a] 1.8[b] 1.2[c] 2.2[d]
Young's modulus [GPa]	197.2[a] 196.5[b] 195.0[c] 199.0[d]	1.8[a] 1.9[b] 2.1[c] 2.1[d]
Ultimate strain [%]	5.07[a] 4.15[c] 3.99[d]	4.2[a] 8.7[c] 13.8[d]

[a] 0.987 cm^2 section; [b] 1.40 cm^2 section; [c] 150 mm^2 section; [d] 140 mm^2 section

Table 5.32. *Statistical properties of prestressing strands (seven wires)*

Studies led by Wisniewski [WIS 07] show a high correlation between the stresses at 0.1%, 1.0% and the ultimate strength, which justifies the relationships in Table 5.32. There is no noticeable correlation between the other properties.

5.2.3. *Properties of structural steel*

The variability in resistance of steel elements is similar to that of reinforcing steel. A lognormal distribution seems suitable, particularly for low resistance. The coefficient of variation will be lower for steels with a high ultimate strength [BYF 97]. Steel characteristics are generally described by nominal values using standard tests. Table 5.33 proposes values for yield strengths which reach up to 380 MPa. The distributions are lognormal. The coefficient a is a spatial position factor (1.05 for webs of hot rolled sections, 1.00 otherwise). Factor k is relative to the fractile which describes the difference between the nominal value and the mean value (values from –1.5 to 2.0 are generally used in accordance with European standards). c is a constant reducing the yield strength; a value of 20 MPa is recommended but care must be given to the loading rate for tensile tests. Lastly, the constant B represents the relationship between ultimate and yield strengths. It varies from 1.1 to 1.5

depending on the steel grade (1.5 for carbon steels while 1.1 is for tempered steel) [AGO 94, MAN 92].

The data in Table 5.33 can be used as *a priori* data. If other measurement data is available, the size of the sample to be considered for the *a priori* data (see Chapter 1) is $n' = 50$.

Variable	Mean	COV	Correlation				
Yield strength [MPa]	$f_{y,nom} = a\exp\left(-k\mathbb{COV}\left[f_y\right]\right) - c$	7%	1	0.75	0	0	−0.45
Ultimate strength [MPa]	$B\mathbb{E}\left[f_y\right]$	4%		1.0	0	0	−0.60
Young's modulus [GPa]	E_{nom}	3%			1	0	0
Poisson coefficient	v_{nom}	3%				1	0
Ultimate strain [%]	ε_{nom}	6%					1

Table 5.33. *Correlations between properties for structural steel*

5.2.4. *Properties of concrete*

Although the statistical distribution of the concrete's compressive strength has been studied for a long time, it has a much more limited influence over the structural resistance at the ultimate limit state than the reinforcement properties. This is explained by the design philosophy which attempts to realize reinforcement yielding. Nonetheless, it is important to have a correct estimation when studying column reliability and serviceability. By using many test results, Table 5.34 gives the value of the coefficient of variation or the standard deviation for concretes issued from various references ($f_{c,k}$ is the characteristic value for the compressive strength). Quality control remains an important parameter for limiting this dispersion.

Quality control	Variation coefficient $\left(f_{c,k} < 28\text{MPa}\right)$	Standard deviation (Mpa) $\left(28 < f_{c,k} < 50\text{MPa}\right)$
Excellent	0.10	2.8
Average	0.15	4.2
Low	0.20	5.6

Table 5.34. *Variation in the compressive strength*

When assessing existing structures, the *in situ* resistance is more relevant than the resistance obtained from cores. The relation between these compressive strengths f_c and $f_{c,k}$ can be taken as:

$$\mathbb{E}[f_c] = 0.675\, f_{c,k} + 7.7 \leq 1.15\, f_{c,k}$$
$$\mathbb{COV}_{f_c}^2 = \mathbb{COV}_{f_{cc}}^2 + 0.0084$$

[5.126]

where $\mathbb{E}(f_c)$ is the mean *in situ* strength, $\mathbb{COV}_{f_{cc}}$ is the coefficient of variation for the results on core samples taken on site, and the value 0.0084 is a constant which expresses the variation between the *in situ* strength and the tests carried out on the cores. The relation [5.126] can be analyzed by examining the influences between f_c and the compressive strength based on specimens. Based on Canadian studies, Bartlett and McGregor [BAR 96] suggested modifying it in the following way:

$$\mathbb{E}[f_c] = F_1 F_2 f_{c,k}$$
$$\mathbb{COV}_{f_c}^2 = \mathbb{COV}_{F_1}^2 + \mathbb{COV}_{F_2}^2$$

[5.127]

where $F_1 f_{c,k}$ is the concrete's strength f_{cc} from cores. This takes into account the material variations, batch treatment, etc. Typically, for concrete cast *in situ,* $\mathbb{E}[F_1] = 1.25$ and $\mathbb{COV}_{F_1} = 0.05$, whereas for precast concrete, $\mathbb{E}[F_1] = 1.19$ ($\mathbb{COV}_{F_1} = 0.05$). Normal or lognormal distributions have been fitted. Factor F_2 converts the strength f_{cc} into the average compressive strength of the *in situ* concrete. After 28 days, the compressive strength had an average value ranging from 0.95 and 1.03. After a year, these values increase by 25%. The coefficient of variation in each case is 0.14. A lognormal distribution seems to be the best probabilistic model for F_2. The spatial variation of the compressive strength in a given structure is also an important data. Canadian studies proposed an approximate coefficient of variation of 7% for an element cast from one batch, to 13% for elements cast from multiple batches.

Stewart [SRE 95] proposes a probabilistic model linking the concrete's compression resistance to the material's curing qualities and implementation:

$$f_c = f_{c,k} F k_{cr} k_{cp}$$

[5.128]

Stewart suggests taking a normal or lognormal model for the compaction coefficient k_{cp}, with a mean value between 0.8 and 1.0 according to the quality of the concrete. The coefficient of variation varies between 6% and 0% according to the compaction quality. For the curing coefficient k_{cr}, Stewart recommends using the same distributions with mean values between 0.66 and 1.0 according to the quality of the concrete and coefficients of variations in the order of 5% to 0%. Parameter F ensures the transfer *in situ* strength to real cylinder strength: a normal distribution may be used with a mean varying from 0.74 to 0.96, and a 10% coefficient of variation [MIR 79a].

Outside these most frequently quoted probablistic models for describing a concrete's compression resistance, other works have been completed over the last 10 years. These are given in Tables 5.35 to 5.37.

Reference	$f_{c,k}$ [MPa]	Bias	Standard deviation [MPa]	Variation coefficient [%]
USA, Canada, Europe [MIR 79a]	< 27	-	-	10 – 20
	≥ 27	-	2.7 – 5.4	-
Sweden [THE 04]	35	1.24	-	8.5
Germany [SOB 93]	25 – 45	-	-	9 – 20
Spain [SOB 93]	25 – 40	1.09 – 1.39	2.6 – 4.2	6 – 11
Canada [BAR 96]	≤ 55	1.25	-	10
Portugal [HEN 98]	20 – 35	1.23 – 1.55	3.9 – 6.6	9 – 17
Austria [STR 03]	25 – 50	1.02 – 2.04	1.0 – 2.7	2 – 6
Portugal [WIS 07]	25	1.26	2.9	7.7
	30	1.18	3.3	7.5
	40	1.18	3.4	5.8
USA [NOW 03]	21 – 41	1.12 – 1.35	1.5 – 4.9	4 – 15
	48 – 83	1.04 – 1.19	5.4 – 9.0	9 – 12

The lowest biases, standard deviations and coefficients of variations generally relate to concretes with better strength.

Table 5.35. *Statistical parameters for in situ cast concrete*

Outside the concrete's compressive strength, there are other parameters of interest for calculating concrete structures: tensile strength f_t, Young's modulus E_c and ultimate compressive strain ε_{cu}. The references [MIR 79a, SPA 92] offer a few models which can be used.

344 Structural Performance

Reference	$f_{c,k}$ [MPa]	Bias	Standard deviation [MPa]	Coefficient of variation [%]
Germany [SOB 93]	50 – 65	-	-	6
Europe [PCS 02]	40 – 90	-	2.0 – 6.0	3 – 11
Portugal [WIS 07]	30	1.23	4.0	8.8
	35	1.08	2.3	4.7
	40	1.08	2.4	4.5
	45	1.00 – 1.02	2.2 – 2.9	3.9 – 5.2
Canada [BAR 96]	≤ 55	1.19	-	5
USA [NOW 03]	34 – 45	1.14 – 1.38	4.1 – 5.7	8 – 12

The lowest biases, standard deviations and coefficients of variations generally relate to concretes with better strength.

Table 5.36. *Statistical parameters for precast concrete*

Reference	Distribution	Mean	Coefficient of variation [%]
USA, Canada, Europe [MIR 79a]	-	0.74 – 0.96	10
			-
Europe [PCS 02]	Determinstic	0.86	-
Spain [SOB 93]			6 – 11
Canada [BAR 96]	Lognormal	0.95 – 1.03	14
Portugal [GON 87]	-	0.69 – 1.02	3 – 14

Table 5.37. *Relations between in situ strength and real cylinder strength*

In [MIR 79a], the tensile strength is modeled by a normal distribution with a mean calculated by the following equation and a coefficient of variation of 13%:

$$f_t = 0.53 \, \mathbb{E}\left(f_c\right)^{1/2} \text{ [MPa]} \tag{5.129}$$

whereas Spaethe [SPA 92] proposes a normal or lognormal distribution with a coefficient of variation between 18% and 20% with the following mean:

$$f_t = 0.30 \, f_{c,k}^{2/3} \text{ [MPa]} \tag{5.130}$$

For high strength concrete (>C50/60), Eurocode 2 [EN 04b] however proposes the following expression:

$$f_t = 2.12 \ln\left(1 + \frac{\mathbb{E}(f_c)}{10}\right) \text{ [MPa]} \tag{5.131}$$

The concrete initial tangent elasticity modulus [MIR 79a] can be modeled by a normal distribution with a coefficient of variation of 8% and with a mean of:

$$E_{ct} = 5{,}015 \sqrt{\mathbb{E}[f_c]} \text{ [MPa]} \tag{5.132}$$

The secant modulus (30% of the maximum stress) is also modeled by a normal distribution with a coefficient of variation of 12% and a mean given by the following:

$$E_{cs} = 4{,}600 \sqrt{\mathbb{E}[f_c]} \text{ [MPa]} \tag{5.133}$$

Eurocode 2 [EN 04b] proposes modeling the secant modulus in the following way:

$$E_{cs} = 22{,}000 \, \mathbb{E}[f_c]^{0.3} \text{ [MPa]} \tag{5.134}$$

This expression must be reduced by 10–30% for sandstone aggregates, but must be increased by 20% for basalt aggregates.

A useful probabilistic model for describing the set of relevant variables which characterize the "concrete" material consists of introducing random variables into the reference data. Thus, according to ISO 2736 or 3893, the concrete's reference compressive strength is obtained by standard tests (cylinders 300 mm high, and 150 mm in diameter) over 28 days:

$$f_c = \alpha(t,\tau) f_{c0}^{\lambda} \tag{5.135}$$

The tensile strength, the elasticity modulus and the ultimate compressive strain are deduced from this variable:

$$\begin{aligned}
f_t &= 0.3 \, f_c^{2/3} \\
E_c &= 10.5 \, f_c^{1/3} \frac{1}{1 + \beta_d \, \varphi(t,\tau)} \\
\varepsilon_u &= 6.010^{-3} f_c^{-1/6} \left(1 + \beta_d \, \varphi(t,\tau)\right)
\end{aligned} \tag{5.136}$$

with:

– λ as a factor which takes into account the variation between the compressive strength in a structural element and that which is given by tests;

– $\alpha(t, \tau)$ as a deterministic function which takes the age of the concrete and the loading period into account:

$$\alpha(t,\tau) = \alpha_1(\tau)\alpha_2(t)$$
$$\alpha_1(\tau) = \alpha_3(\infty) + \left[1 - \alpha_3(\infty)\right]\exp\left(-a_\tau \tau\right)$$
$$\alpha_2(\tau) = a + b\ln t$$

with $\alpha_3(\infty) \approx 0.8, a_\tau \approx 0.04$. Coefficients a and b depend on the type of cement and the weather conditions. Under normal conditions, we would normally take $a = 0.6$ and $b = 0.12$;

– $\varphi(t,\tau)$ is a creep coefficient which can be chosen as deterministic;

– β_d is the ratio between permanent loads and total loads. It depends on the type of structure and is generally ranged between 0.6 and 0.8.

A probabilistic model [KER 91, MAD 83, MIR 79a] of the compressive strength at a point i in a structural element j is built on the basis of this reference strength by introducing additional random variables:

$$f_{c,ij} = \alpha(t,\tau)\left(f_{c0,ij}\right)^{\lambda} Y_{1,j}$$
$$f_{c0,ij} = \exp\left(U_{ij} \Sigma_j + M_j\right)$$

[5.137]

with:

– $f_{c0,ij}$: lognormal variable independent from $Y_{1,j}$ of parameters M_j and Σ_j;

– M_j: logarithmic mean on element j;

– Σ_j: logarithmic standard deviation on element j;

– $Y_{1,j}$: lognormal variable which represents the variations related to curing, hardening conditions, etc., on element j;

– U_{ij}: normal variable representing the variability within element j;

– λ: lognormal variable with a mean of 0.96 and a coefficient of variation of 0.5%.

The other material properties are written identically:

$$f_{t,ij} = 0.3 f_c^{2/3} Y_{2,j}$$

$$E_{c,ij} = 10.5 f_c^{1/3} \frac{1}{1 + \beta_d \, \varphi(t,\tau)} Y_{3,j} \qquad [5.138]$$

$$\varepsilon_{u,ij} = 6.0 \; 10^{-3} f_c^{-1/6} \left(1 + \beta_d \, \varphi(t,\tau)\right) Y_{4,j}$$

where the different random variables $Y_{2,j}$, $Y_{3,j}$ and $Y_{4,j}$ represent the variations of the quantities related to various factors (type and size of aggregates, chemical composition of the cement, weather conditions, etc.). Variables U_{ij} and U_{kj} within element j are correlated by:

$$\rho\left(U_{ij}, U_{kj}\right) = \rho_0 + \left(1 - \rho_0\right) \exp\left(-\frac{\left(r_{ij} - r_{kj}\right)^2}{d_c^2}\right) \qquad [5.139]$$

with $d_c = 5.0$ m and $\rho_0 = 0.5$. For different elements, these variables are taken to be independent.

If measurements are available, the parameters $Y_{k,j}$ from Table 5.38 can be used as *a priori* distribution parameters. These variables are assumed to be lognormal. If additional measurements are available, the parameters in Table 5.38 are updated by taking a *a priori* sample size equal to $n' = 10$. In this case, the updated expressions of Chapter 1 can be applied to the logarithm of the variables.

Variable	Distribution	Mean	Coefficient of variation
$Y_{1,j}$ (compressive strength)	Lognormal	1.0	0.06
$Y_{2,j}$ (tensile strength)	Lognormal	1.0	0.30
$Y_{3,j}$ (Young's modulus)	Lognormal	1.0	0.15
$Y_{4,j}$ (ulitmate strain)	Lognormal	1.0	0.15

Table 5.38. *Parameters of the probabilistic model for concrete [JCS 01]*

$\ln\left(f_{c0,ij}\right)$ is normally distributed provided that the parameters M, Σ are obtained from an ideal infinite sample. In general, concrete production varies from one production site to another, within the same production, and the samples are limited in size. The parameters M, Σ must be treated as random variables. Thus $\ln\left(f_{c0,ij}\right)$ (see Chapter 1) is Student distributed:

$$F_{\ln(f_{c0,ij})}(x) = F_{t_{v''}}\left(\frac{\ln\left(\dfrac{x}{m''}\right)}{s''}\left(1+\frac{1}{n''}\right)^{-1/2}\right)$$ [5.140]

where $F_{t_{v''}}(.)$ is the Student distribution function with v'' degrees of freedom. $f_{c0,ij}$ is then represented by:

$$f_{c0,ij} = \exp\left(m''+t_{v''}s''\sqrt{1+\frac{1}{n''}}\right)$$ [5.141]

The values of m'', s'', n'' and v'' depend on the information available. In the absence of data, the values from Table 5.39 can be applied as *a priori* distribution for $f_{c0,ij}$ [6].

Concrete	Type	Parameters			
		M'	n'	s'	v'
Ready to use	C15	3.40	3.0	0.14	10
	C25	3.65	3.0	0.12	10
	C35	3.85	3.0	0.09	10
	C45	3.98	3.0	0.07	10
Precast	C25	3.80	3.0	0.09	10
	C35	3.95	3.0	0.08	10
	C45	4.08	4.0	0.07	10
	C55	4.15	4.0	0.05	10

Table 5.39. *Parameters of the a priori distribution for $f_{c0,ij}$*

5.3. Geometric variability

Studies in geometric variability are generally led by studying the deviation between the nominal dimension and the real dimension, i.e. $\Delta = X - X_{nom}$. There is a large amount of information regarding the external dimensions of concrete elements (beams, slab) [CAS 91, TIC 79a, TIC 79b]. We should point out here that the element type (reinforced, prestressed), the shape (rectangular, I, T, etc.), the

---

6 If $n''v'' > 10$, a good approximation consists of using a lognormal distribution with mean m'' and standard deviation $s''\sqrt{\dfrac{n''}{n''-1}\dfrac{v''}{v''-2}}$.

material grade, the thickness and position of the element (half-span, abutment) play virtually no role at all; the main essential component is the production method (cast or precast). The expressions in Table 5.40 are valid for dimensions smaller than 1,000 mm. Studying geometric variability on structures seems to show that the real thickness of slabs is bigger than the nominal thickness with a bias reaching up to 1.06, and a coefficient of variation of 0.08. These values are reduced in good quality elements (1.005 and 0.02) or in precast slabs.

However, the useful reinforcement depth seems to be generally worse with a bias between 0.93 and 0.99 [CAS 91] according to whether it is top or bottom reinforcement (Table 5.41). It also differs according to the type of structural element (beam, slab or column).

For metal parts [FAJ 99], the mean of the deviation between real and nominal geometric data (thickness, height, width) does not exceed ±1 mm.

	Mean	Standard deviation
Concrete elements	$0 \leq \mu_\Delta = 0.003\, X_{nom}$ $\leq 3\,\text{mm}$	$\sigma_\Delta = 4\ \text{mm} + 0.006\, X_{nom}$ $\leq 10\ \text{mm}$
Columns and walls	$0 - 5$	$5 - 10$
Slabs (bottom steel)	$0 - 10$	$5 - 10$
Slabs (top steel)	$0 - 10$	$10 - 15$
Beams (bottom steel)	$-10 - 0$	$5 - 10$
Beams (to steel)	$0 - 10$	$10 - 15$

Table 5.40. *Statistical properties of the concrete cover*

Normal distributions are used to characterize the deviation Δ for external dimensions. For the concrete cover, various models have been introduced [TIC 79a]: truncated Beta, translated lognormal or Gamma distributions. The use of these models can be explained by the requirement of truncated distributions on the (lower) tail, or on both tails. Tichy [TIC 79b] has shown that the correlation remains low between different elements, with this correlation being more or less prominent according to the production method (*in situ* cast or precast).

Studies by Mirza and Macgregor [MIR 79b] however, are still the most exhaustive in terms of the statistical characterization of the geometric dimensions of concrete elements, either *in situ* cast or precast. Tables 5.41 to 5.43 summarize this data for slabs, beams, and columns. Mirza and Macgregor recommend normal distributions for the set of properties, with the exception of covers where normal truncated distributions are suggested so as to avoid negative values.

Dimension	Technology	Nominal value [mm]	Deviation [mm]	Standard deviation [mm]
Thickness	*In situ*	100 – 230	0.8	12
	Precast	100 – 230	0.0	5
Depth of top reinforcement	*In situ*	100 – 200	±20	15 – 20
	Precast	100 – 200	±20	3 – 6
Depth of bottom reinforcement	*In situ*	100 – 200	-8 – 9	10 – 15
	Precast	100 – 200	0	3 – 6

Table 5.41. *Variability in geometric properties of slabs*

Dimension	Technology	Nominal value [mm]	Deviation [mm]	Standard deviation [mm]
Overal depth	*In situ*	460 – 960	2.5	5
	Precast	530 – 990	3.2	4
Rib width	*In situ*	280 – 305	2.8	4.8
	Precast	480 – 610	4	6.5
Flange width	Precast	280 – 305	2.8	4.8
Concrete cover (bottom)	*In situ*	12 – 25	–3 – 6	16 – 18
	Precast	50 – 60	3	8 – 9
Concrete cover (top)	*In situ*	19 – 25	–5 – 2	11 – 13
	Precast	19	3	8 – 9

Table 5.42. *Variability in geometric properties of beams*

Dimension	Technology	Nominal value [mm]	Deviation [mm]	Standard deviation [mm]
Rectangular (thickness)	*In situ*	280 – 760	1.6	6.4
	Precast	180 – 410	0.8	3.2
Circular (diameter)	*In situ*	280 – 330	0	4.8
	Precast	280 – 330	0	2.4

Table 5.43. *Variability of geometric properties of columns*

Another study [SOB 93] has enabled the collection data on Spanish production between 1990 and 1993 for bridge elements (precast or not, Table 5.44), whereas Wisniewski [WIS 07] analyzed data on U or I profiles from two precasting sites (Table 5.45).

Dimension	Technology	Nominal value [mm]	Deviation [mm]	Standard deviation [mm]
Horizontal	*In situ*	< 600	-	-
		≥ 600	0.2	3
	Precast	≤ 250	–2 – 0	–5 – 2
		250 – 600	10 – 5	12 – 3.5
		≥ 600	–2 – 1	-
Vertical	*In situ*	≤ 250	0 – 2	2 – 3
		250 – 600	–2	8 – 10
		≥ 600	40 – 20	15 – 22
	Precast	≤ 250	5 – 2.5	5 – 9
		250 – 600	12 – 6	5 – 3.5
		≥ 600	-	-

Table 5.44. *Variability in geometric properties of bridges [SOB 93]*

Dimension	Nominal value [mm]	Deviation [mm]	Standard deviation [mm]
Height (U)	900 – 1900	14 – 18	4.3 – 6.9
Width (U)	2800 – 3500	–10 – 18	0 – 8
Thickness (U)	200	6 – 22	5.3 – 14.1
Height (I)	600 – 1200	–1 – 12	3.4 – 4.2
Width (I)	350 – 440	0 – 2	3.3 – 4.4
	450 – 800	1 – 5	1.6 – 2.5
Thickness (I)	75 – 100	–1 – 2	2.0
	100 – 150	1 – 8	2.2 – 2.4

Table 5.45. *Variability in geometric properties of precast elements [WIS 07]*

5.4. Scale effects

Probabilistic descriptions relating to the resistance of structural components is based on the probabilistic description of the elements' properties, i.e. their geometric dimension and material strength. When the mechanical properties for these elements are calculated based on the mathematical relations between these properties, we notice remarkable differences with the experimental results. This can be explained by the variability induced by experimental observations. In truth, most of these differences come from a simplification of the mathematical model which links the materials' and geometric characteristics to the element's structural behavior. For example, by deriving an expression for a section's resisting capacity, some hypotheses have been made on stress distribution, on the constitutive law for reinforcing steels, on the concrete's tensile strength, etc. These hypotheses are usually conservative. However, they do add a degree of uncertainty to the transfer of different parameters to the element's resistance. We obviously do not see this uncertainty when the statistical properties of a structural part are directly obtained from experimental observations on the same element. However, such tests are not always accessible.

Let R be the property of a structural element. The uncertainty on this property may be assessed by various structural tests on a real scale. However, this requires a significant number of costly tests. This explains why the behavior of structural elements is often analyzed using numerical models and simulations. Nowak and Collins [NOW 00] propose to break down the uncertainty on R according to many categories:

– material properties M;

– manufacture/implementation F;

– behavior modeling P.

This leads us to write [SCH 97]:

$$R = R_d \, M \, F \, P \qquad\qquad [5.142]$$

Factors P, M and F are the ratios in relation to the nominal properties. Whilst taking these factors to be independent and normally distributed, the mean value of R and its coefficient of variation are given by:

$$\mathbb{E}(R) = R_d \, \mathbb{E}(M)\mathbb{E}(F)\mathbb{E}(P)$$
$$\mathbb{COV}(R) = \sqrt{\mathbb{COV}(M)^2 + \mathbb{COV}(F)^2 + \mathbb{COV}(P)^2} \qquad [5.143]$$

The probabilistic model for factor P is generally obtained by comparing the experimental results R_e with the values predicted by the model $\hat{R}_d$ by using the geometry and properties of real materials [SCH 97]:

$$P = \frac{R_e}{\hat{R}_d} \qquad\qquad [5.144]$$

Wisniewski [WIS 07] proposes that we take $\mathbb{E}(P) \approx 1$ for the mean value and a coefficient of variation of only several percent, if advanced calculation models are used. For simple and conservative models, the mean value is higher than 1, and the coefficient of variation may reach 10 to 20%. Melchers [MEL 99] emphasizes that coefficient P does not only depend on the model's uncertainties, but also on errors related to the testing procedure or variability within batches.

The aforementioned values may be adjusted according to whether we are interested in the shear effect or bending moment, particularly in concrete structures. These structural elements demonstrate structural behaviors which are more or less well apprehended by calculation models, which, therefore, justifies many studies on the subject [WIS 07]. Thus, Melchers [MEL 99] proposes taking $\mathbb{E}(P) \approx 1.02$ as the mean value, with coefficients of variation of 3-5%. These values are only relevant in relation to the numerical model used; in the present case, Melchers uses non-linear behavior models with real constitutive behavior laws. Nowak [NOW 99] has obtained similar values for reinforced and prestressed concrete. The JCSS [JCS 01] gives higher means and coefficients of variation (1.20/15%). The JCSS, however, uses linear behavior models based on simplified behavior laws. In each case, these authors recommend lognormal distributions (or truncated normal distributions).

With regard to shear effect, based on more than 300 tests on high strength concrete beams, Bohigas [BOH 02] emphasizes the unsatisfactory nature of models for (reinforced or prestressed) concrete structures. Also, larger errors must be introduced with wider coefficients of variation. Table 5.46 summarizes some of these results.

Model	No reinforcement		Reinforcement	
	Mean	Coefficient of variation (%)	Mean	Coefficient of variation (%)
Eurocode 2	1.02	22.03	1.83	40.29
LRFD – AASHTO	1.28	16.80	1.18	19.23
ACI	1.29	31.21	1.41	26.70
MCFT	1.13	20.00	1.07	17.39

MCFT: Modified Compression Field Theory

Table 5.46. *Modeling shear effect uncertainties [BOH 02]*

Chapter 6

Principles of Decision Theory

6.1. Introduction

Decision making in design and management of structures is an activity in its own right. This process of decision making is often based on predictions and information which invariably contain uncertainties. Risk is inevitable under these conditions.

Problems to do with decision making mainly consist of identifying not only technical risks, but also societal and environmental risks, and optimizing these in relation to different variables. A systematic framework which enables us to take all the aspects of a *decision problem* into account, constitutes a *decision model.*

In any decision analysis, the set of *decision variables* must be identified and defined. In order to test out different alternatives, we find that it is necessary to define an *objective function*, expressed in terms of decision variables. This function is generally expressed in monetary terms, which represent a cost or a gain. Minimizing or maximizing this objective function thus enables us to estimate the *optimal decision variables.* Formally, if $F(., \cdots,.)$ represents this objective function, and if $X_1, \cdots, X_n$ are the decision variables, then the optimal solution is given by:

$$\frac{\partial F\left(X_1, \cdots, X_n\right)}{\partial X_i} = 0, \quad i = 1, \cdots, n \qquad [6.1]$$

The second derivative enables us to verify whether we are dealing with a minimum or a maximum. However, the applicability of the previous optimization procedure is limited: firstly, the "objective" function is rarely expressed as a continuous variable function. Secondly, a decision cannot only be based on available information. In most engineering problems, additional information is needed and may be in the form of controls, inspections, *in situ* or laboratory tests. Since there are a large variety of approaches for collecting this additional information, the range of alternatives is large for the decision problem. To summarize, the general analysis framework for the decision problem must include the following elements:

– a list of possible alternatives, including the acquisition of additional data if necessary;

– a list of possible results for each alternative;

– an estimate of the probability related to each result;

– an assessment of the consequences associated with each combination of alternative and result;

– the decision criterion ("objective" function);

– a systematic assessment of the alternatives.

6.2. The decision model

6.2.1. *Decision tree*

The various elements of the decision problem may be integrated in a *decision tree*. This tree consists of listing the set of alternatives, the possible results, the occurrence probabilities and the consequences.

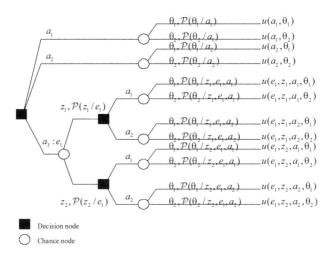

Figure 6.1. *Decision tree*

Figure 6.1 gives an example of a decision tree where $a_i, \theta_j, e_k, z_l, u(a_i, \theta_j)$ respectively denote the alternative i, the result j, the test k designed for obtaining additional information, the test results l, the utility value relating to alternative i and result j. If the utility depends on a test k which gives a result l, then this function is written $u(e_k, z_l, a_i, \theta_j)$. The decision tree starts with a *decision node*, a source of many alternatives. At each alternative, there are many possible results or *branches* at the base of a *chance node*. The results from a chance node are mutually exclusive and collectively exhaustive, which implies that the sum of conditional probabilities is equal to 1.

EXAMPLE 6.1.– Let us consider two alternative projects for building a structure. Project A is based on a standard design for which the degree of performance is estimated at 99%, and costs at €1,500,000. Project B is an innovative design which reduces the cost to €1,000,000. The project's degree of reliability over the degree of performance is unknown, because it is new. However, the design office considers that if the hypotheses are correct, this degree of reliability would be 99%. In the opposite case, the degree of reliability is lowered to 90%.

According to the judgment of the research body, there is a 50% chance that these hypotheses are incorrect. By taking the cost of an unsuitable performance as being €10,000,000, then the decision tree is given by the following figure. A, B, T, $\overline{T}$, S and $\overline{S}$ respectively denote projects A and B, valid and invalid hypotheses, and satisfactory and unsatisfactory performances. u is the utility function expressed in costs.

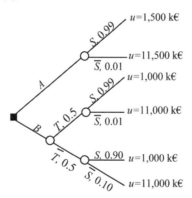

A decision tree is generated in many steps. Firstly, a *reliability tree* (see Chapter 3) is constructed. The *event tree* or *consequence tree* is deduced from this first tree, and the *event tree* or *consequence tree* combines all the possible consequences. Finally, the decision tree is built based on the event tree, combining all the alternatives with it.

EXAMPLE 6.2.– A construction manager is thinking about the performance of one of his structures, in terms of carrying current loads. In order to quantify the risks of failure/collapse, but also to assess the various alternatives, he/she tries to develop a reliability tree, an event tree and a decision tree.

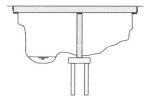

The structure considered is supported by two abutments and an intermediate pile. The pile rests on two groups of piles. The deck encounters degradation problems due to corrosion. We presume that the structure is kept intact if one of the pile groups survives. The probability of a pile group surviving is 5% for the structure's remaining service period. The deck's failure probability is assessed at 1% in the case of heavy corrosion, but only at 0.1% if the degree of corrosion is judged to be low after inspection. The probability of the corrosion developing from a low level to a high level over the remaining service period is 10%.

For building a reliability tree, we will presume that the events are independent. The tree is then given by the following figure. The failure probability is 3.6%.

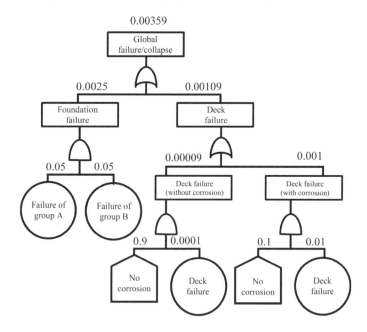

The event tree summarizes the various possible consequences. For this structure, the probability of detecting corrosion early and repairing it is assessed at being 70%. The cost of this operation is valued at €300,000. The cost of failure alone on the structure without alarm is valued at €10,000,000. Based on traffic data, the structure does not carry traffic for 70% of the time. In the case of collapse under traffic, the additional cost related to loss of human lives is around €3,000,000. The event tree may therefore be built on this data.

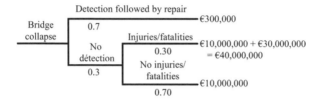

From the previous diagram, we can deduce the average cost for structure failure:

$$C_f = (0.7 \times 0.3 + 0.3 \times 0.3 \times 40 + 0.7 \times 10)10^6$$
$$= (0.21 + 3.6 + 7)10^6 = €10,810,000$$

We may now envisage various intervention measures: to build a new bridge, to restrict weight or to do nothing. In the first case, the cost is €5,000,000 and the reliability is assumed to be fulfilled. In the second case, the failure probability is reduced by 50%. The delay cost due to the weight restriction is fixed to €2,000,000. The decision tree is therefore:

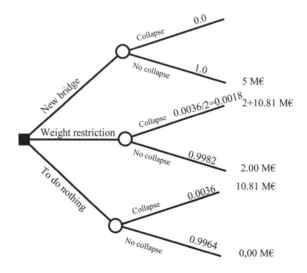

□

6.2.2. *Decision criterion*

The aim of a decision analysis is to make the "right" decision. Most of the time, the comparison between alternatives is made in a probabilistic context, where the probabilities are related to the consequences. When the consequences are expressed in monetary values, the most commonly used criterion is the average cost:

$$\mathbb{E}(a_i) = \sum_j p_{ij} \; c_{ij} \qquad\qquad [6.2]$$

for the alternative i leading to consequences j. p_{ij}, and c_{ij} indicate the probabilities and the costs associated with each result j relating to alternative i. The optimal alternative is then given by:

$$c(a_{opt}) = \max_i \left(\sum_j p_{ij} \; c_{ij} \right) \qquad\qquad [6.3]$$

EXAMPLE 6.3.– We will take example 6.1. The average costs for each project are:

$$\mathbb{E}(A) = 0.99 \times 1,500 + 0.01 \times 11,500 = 1,600 \text{ k€}$$
$$\mathbb{E}(B) = 0.5 \times 0.99 \times 1,000 + 0.5 \times 0.01 \times 11,000$$
$$\qquad + 0.5 \times 0.9 \times 1,000 + 0.5 \times 0.1 \times 11,000$$
$$\qquad = 1,550 \text{ k€}$$

which gives Project B as the optimum solution for the average cost criterion.   □

EXAMPLE 6.4.– Let us take example 6.2. The average costs for each project are:

$$C_{f1} = 0 + 5 = 5 \text{ M€}$$
$$C_{f2} = 0.0018 \times 12.81 + 0.9982 \times 2 = 2.02 \text{ M€}$$
$$C_{f3} = 0.0036 \times 10.81 + 0.9964 \times 0 = 0.039 \text{ M€}$$

The most economical situation is therefore "do nothing".   □

As soon as the decision tree has been created, and the probabilities of possible results through the chance nodes and the consequences have been assessed, the average cost for each alternative can be calculated. A decision analysis based on the initial information is called *a priori decision analysis.* It is not rare for a chance node to be at the bottom of a series of other chance nodes, as shown by Figure 6.2.

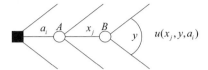

Figure 6.2. *Decision tree with a series of chance nodes*

In Figure 6.2, variable X is discrete and variable Y is continuous. The expected value at point B is therefore given by:

$$\mathbb{E}(a_i / x_j) = \int_{-\infty}^{+\infty} \mathbb{E}(a_i / x_j, y) \, f_Y(y) \, dy = \int_{-\infty}^{+\infty} \sum_j \mathbb{E}(a_i / x_j) \, p_j \, f_Y(y) \, dy \quad [6.4]$$

6.2.3. *Terminal decision analysis*

In order to improve the level of information, and to reduce uncertainties, it is current practice to perform additional tests before making the final decision. This requires additional costs, but may reduce or eliminate uncertainties for some problems. To do so, the decision tree will introduce conditional probabilities. If the decision analysis is adjusted according to new data, then the analysis will be called *terminal decision analysis*.

EXAMPLE 6.5.– Let us use example 6.1. We assume that the design office orders additional tests which cost €50,000. If the hypotheses for project B are erroneous, then the test will be negative (F) in 90% of cases. In the opposite case ($\bar{F}$), the probability of the test failing is only 30%. The new decision tree is given in the following figure:

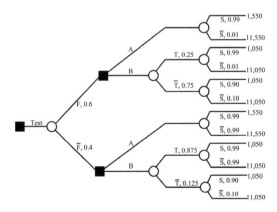

These hypotheses give the following probabilities:

$$\mathcal{P}(F/T) = 0.3$$
$$\mathcal{P}(F/\bar{T}) = 0.9$$
$$\mathcal{P}(\bar{F}/T) = 0.7$$
$$\mathcal{P}(\bar{F}/\bar{T}) = 0.1$$

We then obtain:

$$\mathcal{P}(F) = \mathcal{P}(F/T)\,\mathcal{P}(T) + \mathcal{P}(F/\bar{T})\,\mathcal{P}(\bar{T}) = 0.3\times0.5 + 0.9\times0.5 = 0.6$$

$$\mathcal{P}(T/F) = \frac{\mathcal{P}(F/T)\,\mathcal{P}(T)}{\mathcal{P}(F)} = \frac{0.3\times0.5}{0.6} = 0.25$$

$$\mathcal{P}(\bar{T}/F) = 0.75$$

$$\mathcal{P}(\bar{F}) = 0.4$$

This enables us to express conditional probabilities:

$$\mathcal{P}(V/\bar{F}) = \frac{\mathcal{P}(\bar{F}/T)\,\mathcal{P}(T)}{\mathcal{P}(\bar{F})} = \frac{0.7\times0.5}{0.4} = 0.875$$

$$\mathcal{P}(\bar{T}/\bar{F}) = 0.125$$

If the test is negative, the average costs for the two projects are:

$$\mathbb{E}(A/F) = 0.99\times1,550 + 0.01\times11,550 = 1,650 \text{ k€}$$

$$\mathbb{E}(B/F) = 0.25\times(0.99\times1,050 + 0.01\times11,050)$$
$$+ 0.75\times(0.90\times1,050 + 0.1\times11,050)$$
$$= 1,825 \text{ k€}$$

In the case of test failure, solution A is the best project.

If the test is conclusive, we find:

$$\mathbb{E}(A/\bar{F}) = 1,650 \text{ k€}$$

$$\mathbb{E}(B/\bar{F}) = 1,262.5 \text{ k€}$$

Solution B is then the better project.

The total average cost with the addition test is then given:

$$\mathbb{E}(T_I) = \mathbb{E}(A/F)\mathcal{P}(F) + \mathbb{E}(B/\bar{F})\mathcal{P}(\bar{F}) = 1,455 \text{ k€}$$

Compared to the *a priori* optimum cost (B), we notice that the test is interesting in terms of economics.   □

6.2.4. *Information value*

We have seen how additional information could contribute important decisional elements. But can we give a figure to its value? One of the definitions is to take the following:

$$V_I = \mathbb{E}(T_{-I}) - \mathbb{E}(T_{+I}) \qquad\qquad\qquad [6.5]$$

where $\mathbb{E}(T_{-I})$ and $\mathbb{E}(T_{+I})$ indicate the average costs of alternatives, with and without additional information I. Thus, if V_I exceeds the cost of the test, then the test can be retained. This value is, however, limited by a *value of perfect information* which is simply the difference between the cost of the alternative without a test, and the average cost of an alternative with a 100% reliable test. This cost, written V_{PI} relates to the maximum cost that the decision maker must invest in order to acquire additional information.

EXAMPLE 6.6.– Let us take example 6.5. To assess the total average cost for a test which is 100% reliable, it necessary to use some calculations from example 6.5 with a slightly different decision tree. The calculation for average costs is therefore:

$$\mathbb{E}(A/V) = 0.99 \times 1,500 + 0.01 \times 11,500 = 1,600 \text{ k€}$$
$$\mathbb{E}(B/V) = 0.99 \times 1,000 + 0.01 \times 11,000 = 1,100 \text{ k€}$$

$$\mathbb{E}(A/\bar{T}) = 0.99 \times 1,500 + 0.01 \times 11,500 = 1,600 \text{ k€}$$
$$\mathbb{E}(B/\bar{T}) = 0.90 \times 1,000 + 0.10 \times 11,000 = 2,000 \text{ k€}$$

which gives:

$$\mathbb{E}(PT) = \mathbb{E}(B/T)\mathcal{P}(T) + \mathbb{E}(A/\bar{T})\mathcal{P}(\bar{T}) = 1,350 \text{ k€}$$

The cost of perfect information is given simply by the difference between the cost $\mathbb{E}(PT)$ and the cost of solution B without testing, which was the *a priori* optimum alternative:

$$V_{PI} = \mathbb{E}(B) - \mathbb{E}(PT) = 1,550 - 1,350 = 200 \text{ k€}$$

€200,000 is the maximum sum which can be spent verifying the validity of the hypotheses.

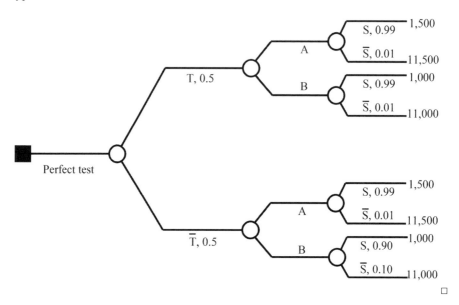

6.3. Controls and inspections

In Chapter 4, we saw how the results from an inspection could be of two different types; either qualitative or quantitative. The first type is for visual inspections or rough measurements. In this case, the information is sometimes even treated afterwards, in order to be provided as a global rating (e.g. 1 for "good", 5 for "bad"). The second type of information is acquired via more thorough investigations, making use of measurements. Both sets of information display advantages and disadvantages, to be used in a reliability reassessment. In some cases, qualitative information is very difficult to use in terms of interpretation and introduction for reliability calculations. This very quickly leads us to note its low influence when carrying out reassessment, moreover if it is associated with a high coefficient of variation (for example, detection of a fatigue crack without measurement, knowing that the detection threshold is widely inaccurate). To a lesser extent, this is also the case for quantitative information which nevertheless has the advantage of being able to be directly connected to a variable or a set of variable's measurements. The quality of the inspection results is therefore as important as the quality of the initial data on the variables.

EXAMPLE 6.7.– (adapted from [CRE 96; LUK 01a]) A welded joint is studied with regard to its fatigue damage. Failure occurs when the defect size $a(t)$ exceeds a critical size a_c. The intuitive safety margin is therefore $M = a_c - a(t)$. A more appropriate safety margin can be used instead, introducing a function which characterizes the damage $\psi(a_1, a_2)$ between two crack sizes a_1 and a_2. Presuming that the joint is subjected to variable amplitude load cycles, a simplified version of the safety margin is given by:

$$M = \int_{a_0}^{a_c} \frac{dx}{\left(Y_0 \sqrt{\pi x}\right)^m} - C\, N(t) \left(\frac{2\, \mathbb{E}(S)}{\sqrt{\pi}}\right)^m \Gamma\left(1 + \frac{m}{2}\right)$$

where x is the crack depth, N is the number of cycles, S is the stress range at the hotspot for the non-cracked section (effective stress range), Y_0 is a stress intensity correction factor, C and m are two material parameters. a_0 and a_c respectively indicate the sizes of the initial defect and the critical defect judged to be unacceptable (or leading the joint rupture). The previous margin may be more simply clarified when noting that:

$$\int_{a_0}^{a_c} \frac{dx}{\left(Y_0 \sqrt{\pi x}\right)^m} = \left(\frac{1}{Y_0 \sqrt{\pi}}\right)^m \frac{1}{\frac{m}{2} - 1} \left[a_0^{-\frac{m}{2}+1} - a_c^{-\frac{m}{2}+1}\right]$$

Therefore, the margin becomes ($\sqrt{\pi}$ has been eliminated for simplification reasons because it appears in the two margin members):

$$M = \left(\frac{1}{Y_0}\right)^m \frac{1}{\frac{m}{2} - 1} \left[a_0^{-\frac{m}{2}+1} - a_c^{-\frac{m}{2}+1}\right] - C\, N(t) \left(2\, \mathbb{E}(S)\right)^m \Gamma\left(1 + \frac{m}{2}\right)$$

The critical defect is conventionally fixed on a size that must not be exceeded. The following table gives the characteristics for these variables.

Variable	Distribution	Mean	Coefficient of variation (%)
m	Deterministic	3.00	-
$C\,[\dfrac{10^{-13}\ \mathrm{mm}}{\mathrm{cycles}(\mathrm{N/mm^{3/2}})^m}\,]$	Lognormal	2.50	10
a_0 [mm]	Lognormal	0.13	20
$\mathbb{E}(S)$ [MPa]	Deterministic	7.80	-
v_0 [10^6 cycles/year]	Deterministic	3.88	-
a_c [mm]	Deterministic	30.00	-
Y_0	Lognormal	1.10	20

By using methods from Chapter 3, we can then calculate the reliability indexes versus time.

Let us now consider that this joint is inspected by a non-destructive test whose detection threshold a_d is marred with uncertainties but modeled by a lognormal distribution with a mean μ_{a_d} and standard deviation σ_{a_d}. We will fix a 20% coefficient of variation for this test, and vary the mean value (0.2 mm, 0.5 mm, 1.0 mm, 2.0 mm, 5.00 mm). The detection probability can be expressed by the event margin $H = a_d - a(t)$. The detection probability at time t is therefore given by $\mathcal{P}\big(a(t) \geq a_d\big) = \mathcal{P}\big(H \leq 0\big)$.

As with the safety margin, an alternative margin is proposed:

$$\hat{H} \approx \left(\frac{1}{Y_0}\right)^m \frac{1}{\dfrac{m}{2}-1}\left[a_0^{-\frac{m}{2}+1} - a_d^{-\frac{m}{2}+1}\right] - C\,N(t)\big(2\,\mathbb{E}(S)\big)^m\,\Gamma\left(1+\frac{m}{2}\right)$$

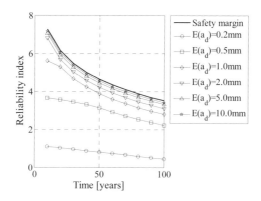

The evolution of the reliability indexes for this event margin may be compared to the reliability index calculated for the safety margin. When the detection threshold increases, the reliability index curve for the event margin comes closer to the curve for the safety margin. This indicates that the probability of detecting a crack is the same as the failure probability! Therefore it is necessary to have low detection thresholds in order to have higher detection probabilities than failure probabilities.

□

It is therefore essential to take into account the reliability of inspection results. In fact, developing *non-destructive inspection techniques* aims to detect early defects and thus to readjust the model used to predict the performance of a structural element. Mediocre quality results may lead us to make inappropriate decisions. If the non-destructive test methods and techniques are essential for structure maintenance, we should pay attention to the fact that they bring in additional uncertainties and hazards. The aim of a non-destructive inspection method is, in fact, to improve the reliability of a component by facilitating the detection of defects or cracks which could reduced its strength. Based on inspection results, maintenance can then be carried out, hence the need to have high performance and reliable techniques on hand, so as to be able to make correct decisions.

In general, the tests are based on the application of stimuli on the component, and on the other hand, on the interpretation of this component's response to these stimuli. Thus, inspections which are repeated on a given crack may lead to different responses due to variations in the initialization and calibration of the measurement system. This variability is inherent in the measurement process. Defects of the same size can also produce different responses to the same stimuli: this variability here will be due to the material's different properties and to the crack's different directions and geometries. Lastly, interpreting the response will be largely influenced by the interpreter's performance (manual or automatic), the inspector's acuity, the ease of access to the element, and by environmental conditions. All these factors contribute to the uncertainties in the inspection technique results: therefore a probabilistic characterization is vital in order to correctly apprehend the capacity of a non-destructive inspection technique, and therefore to decide on relevant maintenance strategies.

6.3.1. *Detection probability: a discrete case*

The result of an inspection or test is the outcome of an event which has or has not been conditioned by the effective or non-effective presence of that which we wish to measure (defect, damage, etc.). The result may be analyzed in four different ways (Figure 6.3):

– result E_1: there is no defect, and nothing is detected;

– result E_2: there is no defect but a defect has been detected;

– result E_3: there is a defect and it has been detected;

– result E_4: there is a defect and it has not been detected.

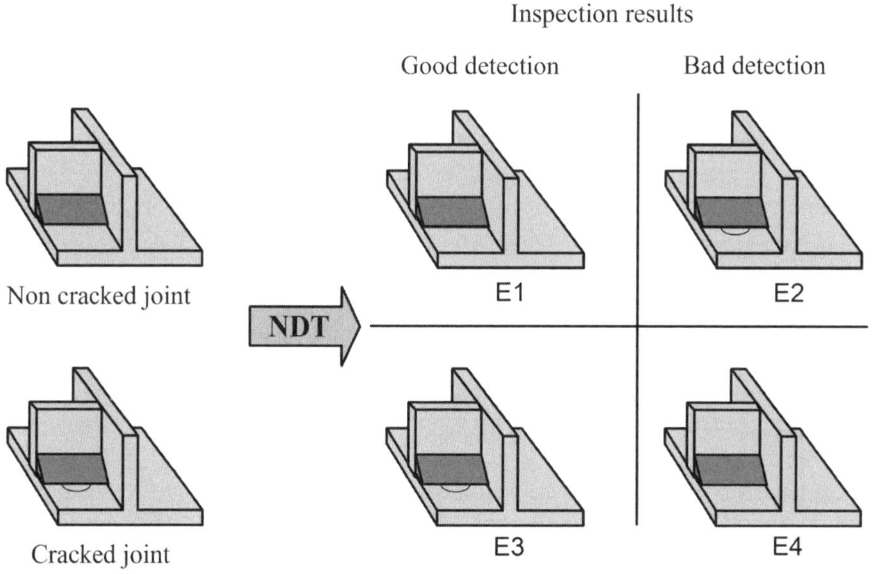

Figure 6.3. *Possible results given by a non-destructive test (NDT) when controlling a joint*

When defining these events, we are interested in the presence or absence of defects after inspection: the aim is to know whether this defect actually exists. From a probabilistic point of view, we consider a binary random variable "presence of defect A", worth 0 in the absence of a defect, and 1 otherwise. We write the random inspection function as $d(.)$ (decision function) which is worth 0 if the test decides that the defect is absent, and 1 otherwise. Thus, the *probability of a false alarm* P_{FA} and the *detection probability* P_D can be written as:

$$P_D(A) = P\big(d(A) = 1 \ \big| A = 1\big)$$
$$P_{FA}(A) = P\big(d(A) = 1 \ \big| A = 0\big)$$

[6.6]

The detection probability P_D defines the probability of deciding on the presence of a crack (detection), knowing that, effectively, there is a crack; the probability of a false alarm P_{FA} is the probability of deciding on the presence of a crack (detection), knowing that, effectively, there is *no* crack.

Let us note that the inspection results may be entirely characterized by the data of a pair (P_D, P_{FA}). The events E_i, $i = 1, 2, 3, 4$ can, in fact, be expressed according to this pair. For E_1, we obtain:

$$P(E_1) = P(A = 0 \mid d(A) = 0) = \frac{P(d(A) = 0 \mid A = 0)}{P(d(A) = 0)} P(A = 0)$$ [6.7]

Let us write P_1 as the probability of there being a crack (*a priori* probability) in the inspected area:

$$P(A = 1) = P_1; \quad P(A = 0) = P_0 = 1 - P_1$$ [6.8]

Finally, the probabilities of various events are given by:

$$P(E_1) = P(A = 0 \mid d(A) = 0) = \frac{\left(1 - P_{FA}(A)\right)\left(1 - P_1\right)}{\left(1 - P_D(A)\right)P_1 + \left(1 - P_{FA}(A)\right)\left(1 - P_1\right)}$$

$$P(E_2) = P(A = 0 \mid d(A) = 1) = \frac{P_{FA}(A)\left(1 - P_1\right)}{P_D(A)P_1 + P_{FA}(A)\left(1 - P_1\right)}$$

$$P(E_3) = P(A = 1 \mid d(A) = 0) = \frac{\left(1 - P_D(A)\right)\left(1 - P_1\right)}{\left(1 - P_D(A)\right)P_1 + \left(1 - P_{FA}(A)\right)\left(1 - P_1\right)}$$ [6.9]

$$P(E_4) = P(A = 1 \mid d(A) = 1) = \frac{P_D(A)\left(1 - P_1\right)}{P_D(A)\,P_1 + P_{FA}(A)\left(1 - P_1\right)}$$

We may also point out that some events are related by the following property:

$$P(E_1) + P(E_2) = 1; \quad P(E_2) + P(E_4) = 1$$ [6.10]

This means that a single pair of events may be considered in order to completely define the inspection. This is due to the fact that the pair (P_D, P_{FA}) is sufficient for representing a test's detection capacities.

EXAMPLE 6.8.– The lower cover of a reinforced concrete beam is tested by radar. Before the inspection, the testing laboratory assumes that there is a probability of 0.60 that the cover will be larger than or equal to 3 cm (event A). We take the reliability of the radar test for this cover thickness to be 0.70 (event *B*), i.e. the probability of obtaining a result for the estimated cover being more than 3 cm is 0.70. By applying the Bayes theorem (see Chapter 1), we obtain:

$$P(A/B) = \frac{P(B/A)P(A)}{P(B/A)P(A) + P(B/\overline{A})P(\overline{A})}$$

but:

$$P(A) = 0.6, \ \ P(\overline{A}) = 1 - P(A) = 0.4$$

$$P(B/A) = 0.7, \ \ P(B/\overline{A}) = 1 - P(\overline{B}/\overline{A}) = 0.3$$

After a test showing that the coating is more than 3 cm, the *a posteriori* probability of this cover being larger than 3 cm is:

$$P(A/B) = \frac{0.6 \times 0.8}{0.6 \times 0.8 + 0.4 \times 0.2} \approx 0.85$$

which changes the confidence probability that the cover thickness is at least 3 cm from 60% to 85%. □

EXAMPLE 6.9.– Let us assume that a new test for the cover thickness from example 6.8 gives a result which shows that the cover is at least 3 cm. By using the *a posteriori* probability calculated in example 6.8 as the *a priori* probability, we obtain:

$$P(A) = 0.85, \ \ P(\overline{A}) = 0.15$$

$$P(A/B_2) = \frac{0.85 \times 0.8}{0.85 \times 0.8 + 0.15 \times 0.2} \approx 0.96$$

Let us take the calculations from example 6.1 whilst changing the *a priori* probability from 0.6 to 0.8:

$$P(A) = 0.8, \ \ P(\overline{A}) = 0.2$$

With the first test, we obtain:

$$P(A/B) = \frac{0.8 \times 0.8}{0.8 \times 0.8 + 0.2 \times 0.2} \approx 0.94$$

With the second test, we obtain:

$$P(A/B_2) = \frac{0.94 \times 0.8}{0.94 \times 0.8 + 0.06 \times 0.2} \approx 0.98$$

We note here that a series of independent (positive) tests leads to an attenuation of the *a priori* probability effect. This property must be connected to the property of ergodic processes (memory loss of initial conditions).   □

EXAMPLE 6.10.– We will take the example of the joint in example 6.7, presuming here that the stress range is 23 MPa and that the number of annual cycles is 2 million. This joint is inspected by a non-destructive test which displays the following reliability: the probability of detection is 5% for cracks smaller than 0.2 mm, 70% for cracks between 0.2 mm and 0.3 mm, and 90% for cracks larger than 0.3 mm. The probability of missing a defect is 95% if the crack is smaller than 0.2 mm, 30% if it is between 0.2 and 0.3 mm, and 10% if it is larger than 0.3 mm. If C_f is the cost of failure, the inspection and repair costs are $0.001\,C_f$ and $0.02\,C_f$ respectively.

The event tree is given by the following diagram:

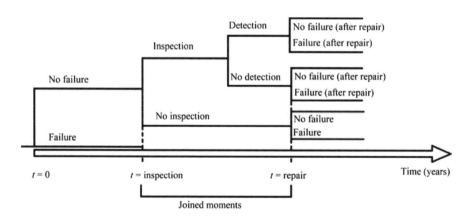

We only want to inspect this joint once over a period of 25 years. Repair quickly follows at this instant in order to join the two moments of intervention.

Repairing enables us to bring the joint back to its initial state. The decision tree is therefore limited to testing two alternatives of either inspection/no inspection at a given moment, knowing that the join is not failing at the moment of inspection. The following figure shows this decision tree.

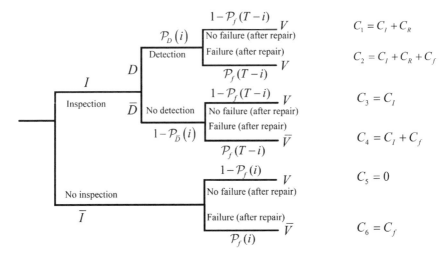

Let i be the moment of inspection. The different costs are written as:

$$\mathbb{E}(C_1) = P_D(i)\big(1 - P_f(T-i)\big)\big(C_I + C_R\big)$$
$$\mathbb{E}(C_2) = P_D(i)\big(P_f(T-i)\big)\big(C_I + C_R + C_f\big)$$
$$\mathbb{E}(C_3) = P_{\bar{D}}(i)\big(1 - P_f(T)\big)\big(C_I\big)$$
$$\mathbb{E}(C_4) = P_{\bar{D}}(i)\big(P_f(T)\big)\big(C_I + C_f\big)$$
$$\mathbb{E}(C_5) = 0$$
$$\mathbb{E}(C_6) = \big(P_f(T)\big)\big(C_f\big)$$

The probability of detection is calculated using the equation (see example 6.7):

$$P(a(t) \geq a_d) \quad \text{with} \quad a_d = 0.2 - 0.3 \text{ mm}$$

$$P_D(i) = 0.05 \times \big(1 - P(a(i) \geq 0.2)\big)$$
$$\qquad + 0.7 \times \big(P(a(i) \geq 0.2) - P(a(i) \geq 0.3)\big)$$
$$\qquad + 0.9 \times P(a(i) \geq 0.3)$$
$$P_{\bar{D}}(i) = 1 - P_D(i)$$

The following table summarizes these results.

T [yrs]	$P_f(i)$	$P_D(i)$	$P_{\bar{D}}(i)$	$\dfrac{E(C_1)}{C_f}$	$\dfrac{E(C_2)}{C_f}$	$\dfrac{E(C_3)}{C_f}$	$\dfrac{E(C_4)}{C_f}$	$\dfrac{E(C_5)}{C_f}$	$\dfrac{E(C_6)}{C_f}$	$\dfrac{E(C_{IR})}{C_f}$	$\dfrac{E(C_{RF})}{C_f}$
1	4.59E-12	6.22E-02	9.38E-01	1.23E-03	3.79E-03	8.82E-04	5.60E-02	0.00E+00	5.97E-02	6.19E-02	5.97E-02
2	6.50E-09	7.04E-02	9.30E-01	1.40E-03	3.75E-03	8.74E-04	5.55E-02	0.00E+00	5.97E-02	6.16E-02	5.97E-02
3	2.54E-07	8.37E-02	9.16E-01	1.68E-03	3.86E-03	8.62E-04	5.47E-02	0.00E+00	5.97E-02	6.11E-02	5.97E-02
4	2.64E-06	1.03E-01	8.97E-01	2.08E-03	4.06E-03	8.44E-04	5.36E-02	0.00E+00	5.97E-02	6.06E-02	5.97E-02
5	1.41E-05	1.27E-01	8.73E-01	2.59E-03	4.25E-03	8.21E-04	5.21E-02	0.00E+00	5.97E-02	5.98E-02	5.97E-02
6	5.01E-05	1.56E-01	8.44E-01	3.18E-03	4.34E-03	7.94E-04	5.04E-02	0.00E+00	5.97E-02	5.88E-02	5.97E-02
7	1.37E-04	1.87E-01	8.13E-01	3.84E-03	4.28E-03	7.64E-04	4.86E-02	0.00E+00	5.97E-02	5.75E-02	5.97E-02
8	3.13E-04	2.20E-01	7.80E-01	4.54E-03	4.07E-03	7.33E-04	4.66E-02	0.00E+00	5.97E-02	5.59E-02	5.97E-02
9	6.24E-04	2.54E-01	7.46E-01	5.26E-03	3.72E-03	7.01E-04	4.46E-02	0.00E+00	5.97E-02	5.42E-02	5.97E-02
10	1.12E-03	2.88E-01	7.12E-01	5.98E-03	3.26E-03	6.69E-04	4.25E-02	0.00E+00	5.97E-02	5.24E-02	5.97E-02
11	1.87E-03	3.22E-01	6.78E-01	6.70E-03	2.75E-03	6.38E-04	4.05E-02	0.00E+00	5.97E-02	5.06E-02	5.97E-02
12	2.91E-03	3.55E-01	6.45E-01	7.40E-03	2.22E-03	6.07E-04	3.86E-02	0.00E+00	5.97E-02	4.88E-02	5.97E-02
13	4.31E-03	3.87E-01	6.13E-01	8.08E-03	1.70E-03	5.77E-04	3.66E-02	0.00E+00	5.97E-02	4.70E-02	5.97E-02
14	6.12E-03	4.17E-01	5.83E-01	8.73E-03	1.24E-03	5.48E-04	3.48E-02	0.00E+00	5.97E-02	4.53E-02	5.97E-02
15	8.37E-03	4.46E-01	5.54E-01	9.35E-03	8.51E-04	5.21E-04	3.31E-02	0.00E+00	5.97E-02	4.38E-02	5.97E-02
16	1.11E-02	4.74E-01	5.26E-01	9.94E-03	5.44E-04	4.95E-04	3.14E-02	0.00E+00	5.97E-02	4.24E-02	5.97E-02
17	1.43E-02	5.00E-01	5.00E-01	1.05E-02	3.19E-04	4.70E-04	2.99E-02	0.00E+00	5.97E-02	4.11E-02	5.97E-02
18	1.81E-02	5.25E-01	4.75E-01	1.10E-02	1.68E-04	4.47E-04	2.84E-02	0.00E+00	5.97E-02	4.00E-02	5.97E-02
19	2.24E-02	5.48E-01	4.52E-01	1.15E-02	7.68E-05	4.25E-04	2.70E-02	0.00E+00	5.97E-02	3.90E-02	5.97E-02
20	2.73E-02	5.70E-01	4.30E-01	1.20E-02	2.91E-05	4.04E-04	2.57E-02	0.00E+00	5.97E-02	3.81E-02	5.97E-02
21	3.27E-02	5.91E-01	4.09E-01	1.24E-02	8.48E-06	3.85E-04	2.44E-02	0.00E+00	5.97E-02	3.72E-02	5.97E-02
22	3.87E-02	6.10E-01	3.90E-01	1.28E-02	1.65E-06	3.66E-04	2.33E-02	0.00E+00	5.97E-02	3.65E-02	5.97E-02
23	4.52E-02	6.28E-01	3.72E-01	1.32E-02	1.63E-07	3.50E-04	2.22E-02	0.00E+00	5.97E-02	3.57E-02	5.97E-02
24	5.22E-02	6.45E-01	3.55E-01	1.35E-02	4.28E-09	3.34E-04	2.12E-02	0.00E+00	5.97E-02	3.51E-02	5.97E-02
25	5.97E-02	6.61E-01	3.39E-01	1.39E-02	3.09E-12	3.19E-04	2.03E-02	0.00E+00	5.97E-02	3.45E-02	5.97E-02

Probabilities $P_f(i)$ and $P(a(t) \ge a_d)$ are calculated using the following safety and event margins by a first order method (FORM):

$$M = \left(\frac{1}{Y_0}\right)^m \frac{1}{\frac{m}{2}-1}\left[a_0^{-\frac{m}{2}+1} - a_c^{-\frac{m}{2}+1}\right] - C\,N(t)\left(2\,E(S)\right)^m \Gamma\left(1+\frac{m}{2}\right)$$

$$\hat{H} \approx \left(\frac{1}{Y_0}\right)^m \frac{1}{\frac{m}{2}-1}\left[a_0^{-\frac{m}{2}+1} - a_d^{-\frac{m}{2}+1}\right] - C\,N(t)\left(2\,E(S)\right)^m \Gamma\left(1+\frac{m}{2}\right)$$

$E(C_{IR}) = \displaystyle\sum_{i=1}^{4} E(C_i)$ is the total average cost related to the inspection/ repair strategy; $E(C_{DN}) = \displaystyle\sum_{i=5}^{6} E(C_i)$ is the cost when there is no inspection

("Do Nothing"). The optimum moment of inspection relates to the time when the average inspection/repair cost falls lower than the average "Do Nothing" cost, which is none other than the cost of failure.

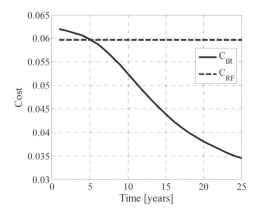

The previous result is not general, and according to the ratio between the cost of inspection/repair with the failure cost, the strategy of "Do Nothing" may prove to be better than the "inspection/repair" strategy. Let us presume that the cost of repair is 10% of the failure cost, and 1% of the inspection cost. In this case, the joint evolutions of the costs (brought back to the cost of failure) do not cross over, and over 25 years, the "Do Nothing" strategy is the most relevant (we will point out that the probabilities of detection, no detection and failure are not modified, which shows that only the costs are influential).

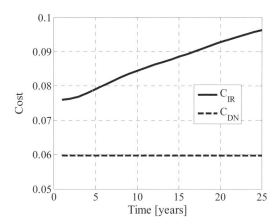

The reliability of a non-destructive inspection technique is assessed on a series of tests on specimens displaying defects, and damages with known positions and sizes. The specimen must be inspected in similar conditions to those found in practice. The first aim of these calibration tests is to demonstrate the efficiency of the non-destructive inspection technique for a particular application by estimating the probability of detecting a defect or damage A (size, type, etc.), $P_D(A)$. To demonstrate this usefulness, the defect inspection protocol must be defined correctly and all the factors which introduce variability within the inspections must represent what is found in practice. For example, k inspections over n defects is not equivalent to one inspection on $k\,n$ defects, even if the inspections are independent. The most important factors which introduce variability during the tests are:

– the differences in the physical properties of cracks with the same nominal sizes;

– the repeatability of the technique responses when the same crack is inspected by the same inspector with the same equipment;

– the human factors.

These factors must be taken into consideration for assessing the test efficiency and therefore in the design of an experiment protocol in order to quantify it.

6.3.2. *Detection probability: a continuous case*

To inspect a structure, an ideal test would be one which enables us to detect an anomaly, or a defect with certainty, when this defect exceeds a given size. This threshold a_d is called the *detection threshold* (Figure 6.4). In truth, this detection threshold is not known with certainty, and therefore must be characterized by a distribution function $F_{a_d}(.)$.

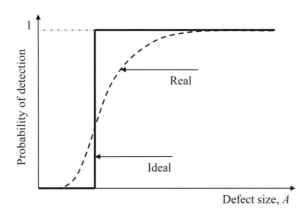

Figure 6.4. *Ideal and real detection thresholds*

The previous section characterized the reliability of an inspection technique from a discrete point of view: detection probability (false alarm, etc.) is given for states or defect size intervals. In the continuous case, the discrete distribution for the probability of detection is replaced by a continuous distribution which is represented by the detection threshold:

$$P_D(a) = F_{a_d}(a) = \mathcal{P}(a_d \leq a)$$ [6.11]

The *probability of detection* $P_D(a)$ expresses the probability of detecting a defect or damage of a given size larger than or equal to a. This measurement is the most commonly used for characterizing the performance of an inspection technique. Such a curve is generally obtained by tests on specimens showing defects of various sizes, while calculating the ratio between the number of defects detected and the number of defects which are actually present. In order to calibrate the functions $P_D(a)$ correctly, we must carry out a large number of tests.

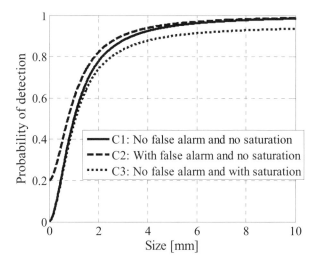

Figure 6.5. *Typical behavior of function $P_D(a)$*

It is possible to highlight three behavior types (Figure 6.5). In principle, the probability of detecting a defect which does not exist, must be zero: this is the same as saying that the probability of false alarm is zero. In the same way, the bigger the defect is, the higher the probability of detection will be, converging asymptotically towards 1. This case is represented by the curve C3 in Figure 6.5. Often, the curve $P_D(a)$ does not tend towards 1 when a tends towards $+\infty$, but tend towards a probability of $1 - p_\infty$ (p_∞ ranging from 0.01 to 0.05). It represents a saturation effect of the method which prevents us from measuring defect sizes which are too large (curve C1). The occurrence of false alarms is characterized by a non-zero probability $1 - p_0$ for $a = 0$ (curve C2). If the curve $P_D(a)$ indicates a test with no false alarm and no saturation, this distribution is easily modified in order to take into account false alarm or non-detection of large cracks:

$$\hat{P}_D(a) = (1 - p_\infty)\left[1 - \frac{1 - p_0 - p_\infty}{1 - p_\infty}(1 - P_D(a))\right] \qquad [6.12]$$

As an example, many distribution models have been proposed in order to mathematically describe the functions $P_D(a)$ which relate to the technique used to detect cracks in materials. Thus, Itagaki and Asada [ITA 77] have suggested taking an exponential model for the inspection technique which uses magnetic particles:

$$P_D(a) = \begin{cases} 0 & a < a_0 \\ 1 - \exp\left[-\dfrac{a - a_0}{\beta}\right] & a_0 \leq a \end{cases} \qquad [6.13]$$

Goranson and Rogers [GOR 83] propose a Weibull distribution for the same technique. But one distribution which is used very often is the lognormal distribution [HAR 77]. It is particularly well adapted for the ultrasound technique:

$$P_D(a) = \begin{cases} 0 & a < a_1 \\ \Phi\left(\dfrac{\ln(a - a_1) - m}{s}\right) & a_1 \leq a \end{cases} \qquad [6.14]$$

m and s are the mean and standard deviation of the normal variable. $\ln a$ $\Phi(.)$ is the cumulative probability function of a standard normal variable. Harris specifies that the lognormal model is rather well adapted for describing the distribution tail of $P_D(a)$, in comparison to the exponential model. This lognormal model is as interesting as it is coherent with the measurement accuracy.

However, the most practical model is the log-odds model [CRE 09a, CRE 10]. Characteristically it is very close to the lognormal model. Two mathematically equivalent forms of the *log-odd* model can be used. The first form is given by:

$$P_D(a) = \frac{\exp(\alpha + \beta \ln a)}{1 + \exp(\alpha + \beta \ln a)} \qquad [6.15]$$

The other form of the *log-odds* model is represented by:

$$P_D(a) = \left\{ 1 + \exp\left[-\frac{\pi}{\sqrt{3}}\left(\frac{\ln(a) - \mu}{\sigma} \right) \right] \right\}^{-1}$$ [6.16]

These two forms are equivalent to $\mu = -\alpha\,\beta^{-1}$ and $\sigma = \pi\,(\beta\,\sqrt{3})^{-1}$.

EXAMPLE 6.11.– Determining the functions $P_D(a)$ for the continuous case (also valid for the discrete case) is based on the defect sizing, since a distribution function must be generated. This is possible for some modern non-destructive inspection techniques, but is not possible for all, or the older measuring systems. The first, and oldest, approach for determining functions of detection probability was based on the sole knowledge of detecting (or not) a defect of a given size (this data being unknown to the inspector). The results from an inspection are therefore formulated in a binary way: detection/non-detection. This kind of data is called "hit/miss" (or HM). Let there be k independent inspections on n independent cracks. Let Z_{ij} be the random variable resulting from the j-th inspection on the i-th crack, with $i = 1,\ldots, n$ and $j = 1,\ldots, k$. We therefore have:

$$\begin{cases} Z_{ij} = 1 \text{ if the result from the } j\text{-th inspection of the } i\text{-th crack is a detection} \\ Z_{ij} = 0 \text{ if the result is no detection.} \end{cases}$$

Z_{ij} follows a Bernoulli distribution with parameter p_i, the probability of detecting a crack with the size a_i. The likelihood function for the sample is given by:

$$\mathcal{L}(\mu, \sigma) = \prod_{i=1}^{n} \prod_{j=1}^{k} p_i^{Z_{ij}} (1 - p_i)^{1 - Z_{ij}}$$

with:

$$p_i = \frac{h(a_i)}{h(a_i) + 1}; \quad h(a_i) = \exp\left[\frac{\pi}{\sqrt{3}}\left(\frac{\ln(a_i) - \mu}{\sigma} \right) \right]$$

assuming that the detection probability is log-odds distributed. The parameter vector to be estimated is $\theta = (\mu, \sigma)$. Thus, we are looking for $\hat{\mu}$ and $\hat{\sigma}$, such that the likelihood function is maximum, i.e. we wish to find $\hat{\theta} = (\hat{\mu}, \hat{\sigma})$ such that $\mathcal{L}(\hat{\mu}, \hat{\sigma}) \geq \mathcal{L}(\mu, \sigma)$, whatever the values of (μ, σ). With the logarithm function strictly increasing, maximizing $\mathcal{L}(\mu,\sigma)$ is equivalent to maximizing $\ln \mathcal{L}(\mu,\sigma)$. The log-likelihood function is:

$$\ln \mathcal{L}(\mu,\sigma) = \sum_{i=1}^{n}\sum_{j=1}^{k}\left[Z_{ij}\ln(p_i)+(1-Z_{ij})\ln(1-p_i)\right]$$

$\hat{\theta}$, which maximizes the log-likelihood function, must fulfill the zero gradient conditions (first derivatives) and the strictly negative Hessian conditions (second derivatives). The first condition gives the following equations:

$$\begin{cases} 0 = \displaystyle\sum_{i=1}^{n}\sum_{j=1}^{k}\frac{Z_{ij}}{k} - \sum_{i=1}^{n}\frac{h(a_i)}{h(a_i)+1} \\[2em] 0 = \displaystyle\sum_{i=1}^{n}\sum_{j=1}^{k}\frac{Z_{ij}}{k}\ln(a_i) - \sum_{i=1}^{n}\frac{\ln(a_i)h(a_i)}{h(a_i)+1} \end{cases}$$

The solutions of these equations may be determined using an iterative Newton-Raphson type method. We can note that this solution may be sensitive to the initial values μ_0 and σ_0, given that these iterative techniques converge towards local maxima or minima, hence the need to use relevant sets of initial values.

As the detection probability function is equivalent to a cumulative probability function, and its parameters have been estimated by the maximum likelihood method, we can put a lower confidence limit on the detection probability [CHE 88]. This lower confidence limit is calculating using the estimators of the variance-covariance matrix. Thus, it reflects the experiment's sensitivity to the number and size of the cracks present in the experimental specimens:

$$[\mathcal{V}]=[\mathcal{I}]^{-1}$$

where $[\mathcal{I}]$ is the Fisher information matrix, whose elements $\mathcal{I}_{lm}$ are:

$$\mathcal{I}_{lm} = -\mathbb{E}\left[\frac{\partial^2 \ln \mathcal{L}}{\partial \theta_l \, \partial \theta_m}(\theta)\right]$$

with $l,\,m = 1,\dots,\,w$ (w = total number of parameters). In applications, we replace μ and σ, known by their estimation with the maximum likelihood method. The Fisher information matrix is therefore given by (taken as the values of the estimators):

$$I(\mu,\sigma) = \frac{n}{\sigma^2}\begin{bmatrix} k_0 & -k_1 \\ -k_1 & k_2 \end{bmatrix}$$

with:

$$I_{11} = \frac{\pi^2 k}{3\sigma^2}\sum_{i=1}^{n}\frac{h(a_i)}{[h(a_i)+1]^2} = \frac{nk_0}{\sigma^2}; \quad I_{22} = \frac{\pi^2 k}{3\sigma^4}\sum_{i=1}^{n}\frac{[\ln(a_i)-\mu]^2 h(a_i)}{[h(a_i)+1]^2} = \frac{nk_2}{\sigma^2}$$

$$I_{12} = I_{21} = \frac{\pi^2 k}{3\sigma^3}\sum_{i=1}^{n}\frac{[\ln(a_i)-\mu]h(a_i)}{[h(a_i)+1]^2} = -\frac{nk_1}{\sigma^2}$$

n being the number of cracks to be detect. As we have said previously, the log-odds model is very similar to the lognormal model, at this stage we will assume that $P_D(a)$ is a lognormal distribution function with parameters $\theta = (\mu,\sigma)$. In this case, the lower confidence limit on function $P_D(a)$ is given by:

$$P_{D,\alpha}(a) = \Phi(\hat{z}-h)$$

where $\Phi(z)$ is the distribution function of the standard normal variable and:

$$\hat{z} = \frac{\ln(a)-\mu}{\sigma}$$

$$h = \left\{ \frac{\gamma}{n\,k_0}\left[1+\frac{(k_0\,\hat{z}+k_1)^2}{(k_0\,k_2-k_1^2)}\right]\right\}^{0,5}$$

where γ is determined according to the following table [CHE 88], the number of cracks analyzed in the sample (sample size) and the desired confidence level $(1-\alpha = 90\%, 95\%$ or $99\%)$.

Sample size	Confidence level			Sample size	Confidence level		
	90%	95%	99%		90%	95%	99%
5	4.197	5.519	8.800	25	3.884	5.222	8.376
6	4.133	5.467	8.692	30	3.871	5.208	8.359
8	4.051	5.393	8.584	40	3.855	5.191	8.338
10	4.002	5.344	8.523	50	3.846	5.180	8.325
12	3.969	5.311	8.482	60	3.839	5.173	8.317
14	3.945	5.287	8.454	80	3.831	5.165	8.306
16	3.928	5.269	8.432	100	3.827	5.159	8.300
18	3.914	5.254	8.415	∞	3.808	5.138	8.273
20	3.903	5.243	8.401				

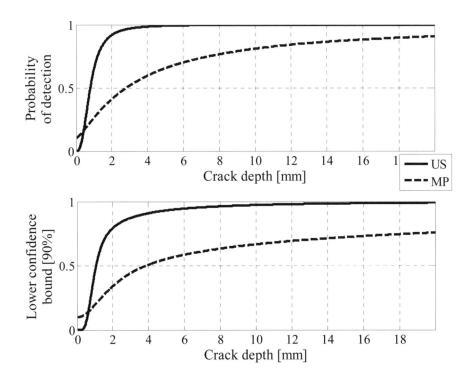

The previous figure gives an application example of this method for calibrating detection probability curves, deduced for ultrasonic and magnetic particle inspection methods on welded joints [CRE 10]; here we will note the presence of the probability of a false alarm for the magnetic particle method, which does not allow a zero probability when there is no defect.   □

6.3.3. *Load tests*

A load test consists of applying known loads to a structure which are lower than the service and design loads in order to limit any risk of failure. Therefore, it is primarily a *performance test*. If a load test is highly correlated with the studied problem, e.g. when measuring structural stiffness so as to assess the load-carrying capacity of a reinforced concrete beam element, it can be very useful in order to modify the probability density function of the load-carrying capacity by truncating it under the test load (Figure 6.6).

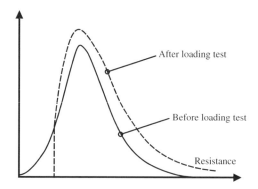

Figure 6.6. *Reduced resistance distribution after load test*

Typically, this raises the reliability, but the test itself may cause a risk of failure in an element or in the structure. Generally, only high loading levels will have a significant impact on the reliability assessment.

Traditionally, the recommendations or regulations for carrying out load tests are rather arbitrary. However, we may try a probabilistic method in order to show the way in which the information obtained can be processed. This analysis is based on a *Bayesian analysis* for calculating the updated failure probability (see Chapter 4). Let us consider, for example, a bridge subjected to a load test. We presume that the limit state is defined by:

$$M_a \leq M_u \tag{6.17}$$

where M_u is the resisting bending moment and M_a the applied bending moment. The failure probability is then given by:

$$P_f = \mathcal{P}(M_u - M_a < 0) \tag{6.18}$$

Let us perform a load test. This test provides a bending moment written M_t. The probability of failure during the test, called the *test risk*, is assessed by:

$$P_t = \mathcal{P}(M_u - M_t < 0) \tag{6.19}$$

If the structure passes the load test, the *residual risk* is then the conditional probability of the structure failing, knowing that is has passed the load test:

$$P_f^{updated} = \mathcal{P}(M_u - M_a < 0 \,|\, M_u - M_t \geq 0) \tag{6.20}$$

EXAMPLE 6.12.– Load tests are carried out on a bridge to check its load-carrying capacity. The moment due to dead loads is evaluated at 3,223 kN.m. We presume that the resisting moment is normally distributed with a bias of 1.1 over the design resistance of 11,395 kN.m and a 12% coefficient of variation. Accepting that the moment due to dead loads is also normally distributed with a bias of 1.0 and a 1% coefficient of variation, the reliability index and the direction cosines with the value of 2,600 kN.m for live loads (20% coefficient of variation and normal distribution) are given by:

$$\beta = \frac{m_R - m_{S_D} - m_{S_L}}{\sqrt{\sigma_R^2 + \sigma_{S_D}^2 + \sigma_{S_L}^2}}$$

$$= \frac{1.1 \times 11,395 - 3,223 - 2,600}{\sqrt{(1.1 \times 11,395 \times 0.12)^2 + (3,223 \times 0.1)^2 + (2,600 \times 0.2)^2}} \approx \frac{6,711.5}{1623.8} = 4.13$$

$$\alpha_R = \frac{\sigma_R}{\sqrt{\sigma_R^2 + \sigma_{S_D}^2 + \sigma_{S_L}^2}} \approx \frac{1.1 \times 11,395 \times 0.12}{1,623.8} = 0.926$$

$$\alpha_{S_D} = -\frac{\sigma_{S_D}}{\sqrt{\sigma_R^2 + \sigma_{S_D}^2 + \sigma_{S_L}^2}} \approx -\frac{3,223 \times 0.1}{1,623.8} = -0.198$$

$$\alpha_{S_L} = -\frac{\sigma_{S_D}}{\sqrt{\sigma_R^2 + \sigma_{S_D}^2 + \sigma_{S_L}^2}} \approx -\frac{2,600 \times 0.2}{1,623.8} = -0.320$$

The values of these direction cosines indicate that the resisting moment and the live load moment are the most important variables when calculating the probability of failure.

Now, assuming that the (deterministic) moment due to load tests is twice the mean value of the live load moment, then the reliability index of the test risk is given by:

$$\beta_H = \frac{m_R - m_{S_D} - m_{S_L}}{\sqrt{\sigma_R^2 + \sigma_{S_D}^2 + \sigma_{S_T}^2}} = \frac{1.1 \times 11,395 - 3,223 - 5,200}{\sqrt{(1.1 \times 11,395 \times 0.12)^2 + (3,223 \times 0.1)^2}}$$

$$\approx \frac{4,111.5}{1,538.3} = 2.673$$

The direction cosines respectively are worth:

$$\alpha_R = \frac{\sigma_R}{\sqrt{\sigma_R^2 + \sigma_{S_D}^2}} \approx \frac{1.1 \times 11,395 \times 0.12}{1,538.3} = 0.977$$

$$\alpha_{S_P} = -\frac{\sigma_{S_P}}{\sqrt{\sigma_R^2 + \sigma_{S_D}^2}} \approx -\frac{3,223 \times 0.1}{1,538.3} = -0.209$$

Chapter 4 recalls that if $M = R - S_1$ and $H = R - S_2$ respectively are the safety and event margins allowing β_M and β_H as reliability indexes, then the reliability index $\beta_M^{residual}$ of the residual risk is given by:

$$\beta_M^{residual} = \frac{\beta_M - \rho_{MH}\beta_H}{\sqrt{1 - \rho_{MH}^2}}$$

where ρ_{MH} is the correlation between margins M and H. This correlation is calculated using the direction cosines of these margins for the variables R and S_D (provided that the variables are independent): $\rho_{MH} = \alpha_{M,R}\,\alpha_{H,R} + \alpha_{M,S_D}\,\alpha_{H,S_D}$.

In the present case, the reliability index for the residual risk is given by:

$$\beta_r = \frac{\beta - \rho\,\beta_H}{\sqrt{1 - \rho^2}} = \frac{4.13 - \rho\,2.673}{\sqrt{1 - \rho^2}}a$$

with $\rho = 0.977 \times 0.926 + 0.209 \times 0.198 = 0.946$. The residual risk has $\beta_r = 4.94$ as the reliability index. This index has to be compared with the initial reliability index; if the test is positive, the structure's reliability must be increased.   □

6.4. Maintenance optimization

Structural performance is generally controlled by institutionalized procedures, such as design regulations, standards and quality assurance systems. However, are they sufficient for assessing the performance and the residual lifetime of a structure when faced with potential hazards (degradations, extreme loading, etc.)?

The core of the structural risk control lies in practice within the safety coefficients or, in modern versions, in partial safety factors. These concepts should be linked together with the concept of failure probability. However, for technical reasons, these concepts are simplified in a calibration process. The result is that the failure probability becomes a nominal value. For design standards, this estimate is based on properties of resistance and predefined loads. This calibration process also leads us to define partial factors which cover a wide variety of structures. Because these coefficients are applied generally and not specifically, the failure probability in standards or regulations does not give a correct estimation of the true failure probability of a particular structure which has been calculated using the design rule. This particularly explains why applying design rules for verifying the performance of a structure which already exists or is in use, tends to be conservative.

The second reason explaining the insufficient nature of current approaches for assessing structural performance lies in the fact that most structures are part of macro-systems, where a structure or a system will therefore essentially ensure one or many functions. Modifying its performance will induce maintenance in order to fix it. It is, then, necessary to systematically and attentively monitor the structural and service conditions and to carry out, in good time, safeguarding, maintenance, or repair procedures which make it possible to maintain serviceability and structural safety.

Techniques such as *probabilistic risk assessment* fulfill these demands and were introduced into the study of chemical and nuclear plant safety in the last 30 years. Taking into account the consequences of the failure of a component (and therefore the very idea of risk analysis) also came about very early, in the construction of offshore platforms particularly [GOY 02]:

– when introducing system analyzes in the structural analysis for justifying new designs (collapse analysis, push over analysis, member removing analysis);

– when optimizing inspection-maintenance-repair planning for assessing existing structures.

In the first case, the consequences are assessed according to the structure's redundancy. In the second case, the consequence was dealt with by introducing costs (inspection costs, repair costs, failure costs) in the optimization models. These optimization models have also been gradually integrated into the formulation of system analyses. Goyet [GOY 02] carefully describes the general methodology for performance assessment by a risk analysis, basing it on five levels:

1. Hazard identification (loading, degradation, brittle rupture, etc.).

2. Failure mode identification (component), their limit states and failure mechanisms (system).

3. Estimating failure probabilities (component and system).

4. Determining consequences of failure and assessing the risks for each mode and mechanism.

5. Determining risk scenarios and risk control (intervention, monitoring, etc.).

Of course, point 3 shows that this methodology is replaced in a probabilistic context. However, it is also desirable to establish optimization models which enable us to identify the best strategies and maintenance calendars in order to reach these objectives (point 5). Various models may be built, and may differ according to the structures or the systems, management issues and the financial budgets available. But, all the points are based on a balance between technical issues and economical consequences. It is by building and assessing the relationship between costs and benefits (called *cost/benefit analyses*) associated with each management alternative, that the optimal strategies can be identified and applied.

Maintenance can be defined as the combination of all the intended administrative and technical activities to keep a component or a system (or to restore it) in a state or condition allowing it to fulfill its expected functions. In practice, inspections represent the most relevant and efficient way of testing and detecting performance evolution. In this context, the inspections look to answer four essential questions:

– What to inspect?

– How to inspect?

– Where to inspect?

– How many times to inspect?

Obviously, the inspection process has a direct cost (intervention costs), but also indirect costs linked to the constraints imposed on the structure being inspected (restricting or stopping the service). For this reason, it is then also necessary to plan inspections in such a way so that the benefits of these inspections balance out the service restrictions caused by the inspections themselves.

Over the last 20 years, research and the implementation of inspection, maintenance and repair techniques have been carried out so as to try to fulfill the four previously requirements. In this light the probabilistic risk assessment (PRA) was developed and extended in order to apprehend the optimization of inspections and interventions. Optimizing inspection calendars has been part of much research which has been thought to be an extension of the usual methods for structural reliability. The approach was to minimize cost objective functions under constraints of reliability; the theoretical corpus generally used is the *decision analysis*. The formulations were not very flexible insofar as some decision analysis parameters were enforced arbitrarily: e.g. the very utility of the inspection was not studied and only one mitigation alternative was considered in the case of damage detection (systematic repair). In this way, the inspection of pressurized systems, containers storing dangerous products, marine structures, etc., were traditionally led on the basis of instructions which fix locations, frequencies and methods of inspection, based on equipment operational knowledge. Although not very flexible, these instructions have acquired adequate levels of reliability and safety in most cases.

Over recent years, the points of view for certain operators have, however, changed in favor of less conservative approaches which take better account of the real behavior of equipment. In fact, the regulatory instructions display certain disadvantages: they do not encourage the analysis of specific dangers, failure consequences (performance loss) and risks created by every component in the equipment. Moreover, they do not take advantage of the gain which comes from the experience of using these systems, equipment and structures correctly, and do not assign limited inspection resources on the most important hotspots. Becoming aware of this has caused a certain number of sectors, including oil chemistry, nuclear engineering, etc., to develop a risk-based procedure using experience, improving understanding of degradation mechanisms, and writing serviceability assessment procedures. In addition, economic stresses and productivity requirements have led operators to focus on *risk-based inspection* (RBI): its objective is to determine the criticality of structural components, to order them based on this criticality and then to determine, for the most critical components, an inspection-maintenance-repair calendar which respects, at best, a predefined socio-economic optimum. Risk analysis as a doctrine with its methods and tools is then turned into a reference framework from which structural inspection methodologies can be established.

6.4.1. *Identification of degradations and failure modes*

Even if certain degradation processes (such as fatigue crack propagation or corrosion) are taken into account from the design phases, they are considered according to given modeling, manufacturing, construction and operation hypotheses (including considering uncertainties on all these stages). But if these hypotheses are not verified, deterioration process(es) may lead to a loss in performance, far beyond being acceptable. In every situation, there is a difference between the theoretical reliability which characterizes the structure in the design phase, and the real reliability which characterizes the structure in the operation phase. Therefore, the question is raised regarding the ability of a structure to fulfill its initial or new performance. We must, therefore, continuously carry on to an assessment of the structure and to discuss the safety/serviceability over a time period. Moving onto this discussion with regard to updated information (advances in knowledge, regulation changes, data from the structure's service conditions, etc.) is the major objective in any inspection strategy. In each case, justifying the ability to fulfill a function will be conditioned to the implementation of an RBI analysis or the obligation to strengthen the structure by appropriate methods.

Inspection, maintenance and repair will try to ensure that the acceptable performance levels are reached, which results in the following objectives:

– ensuring that the risks concerning staff are the lowest possible within reasonable limits (ALARP principle – *as low as reasonable principle*);

– ensuring that the risks related to the environment are maintained within the limits specified over the structure's lifespan;

– ensuring that the structure stays within the design limits, enabling us to continue using it whilst complying to legislation over the structure's lifespan;

– ensuring that the operational objectives are maintained for the structure's lifespan.

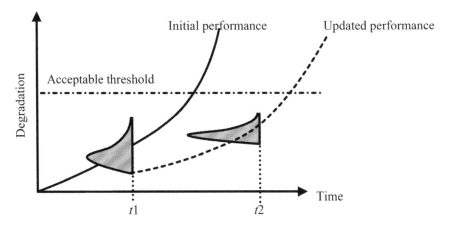

Figure 6.7. *Diagram showing an updated degradation process*

Inspections are tools which aim to reduce uncertainties related to degradations, or their criticality before they get too large. Figure 6.7 shows this result.

Figure 6.8 gives three examples of performance evolution. The first (dashed line) indicates "normal" or accepted evolution of the performance over time, for example, fatigue damage on a joint. The dotted curve denotes an abnormal performance evolution, since the acceptable threshold is reached before the design working life. Lastly, the bold line shows the performance profile with a series of interventions in order to "maintain" the performance level over time. The instant which corresponds to crossing the minimum acceptable performance is, then, the last moment to take action. In fact, beyond this point, the performance level no longer corresponds to the targeted level. This time is, then, not necessarily the optimum instant with regard to maintenance costs.

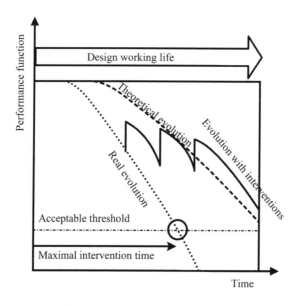

Figure 6.8. *Evolution profile of performance over time*

6.4.2. *Decision process and RBI analysis*

In an RBI analysis, the decision process may be characterized by Figure 6.9. In this figure, the parameters which define the inspection planning are listed in the set $\mathbb{I} = (\mathcal{T}, \mathcal{L}, \mathcal{R})$, where $\mathcal{T} = (t_1, \cdots, t_N)$ are the scheduled inspection times, $\mathcal{L} = (L(t_1), \cdots, L(t_N))$ are the inspected areas for each inspection with $L(t_i) = (l_1, \cdots, l_{M(t_i)})$, and $M(t_i)$ is the number of inspected areas at time t_i. Lastly, $\mathcal{R} = (R(t_1), \cdots, R(t_N))$ represents the inspection quality [SOR 02].

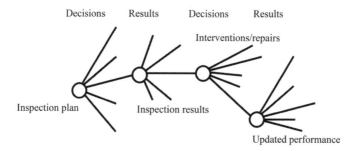

Figure 6.9. *Decision tree for an RBI analysis*

In principle, the inspection results are uncertain, insofar as they not only depend on the inspection's reliability, but also on the level of degradation. In Figure 6.9, theses uncertainties are modeled by a random vector $S = \left(S(t_1), \cdots, S(t_N) \right)$, in which the components refer to the inspection results for the $M(t_i)$ inspected areas. $\mathcal{D}(s)$ is the decision rule which defines the maintenance or repair to be carried out in the case of inspection result s. Θ represents the structure's performance.

If we refer to the concepts from the decision theory, it is appropriate to associate a utility function $\mathcal{U}(\mathbb{I}, \mathcal{S}, \mathcal{D}(\mathcal{S}), \Theta)$. The optimum inspection planning maximizes the desired utility:

$$\mathcal{U}^* = \max_{\mathbb{I}, \mathcal{D}} \mathbb{E}_{\Theta, \mathcal{S} \mid \mathbb{I}} \left[\mathcal{U}(\mathbb{I}, \mathcal{S}, \mathcal{D}(\mathcal{S}), \Theta) \right] \qquad [6.21]$$

In practice, the utility concept is replaced by costs induced by failure and maintenance, including inspections. This changes the problem of maximization into a problem of minimization with an identical level of formalism:

$$C^* = \min_{\mathbb{I}, \mathcal{D}} \mathbb{E}_{\Theta, \mathcal{S} \mid \mathbb{I}} \left[C_T(\mathbb{I}, \mathcal{S}, \mathcal{D}(\mathcal{S}), \Theta) \right] \qquad [6.22]$$

C_T represents the total cost over the design working life (also called the intervention horizon). This cost can be broken down into many costs: inspection, repair, and failure. It is natural to add constraints to this problem of minimization, which corresponds to the acceptability criteria. One of the most frequently used criteria is reliability. This implies that the failure probability $\mathcal{P}_f(T)$ cumulated over the working design life T remains lower than an acceptable failure probability $\mathcal{P}_f^0$:

$$\begin{cases} \min_{\mathbb{I}, \mathcal{D}} \left(\mathbb{E}\left[C_{insp}(\mathbb{I}, \mathcal{D}) \right] + \mathbb{E}\left[C_{rep}(\mathbb{I}, \mathcal{D}) \right] + \mathbb{E}\left[C_f(\mathbb{I}, \mathcal{D}) \right] \right) \\ \mathcal{P}_f(T) \le \mathcal{P}_f^0 \end{cases} \qquad [6.23]$$

where $\mathbb{E}\left[C_{insp}(\mathbb{I}, \mathcal{D}) \right], \mathbb{E}\left[C_{rep}(\mathbb{I}, \mathcal{D}) \right], \mathbb{E}\left[C_f(\mathbb{I}, \mathcal{D}) \right]$ represent the average costs of inspection, repair and failure respectively.

6.4.3. Maintenance types

Two major maintenance strategies may be applied to structural components: *corrective maintenance* and *preventive maintenance*. Using either of these strategies differs according to the element considered, but also according to the structure type and the operation and inspection strategy.

6.4.3.1. *Corrective maintenance*

Corrective maintenance implies that an action only takes place when a loss of performance is noted. We may also refer to *palliative maintenance* (temporary) or *curative maintenance* (permanent). In practice, this maintenance is interesting if the consequences of the loss of performance are not serious, and if preventatively avoiding the loss of performance is costly. Thus, if this loss intervenes, the component will be repaired or replaced. This type of maintenance will generally induce low costs over the first few years, then higher costs thereafter. Also, there is a risk of inducing indirect costs disturbing the infrastructure management, as well as an accumulation of maintenance works.

6.4.3.2. *Preventative maintenance*

Preventative maintenance involves reducing the probability of failure or degradation for the structure. Therefore, we distinguish between *systematic maintenance* and *conditional maintenance*. Within the framework of systematic maintenance, maintenance is regularly carried out, whatever condition the element is in. This type of maintenance may raise costs, but reduces the risk of disturbing the infrastructure's operation. For example, this is the case in cleaning small visible bridge parts (gutters, etc.) or changing mechanical parts.

In the context of conditional maintenance, interventions (number and type) are a function of the element's condition and the inspection results. Repair works may be decided on according to these results. The inspection results can be predetermined, and the time intervals between each of these times may be identical or different.

Preventative maintenance is interesting if the costs induced by a performance loss are high, and if the costs of repair and inspection are relatively low compared to the former.

6.5. Life cycle cost analysis

The challenge faced by engineers (hence the existence of a life cycle cost analysis) is to specify a set of decisions, actions and interventions, and their instants, so that a structure may reach its expected working lifetime.

The cost of a structure is never spent in one single go. A structure is a long term investment. After its design and construction, it must undergo periodic maintenance, repair or rehabilitation, allowing it to maintain durability, serviceability or structural safety. The manager will eventually be able to decide when it is time to replace it, designating its real service lifespan.

The sequence of actions or events (construction, operation, ageing, damage, repair, replacement) are known as *life cycles*. The structures's owner and the manager will have to choose the management strategies to be followed, the repairs to be carried out, based on predictions. Life cycle cost analyses form a set of economic principles and calculation procedures in order to compare current and future costs, so as to identify the technical-economical strategy which will provide the expected services for the structure. Based on technical requirements that the structure must verify, the lowest total cost of these actions will give the best strategy. From this, we can manage the financial resources more efficiently.

The life cycle cost analysis is a relatively old technique, since it dates back to the 19th Century. Its first application was in the road infrastructure sector, going back to the 1960s (*highway cost model* – HCM). This model was already considering the users' and owner's costs, as a basis for roadway management procedures. When user costs are introduced, the life cycle cost analysis becomes an extension of the cost/benefit approach. Where this model is essentially a consideration of current and future costs, the cost/benefit analysis attempts to evaluate a wide set of expenditures and returns on the investment which exceed the simple requirement of setting up a project. As an example, the cost/benefit approach has been widely used in dam construction projects, by also considering the benefits related to creating a lake (touristic area, boating activities, etc.) and not only the contribution to irrigation or electricity. The important metric of a cost/benefit analysis, the ratio of benefit over cost, requires a distinction to be made between the benefits and the costs; but the allocation of costs or benefits is often badly apprehended. The metric used in life cycle cost analysis is based on the discounted sum of all the costs. It is, however, restrictive, because the comparison between strategies sometimes requires some of the benefits.

6.5.1. *Discount calculations*

Life cycle cost concepts may constitute a technical-economic approach to assessing the cost of a structure over its lifetime, from its design to the end of its working life (including its demolition). Therefore, it is a matter of quantifying the discounted costs (i.e. costs expressed as *present* money) of the various actions taken over the structure's lifetime. The sum of these different discounted costs represents the *net present cost value* of the structure. This analysis aims to assess the investments necessary today in order to carry out these actions in the future. This enables us to determine the *global cost* of a structure, including the costs of design, construction, maintenance, repair, disturbance and destruction.

Therefore, such an analysis tries to realistically estimate how much a structure will cost in the long term, and to allow for a rational economic and financial management. It is often used for comparing various solutions for improvement.

Discounting the costs when calculating the net present cost is explained by the need to compare the costs on the same monetary basis. The present value (or present cost) is based on an investment principle. A capital P is invested over a period t (generally expressed in years) at an interest rate of r gives the final value at the end of period t:

$$C = P(1+r)^t \qquad\qquad [6.24]$$

The value P represents the *present cost* related to expenditure C in the year t for updated discount rate r:

$$P = \frac{C}{(1+r)^t} \qquad\qquad [6.25]$$

In practice, many costs may intervene over the structure's use period; if $(t_i)_{1\le i\le n}$ represents the times of intervention and $(C_i)_{1\le i\le n}$ the corresponding costs, then the total present value is given by:

$$\overline{P} = \sum_{i=1}^{n} \frac{C_i}{(1+r)^{t_i}} \qquad\qquad [6.26]$$

This approach favors the comparison of different investment and/or maintenance schemes on an equitable basis, with the scheme providing the minimum present value generally being preferred.

The *whole life cycle cost* concept is defined as the sum of the structure's investment costs and the set of discounted costs $\overline{P}$ related to inspection, maintenance, repair and demolition.

EXAMPLE 6.13.– Let us consider that two options are offered to an authority to replace a structure, either today or in 60 years, however being aware that strengthening will be necessary before it can be replaced for the second solution. These options are shown in the following table, with the discount rate being taken at 3%. The service life is fixed at 120 years.

Option 1	Operation	Cost (€)	Discounted net cost
Year			
0	Strengthening	400,000	400,000
30	Repair (concrete)	10,000	4,120
60	New structure	600,000	101,840
90	Repair (concrete)	10,000	699
120	Waterproof system repair	20,000	576
120	Repair (concrete)	15,000	432
		Total	507,667
Option 2	**Operation**	**Cost (€)**	**Discounted net cost**
Year			
0	New structure	600,000	600,000
30	Repair (concrete)	10,000	4,120
60	Waterproof system repair	20,000	3,395
60	Repair (concrete)	15,000	2,546
90	Repair (concrete)	20,000	1,399
120	None	0	0
		Total	611,459

This example shows that the strengthening solution with delayed replacement is more economically appropriate than immediate replacement.   □

6.5.2. *Discount rate*

The discount is a particularly important parameter when calculating present cost values. It varies according to the adopted investment strategy, or differently according to the financial risks encountered. In developed countries, with the exception of Japan, the rates vary between 2% (Switzerland), 3% (Germany) and 4–6% (USA, France, UK). The latter values are frequently considered to be high, where rates from 2–3% would be more realistic. Japan even uses zero rates, under the pretext that authorities will continue to allocate maintenance budgets rather than giving them to the financial market! The choice of discount rate is, therefore, essential within the framework of researching optimum intervention solutions. Trying to determine the management requirements for structures intended to last for 100 years, for example, may be debatable if the discount rate of 6% is kept. In fact, such rates have a tendency to make any economic calculation insignificant beyond a 30–40 year period of time. This involves distinguishing the *service lifetime* and the *maintenance lifetime*. During this management period, the choices made can be maintained or optimized again in light of past management strategies.

Generally, the exponential character of the relation between present and future values may have a dramatic impact on the consequences of future spending. Any life cycle cost analysis with large discount rates will be less favorable for long projects with high initial costs in relation to the benefits to come. Those defending costly projects recommend very low discount rates in order to make them viable. However, such rates may encourage the creation of projects with unreasonable costs. Figure 6.10 shows the influence of the discount rate; this figure shows that financing the replacement of a structure after 40 years with a rate of 8% are close to $\frac{1}{10}$ for those needed with a rate of 2%.

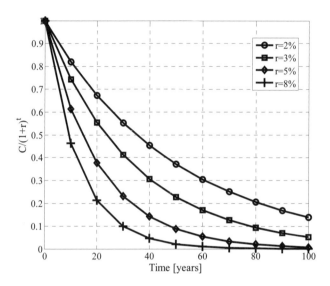

Figure 6.10. *Influence of discount rate*

Discount rates are actually made up of three components which measure:

– the real investment opportunities, or the cost of capital r_{cc};

– the financial risk related to investments r_{fr};

– the anticipated inflation rate r_i;

which gives:

$$r = (1+r_{cc})\times(1+r_{fr})\times(1+r_i)-1 \approx r_{cc}+r_{fr}+r_i \qquad [6.27]$$

If we ignore the financial risk for public organizations, an investment rate of 4% and an inflation rate of 2% then easily leads to a 6% discount rate.

EXAMPLE 6.14.– Expansion joints of a bridge are replaced every 10 years. The structure's service lifetime is 100 years, which is equivalent to replacing the expansion joints 10 times. An expansion joint is estimated to cost €5,000. The service cost is then given by:

$$\overline{P} = 5000 \times \sum_{k=1}^{10} \frac{1}{(1+r)^{10k}}$$

which, for discount rates of 2, 5 and 8%, gives €19,680, €7,890 and €4,312; a difference of more than €15,000!   □

Choosing a discount rate may be difficult. This rate is often considered to be an investment opportunity. In other words, budgets may still be regarded as financially "productive", which implies that $(1 + r) > 1$. A manager, for example, generally does not have the choice of investing part or full budget on financial investments to produce extra revenue, instead of using up his budget on maintenance, restoration, repair, or construction. In fact, the discount rate has two aims: on the one hand to reflect this financial opportunity in terms of returns on the investment (e.g. in the private sector), and on the other hand, to quantify the benefits (if there are any) of the delayed actions. A zero discount rate will imply that maintenance or repair interventions will not be important; only the total cost is a key factor. On the contrary, very high rates will favor short term alternatives. Reality is obviously between these two extremes.

Some analysts defend the idea that discount rates for public spending must be in the same order as for private spending. Others demand lower rates, of at least the market loan interest rate.

6.5.3. *Some results from discounting analysis*

Equation [6.25] shows the basic case for discounting calculations. It is a matter of estimating the present value of a single future financial flow occurring in the year t (Figure 6.11).

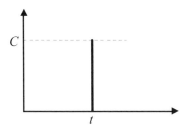

Figure 6.11. *Representation of a future single financial flow*

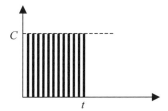

Figure 6.12. *Representation of a constant future financial flow*

Another discounting problem consists of determining the present flow value spent uniformly from year 1 to year t (Figure 6.10). In this case, the discounted value is given by the expression:

$$P = C\frac{(1+r)^t - 1}{r(1+r)^t}$$ [6.28]

Equation [6.28] is particularly useful for representing periodic maintenance spending.

Figure 6.13. *Representation of financial flow in arithmetic progression*

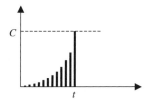

Figure 6.14. *Representation of a financial flow in geometric progression*

There are other equally important expressions used for describing costs; they deal with spending in arithmetic or geometric progression flows. These two cases, shown in Figure 6.13 and 6.14, lead us to the following present values:

– arithmetic progression flow with growth C/t :

$$P = C \left[\frac{1}{r(1+r)^t} \right] \left[\frac{(1+r)^t}{r} - t \right] \qquad [6.29]$$

– exponential progression flow with growth α:

$$P = C \left[\frac{1}{r_c(1+r_c)^t} \right] \left[\frac{(1+r_c)^t}{r_c} - t \right]; \quad r_c = \frac{1+r}{1+\alpha} - 1 \qquad [6.30]$$

where r_c is a composite discount rate.

6.5.4. *Condition, working lifetime and life cycles*

As a long term investment over many years, a structure appears as the product of decisions and actions taken, as well as in design and construction phases as in service (operation) and maintenance phases. One of the main issues in a structure's life cycle cost analysis is, without any doubt, determining this service lifetime.

In practice, this service (or working) lifetime ends when the structure is no longer useful, or when it becomes dangerous, obsolete, or unable to fulfill performance requirements in terms of function, even when resorting to repair. Experience tends to prove that, on average for bridges, this service lifetime is 50 years in developed countries. This value must be taken carefully, because exceptional structures (such as the Millau Viaduct, in France) have a prescribed working lifetime which is much higher, namely 120 years.

In a life cycle cost analysis (LCCA), this service lifetime is rather useless. In fact, the working lifetime of a structure and its elements are, in principle, filled with uncertainty. The life cycle cost analysis then involves choosing an *intervention time horizon* defined by the period over which the costs are estimated. This horizon may differ greatly from the service lifetime. Bridges will always have a working lifetime larger than this intervention horizon in a life cycle cost analysis. For example, the structure will have given years of adequate usage before the analysis is carried out, or will give many years of service well after the period used for the analysis.

The existence of many kinds of elements – structural, electrical, mechanical – introduces very different degradation mechanisms (causes, consequences). Thus, parts of civil engineering frequently have a small number of dominant degradation mechanisms, which cannot be easily predicted or anticipated. As these parts or elements are mostly unique with regard to their design, construction, loading and boundary conditions, inspections are the only means of assessing the existence or non-existence of degradation, and therefore, the structural condition. Mostly, mechanical parts display a very limited number of degradation processes (fatigue, for example). As they are mainly produced in a series, there is more statistical information on the time behavior of these degradations. However, the complexity of some of these mechanical systems makes them more difficult to inspect. Electrical and electronical parts generally have a very limited number of degradation processes. Their production in a series also means that we can estimate the degradation kinetics more accurately. However, their relatively short lifespan makes using regular inspections impractical.

Degradation profiles are generally built to include routine maintenance procedures over the working lifetime. They influence the speed of degradation, and ignoring these would give shorter working lifespans. However, using maintenance alone, without substantial repair or strengthening, means that a structure will actually have a working lifetime which is much lower than the usual values of 100 years.

Independently from a structure's ageing nature, obsolescence is also a parameter which may reduce a structure's working lifetime. It results in a change in the performance requirements (related to changes in regulation or a different use). These changes lead to us to raise the performance requirements and therefore the acceptable threshold. Such a problem is encountered in the increase of live loads which may impose higher or more frequent load effects. We can, however, identify three factors for this obsolescence:

– technological changes influence the service objectives and levels, this is the case for live loads which are higher than those used during the structure's design;

– regulatory changes involves new safety constraints which require changing the way to operate the structure;

– socio-economical changes may substantially alter the requirements, which lead to an increase of live loads in terms of volume and weight.

In most cases, obsolete structures often continue to fulfill their functions, but with performance levels which are lower than the new requirements. In this case, a structure may be preserved but with live load restrictions.

In the majority of cases (and this should even be a principle for good management!), we will not let a structure follow the degradation curve to the point of exhausting all the performance requirements without some form of intervention. From time to time, the authorities responsible for its management will have carried out repair works which will improve its performance level (Figure 6.18). The challenge faced by the manager is, then, to specify in an economically and technically realistic light, these interventions over the working lifetime. This sequence of interventions and actions is often called the life cycle's *activity profile.*

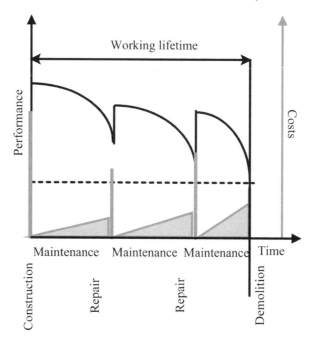

Figure 6.15. *Activity and cost profiles*

The different activities or interventions are generally associated with costs, as indicated in Figure 6.13. This *cost profile* (may also include user costs[1]) shows the importance and the occurrence times.

At the end of its service life, a structure will have a *terminal value* which must be taken into account in a life cycle cost analysis. This value may be positive (when materials or elements can be recycled), or negative (if the destruction costs exceed the cost of reselling the structure). In most cases, the intervention horizon will not coincide with the working lifetime, and a structure will then, during this horizon, have a *residual value,* attributed to the service years that it could gain up until reaching the point of unacceptable performance. This residual value differs from the terminal value.

---

1 User costs may be of different kinds: delayed costs (roads and railways), non-availability costs (energy, water, etc.).

EXAMPLE 6.15.– The cost for building a steel bridge is valued at €2,000,000, carried out over two years after one year of design. The structure's intervention horizon is 80 years. The design cost is assigned at the end of the first year and is valued at €100,000. The cost of this structure is paid at the end of the second and third year. We presume that cost is uncertain, according to the following distribution:

Final cost	10% below the net present value	Net present value	10% above the net present value	20% above the net present value
Probability	0.1	0.6	0.20	0.10

The net discounted cost for building this structure, by accepting a discount rate of 6%, is:

$$C_c = \frac{1,000,000}{(1.06)^2} + \frac{1,000,000}{(1.06)^3} = 1,729,616 \text{ €}$$

which gives the average discounted cost:

$$\overline{C}_c = 1,729,616 \times [0.90 \times 0.10 + 1.00 \times 0.60 + 1.10 \times 0.20 + 1.20 \times 0.10]$$
$$= 1,781,504 \text{ €}$$

We presume that this structure will be inspected every two years for a unitary cost of €1,000. This cost may be considered as uniform annual cost which is charged at the end of the year, at €500 per year, over 80 years. The discounted inspection cost, knowing that the inspections will not start before the 6th year, is given by:

$$C_I = 500 \left[\frac{(1.06)^{80} - 1}{0.06 \times (1.06)^{80}} - \frac{(1.06)^4 - 1}{0.06 \times (1.06)^4} \right] = 6,522 \text{ €}$$

As this structure is a steel bridge, the need for repainting is necessary. The average time between two repaintings is estimated as 15 years, for a cost of €150,000. We estimate that this cost has a 20% change of being lower than 20%, and a 20% change of being higher than 30%. The average present cost of repainting, knowing that the cost if charged at the end of the year, is therefore:

$$\overline{C}_{ep0} = 150,000 \times [0.80 \times 0.20 + 1.00 \times 0.60 + 1.20 \times 0.20] = 153,000 \text{ €}$$

$$\overline{C}_{ep0}(15 \text{ years}) = 153,000 \times \left[\frac{1}{(1.06)^{15}} + \cdots + \frac{1}{(1.06)^{75}} \right] = 108,169 \text{ €}$$

Repainting has a 10% chance of lasting 12 years, a 70% of lasting 15 years, and a 20% chance of lasting 18 years. The average present cost taking into account these three possibilities is then:

$$\overline{C}_{ep} = 0.10 \times \overline{C}_e (12 \text{ years}) + 0.70 \times \overline{C}_e (15 \text{ years}) + 0.20 \times \overline{C}_e (18 \text{ years}) = 106,860 \ €$$

The pavement is replaced every 10 years after construction. Presuming that there is a cost of €25,000 for each replacement charged at the end of every year, the present cost of this replacement is (first pavement in year 2 included in the construction costs):

$$\overline{C}_{ec} = 25,000 \times \left[\frac{1}{(1.06)^{12}} + \cdots + \frac{1}{(1.06)^{72}} \right] = 27,658 \ €$$

Knowing that the structure is located in a flood zone, measured by a rising severity of 0 to 6, the annual cost of vulnerability may be calculated using the following table (in €):

Severity of water rising	0-3	4	5	6
User cost	0	0	10,000	500,000
Agency cost	0	500	4,000	100,000
Probability	0.74	0.2	0.05	0.01

which gives us, assuming the costs to be charged at the end of each year:

$$C_a = 0.74 \times 0 + 0.20 \times 500 + 0.05 \times 4,000 + 0.01 \times 100,000 + 0.002 \times 2,000,000$$
$$= 5,300 \ €$$
$$C_u = 0.74 \times 0 + 0.20 \times 0 + 0.05 \times 10,000 + 0.01 \times 500,000 + 0.002 \times 2,000,000$$
$$= 9,500 \ €$$
$$\overline{C}_v = (5,300 + 9,500) \left[\frac{(1.06)^{80} - 1}{0.06 \times (1.06)^{80}} \right] = 244,335 \ €$$

In the year when the pavement is replaced, the number of delayed vehicles is assessed at 1,200 vehicles for an hourly cost of €8. We take the delay to be 30 min/vehicle. The user cost is therefore:

$$C_U = 1,200 \times 0.50 \times 8.00 = 4,800 \ €$$

This cost increases by 16% every 10 years. The present user cost on the intervention horizon, knowing this cost is chargeable at the end of the year, is therefore:

$$\bar{C}_U = 4,800 \frac{1}{(1.06)^{12}} + 5,571 \frac{1}{(1.06)^{22}} + \cdots + 10,105 \frac{1}{(1.06)^{62}} + 11,727 \frac{1}{(1.06)^{72}}$$

$$= 6,452 \; €$$

The set of previous present average costs allows us to calculate the net present value of the structure, disregarding the possible destruction and recycling costs:

$$\bar{C}_T = C_S + \bar{C}_c + \bar{C}_I + \bar{C}_{ep} + \bar{C}_{ec} + \bar{C}_U + \bar{C}_V$$

$$= 100,000 + 1,781,504 + 6,522 + 106,860 + 27,658 + 6,452 + 255,335$$

$$= 2,273,331 \; €$$ □

6.6. Maintenance strategies

The decisions relating to managing a particular structure are:

– maintenance strategies;

– management methods;

– maintenance application period;

– maintenance application instants.

A *maintenance strategy* is often a political decision. In particular, it is a matter of keeping the appropriate strategy which minimizes costs and maximizes the efficiency of the maintenance. Options for maintenance include [ORC 10]:

– taking no action until the structure becomes unsafe or unsuitable, or until strengthening or service restrictions are necessary;

– carrying out preventative action to reduce the degradation profile by avoiding or delaying repair, strengthening or service restrictions;

– intervening according to the inspection results following predefined or optimized inspection schedules. Following the inspection results, there are many possible choices, from do-nothing to strengthening/replacement.

Replacing a structure when it no longer fulfills its functions (loss of performance) is an alternative to strengthening, and the choice between these two strategies is effectively based on an economic decision.

6.6.1. *Corrective maintenance*

Corrective maintenance means that no action is taken so long as the structure's performance does not reach the acceptance threshold. This means that the direct (interventions and works) and indirect costs (reduced use of the structure) can be pushed back to later dates. The main disadvantage of this strategy is that it may simultaneously lead to interventions for structures in the same age bracket. This strategy may seem relevant for small groups of structures, for which the decision to replace them and let them last as long as possible has been chosen.

Let T be the period of the maintenance strategy (intervention horizon). If the acceptable performance threshold is reached at time t_k, the present (discounted) maintenance cost will be:

$$\hat{C}_M(t_k) = \frac{C_f + C_{int}}{(1+\alpha)^{t_k}} \qquad [6.31]$$

where C_f is the *cost of failure* and C_{int} is the *cost of necessary intervention* (repair, etc.). Expression [6.31] indicates that at $t = 0$, the sum $\hat{C}_M(t)$ must be designated so that the capital C_f is available when $t = t_k$. α is the *discount rate*.

If $\Delta P_f(t_k, t)$ represents the probability of failure $]t_k, t]$, then we obtain:

– for $0 \le t \le t_1$:

$$\hat{P}_f(t) = P(M(t) \le 0) \qquad [6.32]$$

– for $t_1 \le t \le t_2$:

$$\begin{aligned}\hat{P}_f(t) &= P_f(t_1) + \Delta P_f(t_1, t) \\ &= P_f(t_1) + P(M(t_1) > 0 \cap M^1(t) \le 0)\end{aligned} \qquad [6.33]$$

and so on for each inspection time. $M^k(t)$ represents the event margin after intervention at time t_k.

If $P_f(t_k)$ is the probability of failure at time t_k, then the *average present cost* for maintenance at time t_k is:

$$\mathbb{E}\left(\hat{C}_M(t_k)\right)=\frac{C_f+C_{\text{int}}}{\left(1+\alpha\right)^{t_k}}\left(\hat{P}_f(t_k)-\hat{P}_f(t_{k-1})\right) \qquad [6.34]$$

knowing that at t_{k-1} intervention has already been carried out. The total average cost of the corrective maintenance E_{MC} over the period T is then the sum of the total average costs with each intervention:

$$C_{\text{MC}}=\sum_{k=1}^{n}\frac{C_f+C_{\text{int}}}{\left(1+\alpha\right)^{t_k}}\left(\hat{P}_f(t_k)-\hat{P}_f(t_{k-1})\right) \qquad [6.35]$$

with n being the number of interventions. Within the framework of this maintenance strategy, we are looking to minimize this amount of intervention, provided that the acceptable threshold is always verified.

6.6.2. *Systematic maintenance*

The strategy for systematic maintenance consists of taking action over regular time intervals to maintain an acceptable performance. In this case, it is possible to optimize the instants of intervention. If the acceptable performance expressed by the probability of failure P_{fa} is reached, then the intervention will always be enforced, and we will presume that the probability of failure after intervention is low enough to be disregarded.

For a systematic strategy, maintenance is systematically carried out at times t_k. If failure occurs before t_k but after t_{k-1} (date of the last intervention), then the present average cost of failure is:

$$C_f=\frac{C_f}{\left(1+\alpha\right)^{t_k}}\left(\hat{P}_f(t_k)-\hat{P}_f(t_{k-1})\right) \qquad [6.36]$$

with the probability of failure $P_f(t)$ being given by equations [6.32] and [6.33]. At point t_k, the cost of intervention will be:

$$C_{\text{int}}=\frac{C_{\text{int}}}{\left(1+\alpha\right)^{t_k}} \qquad [6.37]$$

The total average cost until t_k is then:

$$C_{MS}(t_k) = \frac{C_f}{(1+\alpha)^{t_k}}\left(\hat{P}_f(t_k) - \hat{P}_f(t_{k-1})\right) + \frac{C_{int}}{(1+\alpha)^{t_k}} \qquad [6.38]$$

If the number of interventions over the service life is n (this is enforced by the manager on the difference of corrective maintenance), the average total cost over T is:

$$C_{MS} = \sum_{k=1}^{n}\left(\frac{C_f}{(1+\alpha)^{t_k}}\left(\hat{P}_f(t_k) - \hat{P}_f(t_{k-1})\right) + \frac{C_{int}}{(1+\alpha)^{t_k}}\right) \qquad [6.39]$$

This cost may be optimized in relation to the times of intervention t_k and to the number of interventions n. We can write this optimization as:

$$\min_{t_1, \cdots, t_k, \cdots t_n} \mathrm{E}_{MS} \qquad [6.40]$$

under constraint $\hat{P}_f(T) \geq P_{fa}$, the probability of failure $\hat{P}_f(t)$ being an increasing function. Here we see again the formalism of section 6.4.2.

The advantage of this strategy is based on the fact that the cost of preventative action is low in relation to costs for repair or strengthening. When there are signs of degradation, we can reduce the degradation profile. However, this, then, involves frequent maintenance. It is also clear that, for certain structures, it may be a matter of costly, useless strategies, whereas for other structures, the kind of intervention should give better results.

6.6.3. Conditional maintenance

For conditional maintenance, interventions are conditioned by the inspection results. Intervention times can either be regularly spaced, or not. Similar to the systematic strategy, a reliability constraint expressed by an acceptable probability of failure must be checked.

Let us assume that the working life is divided into n inspections at time t_k. In the case of conditional maintenance, it is important to define the criteria conditioning the actions which need to be undertaken according to the possibilities encountered (Figure 6.16). If we refer to the concept of the *event margin* introduced in Chapter 4, each intervention type will be conditioned by the occurrence of a $\{H \leq 0\}$ or $\{H = 0\}$ type event. For reasons of simplicity, we will limit ourselves to the formalism of introducing the $\{H \leq 0\}$ type event, the second case being dealt with in a similar way. Each action taken after each inspection influences the event and safety margins at the next inspection. It is necessary to introduce another notation describing the sequence of events. For example, with an intervention of type k at time t_1 and an action of type l at moment t_2, then the safety margin for failure during $t_2 \leq t \leq t_3$, is $M^{k,l}$. The event margin at time t_2 is $H^{k,l}$ (written H^k for t_1). $M(t)$ will define the safety margin before the first inspection [LUK 01b].

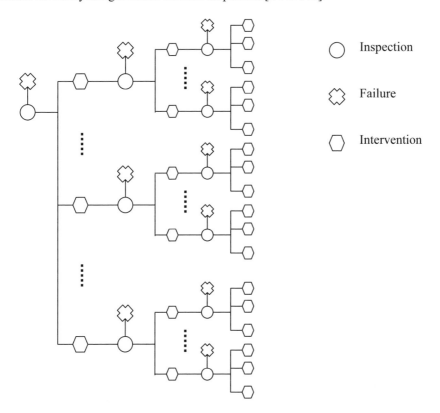

Figure 6.16. *Event tree for conditional maintenance*

If n_{int} defines the number of intervention, then the probability of intervention number r for $r \geq 2$ is also determined by:

$$\hat{P}_{\text{int}}^r\left(t_1\right) = \mathcal{P}\left(M\left(t_1\right) > 0 \cap H^r \leq 0\right) \qquad [6.41]$$

$$\hat{P}_{int}^r(t_2) = P_{int}^{0,r}(t_2) + \cdots + P_{int}^{n_R,r}(t_2)$$

$$= P\left(M(t_1) > 0 \cap H^0 \le 0 \cap M^0(t_2) > 0 \cap H^{0,r} \le 0\right)$$

$$\vdots$$

$$+ P\left(M(t_1) > 0 \cap H^{n_R} \le 0 \cap M^0(t_2) > 0 \cap H^{n_R,r} \le 0\right)$$

[6.42]

and so on for each inspection.

The *average cost of intervention* at time t_k is therefore:

$$C_{int}(t_k) = \sum_{r=1}^{n_{int}} \frac{C_{int}(r)}{(1+\alpha)^{t_k}} \hat{P}_{int}^r(t_k)$$

[6.43]

and the average cost of failure is given by equation [6.36]. Therefore the total average cost is the summation of the different terms over n inspections:

$$C_{MCo} = \sum_{k=1}^{n} \left(C_{int}(t_k) + C_f(t_k)\right)$$

$$= \sum_{k=1}^{n} \sum_{r=1}^{n_{int}} \frac{C_{int}(r)}{(1+\alpha)^{t_k}} \hat{P}_{int}^r(t_k) + \sum_{k=1}^{n} \frac{C_f}{(1+\alpha)^{t_k}} \left(\hat{P}_f(t_k) - \hat{P}_f(t_{k-1})\right)$$

[6.44]

An additional term is often added to this average total cost; it is the *average total inspection cost* where C_{insp} is the *cost of an inspection*:

$$\sum_{k=1}^{n} C_{insp}(t_k) = \sum_{k=1}^{n} \frac{C_{insp}}{(1+\alpha)^{t_k}} \left(1 - \hat{P}_f(t_k)\right)$$

[6.45]

which gives:

$$C_{MCo} = \sum_{k=1}^{n} \left(C_{int}(t_k) + C_f(t_k) + C_{insp}(t_k)\right)$$

[6.46]

In an identical way to the systematic strategy, this cost may be optimized in relation to the times of intervention t_k and the number of n interventions. This *economical optimization* is written as:

$$\min_{t_1, \cdots, t_k, \cdots, t_n} \ \mathrm{E}_{\mathrm{MCo}} \qquad\qquad [6.47]$$

under constraint $\hat{P}_f(T) \geq P_{fa}$. We may also associate constraints with the times of inspection (particularly over the period between two inspections). Once the optimal inspection schedule is obtained, it is then possible to look for scheduling interventions which make it possible to reach the optimal cost [CRE 02, CRE 03a].

EXAMPLE 6.16.– We will use example 6.10 again and analyze it from a conditional maintenance point of view. It has been decided that a detailed inspection over a period of 25 years will be carried out. The discount rate is fixed at 5%. The event tree from example 6.10 is modified as follows:

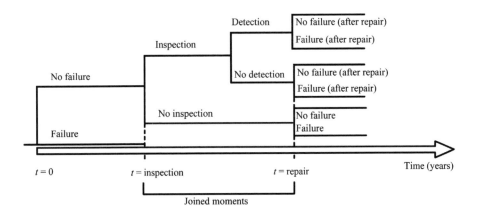

The discounted costs for inspection, repair, and failure are given by:

$$C_1(i) = \frac{C_I}{(1+\alpha)^i}\big(1 - P_f(i)\big)$$

$$C_R(i) = \frac{C_R}{(1+\alpha)^i}\,\mathcal{P}_D(i)\big(1 - P_f(i)\big)$$

$$C_f = \frac{C_f}{(1+\alpha)^i}\big(P_f(i)\big) + \frac{C_f}{(1+\alpha)^T}\big(1 - P_f(i)\big)\mathcal{P}_D(i)\big(P_f(T-i)\big)$$
$$+ \frac{C_f}{(1+\alpha)^T}\big(1 - P_f(i)\big)\mathcal{P}_{\bar{D}}(i)\big(P_f(T)\big)$$

The total cost may then be outlined according to the time of inspection; this is minimal when the inspection is carried out at around 12 years.

A very high discount rate leads to a shift in the optimum intervention time.

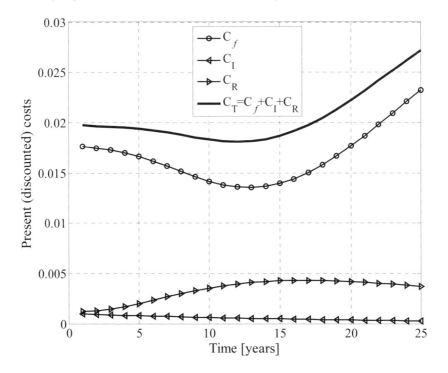

We will notice here that the cost of the "inspection/repair" solution from example 6.10 can be written as:

$$\mathbb{E}\big(C_{IR}(i)\big) = \mathcal{P}_D(i)\big(1 - P_f(T-i)\big)\big(C_I + C_R\big) + \mathcal{P}_D(i)\big(P_f(T-i)\big)\big(C_I + C_R + C_f\big)$$
$$+ \mathcal{P}_{\bar{D}}(i)\big(1 - P_f(T)\big)\big(C_I\big) + \mathcal{P}_{\bar{D}}(i)\big(P_f(T)\big)\big(C_I + C_f\big)$$
$$= \mathcal{P}_D(i)\big(C_I + C_R\big) + \mathcal{P}_D(i)\big(P_f(T-i)\big)\big(C_f\big)$$
$$+ \mathcal{P}_{\bar{D}}(i)\big(C_I\big) + \mathcal{P}_{\bar{D}}(i)\big(P_f(T)\big)\big(C_f\big)$$
$$= \mathcal{P}_D(i)\big(C_I + C_R\big) + \mathcal{P}_D(i)\big(P_f(T-i)\big)\big(C_f\big)$$
$$+ \big(1 - \mathcal{P}_D(i)\big)\big(C_I\big) + \big(1 - \mathcal{P}_D(i)\big)\big(P_f(T)\big)\big(C_f\big)$$
$$= \big(C_I\big) + \mathcal{P}_D(i)\big(C_R\big) + \mathcal{P}_D(i)\big(P_f(T-i) - P_f(T)\big)\big(C_f\big) + P_f(T)\big(C_f\big)$$

For a discount rate of 0%, the total cost is given by:

$$C_T(i) = \left(1 - P_f(i)\right)\left(C_I\right) + \mathcal{P}_D(i)\left(1 - P_f(i)\right)\left(C_R\right)$$
$$+ \left(1 - P_f(i)\right)\mathcal{P}_D(i)\left(P_f(T - i) - P_f(T)\right)\left(C_f\right)$$
$$+ \left(P_f(i)\right)\left(C_f\right) + \left(1 - P_f(i)\right)P_f(T)\left(C_f\right)$$

We notice that the two expressions differ from each other. Firstly, in example 6.10, we were looking to choose between the two alternatives, "do nothing", and "inspection/repair", by giving ourselves a decisional choice at a given time. From this we presumed that the combination was fair. This first analysis is placed, therefore, at the bottom of the decision tree at the time of inspection. The term $\left(1 - P_f(i)\right)$, a common multiplicative factor to the set of costs, was therefore removed. Secondly, the total cost of conditional maintenance involves looking at the set of possible solutions, therefore, failure before inspection (omitted from the first analysis): $\left(1 - P_f(i)\right)P_f(T)\left(C_f\right)$. The two approaches are resolutely different in their assessment of failure trees. □

Bibliography

[AAS 10] AASHTO, *LRFD Bridge Design Specifications*, Washington DC, 2010.

[AFG 04] AFGC GUIDE, *Guide for the Implementation of Performance-type and Predictive Approach Based upon Durability Indicators*, French Association of Civil Engineering, Paris, France, 2004.

[AGO 94] AGOSTONI N., BALLIO G., POGGI C., "Statistical analysis of the mechanical properties of structural steel", *Construzioni Metalliche*, vol. 2, pp. 31–39, 1994.

[ALL 81] ALLEN D.A., "Criteria for design safety factors and quality assurance expenditure", in MOAN T., SHONOZUKA M. (eds), *Structural Safety and Reliability*, Elsevier, Amsterdam, pp. 667–678, 1981.

[AND 04] ANDRIEU-RENAUD C., SUDRET B., LEMAIRE M., "The PHI2 method: A way to compute time-variant reliability", *Reliability Engineering and System Safety*, vol. 84, pp. 75–86, 2004.

[ANG 84] ANG A.H., TANG W.H., *Probability Concepts in Engineering Planning and Design*, Volumes 1 and 2, John Wiley & Sons, New York, 1984.

[BAI 96] BAILEY S.F., Basic principles and load models for the structural safety evaluation of existing road bridges, Thesis, Ecole polytechnique fédérale de Lausanne, 1996.

[BAI 02] BAILEY S.F., ANTILLE S., BEGUIN P., IMHOF D., BRÜHWILER E., "Niveau de sécurité requis pour l'évaluation des ponts-routes existants", *OFROU* 84/99, 2002.

[BAK 08] BAKER J.W., SCHUBERT M., FABER M., "On the assessment of robustness", *Structural Safety*, vol. 30, pp. 253–267, 2008.

[BAR 96] BARTLETT F.M., MACGREGOR J.G., "Statistical analysis of the compressive strength of concrete in structures", *ACI Materials Journal*, vol. 93, no. 2, 1996.

[BD 01] BD 79, Level 4 and level 5 methods of assessment for bridges, Highways Agency, Draft, 2001.

[BEN 68] BENJAMIN J.R., "Probabilistic models for seismic force design", *Journal of the Structural Division, ASCE*, vol. 94, ST5, pp. 1174–1195, 1968.

[BEN 70] BENJAMIN J.R., CORNELL C.A., *Probability, Statistics and Decision for Civil Engineers*, McGraw Hill, New York, 1970.

[BEN 00] BENTZ E.C., Sectional analysis of reinforced concrete members, Thesis, University of Toronto, 2000.

[BER 85] BERGER J.O., *Statistical Decision Theory and Bayesian Analysis*, 2nd edition, Springer-Verlag, New York, 1985.

[BOH 02] BOHIGAS A.C., Shear design of reinforced high-strength concrete beams, Thesis, Technical University of Catalonia, Barcelona, 2002.

[BRE 84] BREITUNG K., "Asymptotic approximations for multinormal integrals", *Engineering Mechanics, ASCE*, vol. 110, no. 3, pp. 357–366, 1984.

[BRI 01] BRIME, Bridge management in Europe, Final report, European project PL97-2220, European Union, 2001.

[BRO 79] BROWN C.B., "A fuzzy safety measure", *ASCE Journal of the Engineering Mechanics Division*, vol. 105, EM5, pp. 855–872, 1979.

[BRU 02] BRÜHWILER E., BAILEY S.F., "Target safety levels for the evaluation of existing bridges", IABMAS Conference, Barcelona, Spain, July 2002.

[BUC 90] BUCHER C.G. *et al.*, "A fast and efficient response surface approach for structural reliability problems", *Structural Safety*, vol. 7, pp. 57–66, 1990.

[BYF 97] BYFIELD M.P., NETHERCOT D.A., "Material and geometrical properties of structural steel for use in design", *Structural Engineer*, vol. 75, pp. 21, 1997.

[CAL 91] CALGARO J.A., "Introduction à la réglementation technique", *Annales des Ponts et Chaussées*, vol. 60, 1991.

[CAL 96] CALGARO J.A., *Introduction aux Eurocodes*, Presses de l'Ecole nationale des Ponts et Chaussées, Paris, 1996.

[CAN 00] CAN/CSA-S6, Canadian bridge design code, Toronto, 2000.

[CAS 91] CASCIATI F., NEGRI I., RACKWITZ R., "Geometrical variability in structural members and systems", *JCSS Working Document*, January 1991.

[CHE 88] CHENG R.C.H., ILES T.C., "One-sided confidence bands for cumulative distribution functions", *Technometrics*, vol. 30, no. 2, pp. 155–159, 1988.

[CIB 89] CIB W81, Actions on structures – live loads in buildings, International council for building, report 116, Rotterdam, 1989.

[CIR 77] CIRIA, Rationalization of safety and serviceability factors in structural codes, Construction Industry Research and Information Association, report no. 63, London, 1977.

[COO 97] COOPER D.I., "Development of short span bridge-specific assessment live loading", in DAS P.C. (ed.), *Safety of Bridges*, Thomas Telford, London, 1997.

[COR 67] CORNELL C.A., Some Thoughts on Maximum Probable Loads, Structural Safety Insurance, Memorandum to ASCE Structural Safety Committee, MIT, Cambridge, 1967.

[COX 09] COX L.A., *Risk Analysis of Complex and Uncertain Systems*, International Series in Operations Research & Management Science, Volume 129, Springer-Verlag, New York, 2009.

[CRE 96] CREMONA C., "Reliability updating of welded joints damaged by fatigue", *International Journal of Fatigue*, vol. 18, no. 8, pp. 567–575, 1996.

[CRE 01] CREMONA C., "Optimal extrapolation of traffic load effects", *Structural Safety*, vol. 23, no. 1, pp. 31–46, 2001.

[CRE 02] CREMONA C., *Applications des notions de fiabilité à la gestion des ouvrages existants*, Presses de l'Ecole nationale des Ponts et Chaussées, Paris, 2002.

[CRE 03a] CREMONA C., FOUCRIAT J.C., *Comportement au vent des ponts*, Presses de l'Ecole nationale des Ponts et Chaussées, Paris, 2003.

[CRE 03b] CREMONA C., "Life cycle cost concepts applied to welded joint inspection in composite bridges", *International Journal of Steel Structures*, vol. 3, no.2, pp. 95–105, 2003.

[CRE 04] CREMONA C., *Aptitude au service des ouvrages*, Presses du LCPC, Paris, 2004.

[CRE 09a] CREMONA C. *et al.*, "Determination of PoD curves for T-joints", *10th International Conference on Structural Safety and Reliability, ICOSSAR 2009*, Osaka, Japan, 13–17 September 2009.

[CRE 09b] CREMONA C., BYRNE G., MARCOTTE C., "Structural reassessment of prestressed concrete beams by bayesian updating", *10th International Conference on Structural Safety and Reliability, ICOSSAR 2009*, Osaka, Japan, 13–17 September 2009.

[CRE 10] CREMONA C. *et al.*, "Reliability of non-destructive testing for crack detection", in LARAVOIRE J. (ed.), *Steel-concrete Composite Bridges*, Ecole nationale des Ponts et Chaussées Press, Paris, 2010.

[DES 95] DESROCHES A., "Concepts et méthodes probabilistes de base de la sécurité", *Tec&Doc*, 1995.

[DEV 92] DEVALPURA R.K., TADROS M.K., "Critical assessment of ACI 318 Eq.(18.2) for prestressing steel stress at ultimate flexure", *ACI Structural Journal*, vol. 89, pp. 538–546, 1992.

[DEV 96] DEVICTOR N., Fiabilité et mécanique: méthodes FORM/SORM et couplages avec des codes d'éléments finis par des surfaces de réponse adaptatives, Thesis, Blaise Pascal University, Clermont-Ferrand, December 1996.

[DIT 79] DITLEVSEN O., "Narrow reliability bounds for structural systems", *Journal of Structural Mechanics*, vol. 7, no. 4, 1979.

[DIT 88] DITLEVSEN O., BJERAGER P., "Plastic reliability analysis by directional simulation", *Journal of Engineering Mechanics, ASCE*, vol. 115, no. 6, pp. 1347–1362, 1988.

[DIT 97] DITLEVSEN O., MADSEN H.O., *Structural Reliability Methods*, Wiley, Chichester, 1997.

[DS 03] DS 409, "Code of practice for the safety of structures", Danish Standards Association, supplement no. 1:1998, 2003.

[EN 02] EN 1990, Eurocode 0: Basis of structural design, CEN, 2002.

[EN 03] EN 1991, Eurocode 1: Actions on Structures – Part 2, Traffic Loads on Bridges, CEN, 2003.

[EN 04a] EN 1998, Eurocode 8: Design of Structures for Earthquake Resistance – Part 1: General Rules, Seismic Actions and Rules for Buildings, CEN, 2004.

[EN 04b] EN 1992, Eurocode 2: Design of Concrete Structures, CEN, 2004.

[EN 05] EN 1991, Eurocode 1: Actions on Structures – Part 1–4: Wind Actions, CEN, 2005.

[EN 06] EN 1991, Eurocode 1: Actions on Structures: Part 1–7: Accidental Actions, CEN, 2006.

[FAJ 99] FAJKUS M. *et al.*, "Random properties of steel elements produced in Czech Republic", *Eurosteel'99*, paper no. 90, Prague, Czech Republic 1999.

[FIB 06] FEDERATION INTERNATIONALE DU BETON, "Model code for service life design", Lausanne, 2006.

[GAY 04] GAYTON N., MOHAMED A., SORENSEN J.D., PENDOLA M., LEMAIRE M., "Calibration methods for reliability-based design codes", *Structural Safety*, vol. 26, pp. 91–121, 2004.

[GON 87] GONÇALVES A.F., "Resitencia do beato nase structuras", *Thesis, National Laboratory of Civil Engineering*, Lisbon, 1987.

[GOR 83] GORANSON U.G., ROGERS J.T., "Elements of damage tolerance verification", *12th ICAF*, pp. 1–20, 1983.

[GOY 02] GOYET J., STRAUB D., FABER M.H., "Risk based inspection planning of offshore installations", *Structural Engineering International*, vol. 3, pp. 200–208, 2002.

[GUL 02] GULVANESSIAN H., CALGARO J.A., HOLICKÝ M., *Designers' Guide to EN 1990 Eurocode: Basis of Structural Design*, Thomas Telford Ltd, London, 2002.

[HAR 77] HARRIS D.O., "A means of assessing the effects of NDT on the reliability of cyclically loading structures", *Materials Evaluation*, vol. 35, no. 7, pp. 57–69, 1977.

[HAS 74] HASOFER A.M., LIND C., "Exact and invariant second-moment code format", *Journal of the Engineering Mechanics division, ASCE*, vol. 100, EM1, 1974.

[HEN 98] HENRIQUES A.A., Aplicaçao de novos conceitos de sugurança no dimensionamenyto do betao estructural, Thesis, University of Porto, 1998.

[HOH 88] HOHENBICHLER M., RACKWITZ R., "Improvement of second-order reliability estimation by importance sampling", *Engineering Mechanics, ASCE*, vol. 114, no. 12, pp. 2195–2199, 1988.

[IAB 01] IABSE/AIPC, Safety, risk and reliability: trends in engineering, Malta, 21-23 March 2001.

[IAB 02] IABSE/AIPC, Safety, risk and reliability: trends in engineering, Malta, March 2002.

[ISO 99] ISO/CD 13822, Bases for design of structures – assessment of existing structures, 1999.

[ITA 77] ITAGAKI H., ASADA H., "Bayesian analysis of inspection-proof loading-regular inspection procedure", *Proceedings of HOPE Symposium*, JSME, pp. 481–487, Tokyo, Japan, 1977.

[JCS 00] JCSS, JOINT COMMITTEE ON STRUCTURAL SAFETY, Probabilistic evaluation of existing structures, RILEM, 2000.

[JCS 01] JCSS, JOINT COMMITTEE ON STRUCTURAL SAFETY, Probabilistic model code, 2001.

[JEN 01] JENSEN F.V., *Bayesian Networks and Decision Graphs*, Springer, New-York, 2001.

[KAP 92] KAPUR J.N., KERASAN H.K., *Entropy Optimization Principles with Applications*, Academic Press, London, 1992.

[KER 91] KERSKEN-BRADLEY M., RACKWITZ R., "Stochastic modeling of material properties and quality control", *JCSS Working Document*, IABSE, March 1991.

[KJE 40] KJELLMANN W., *Säkerhetsproblemet ur principiell och teoretisk synpunkt*, Ingeniörs Vetenskaps Akademien, Stockholm, 1940.

[LEM 09] LEMAIRE M., *Reliability of Structures*, ISTE, London and John Wiley & Sons, New York, 2009.

[LIN 73] LIND N.C., "The design of structural design norms", *Journal of Structural Mechanics*, vol. 1, 1973.

[LIN 76] LIN Y.K., *Probabilistic Theory of Structural Dynamics*, Krieger, Huntington, 1976.

[LUK 01a] LUKIC M., CREMONA C., "Probabilistic assessment of welded joints versus fatigue and fracture", *Journal of Structural Engineering*, vol. 127, no. 2, pp. 211–218, 2001.

[LUK 01b] LUKIC M., CREMONA C., "Probabilistic optimization of welded joints maintenance versus fatigue and fracture", *Reliability Engineering and System Safety*, vol. 72, no. 3, pp. 253–264, 2001.

[MAD 83] MADSEN H.O., BAZANT Z.P., "Uncertainty analysis of creep and shrinkage effects in concrete structures", *ACI Journal*, vol. 80, no. 2, pp. 116–127, 1983.

[MAD 86] MADSEN H.O., KRENK S., LIND N.C., *Methods of Structural Safety*, Prentice-Hall, New York, 1986.

[MAD 89] MADSEN H.O., "Optimal inspection planning fatigue damage of offshore structures", *ICOSSAR89*, pp. 2099–2106, 1989.

[MAD 90] MADSEN H.O., "Probability based optimization of fatigue design inspection and maintenance", *Symposium IOS*, Glasgow, United Kingdom, 1990.

[MAN 05] MANAGEMENT OF BRIDGES, *Anglo-French Liaison*, Thomas Telford, London, 2005.

[MAN 92] MANZOCCHI G.M.E, CHRYSSANTHOPOULOS M.K., ELNASHAI A.S., Statistical analysis of steel tensile test data and implications on seismic design criteria, ESEE report 92-7, Imperial College, London, 1992.

[MAT 95] MATTEO J., DEODATIS G., BILLINGTON D.F., "Safety analysis of suspension-bridge cables: Williamsburg Bridge", *ASCE Journal of Structural Engineering*, vol. 120, no. 11, pp. 3197–3211, 1995.

[MAY 26] MAYER M., *Die Sicherheit der Bauwerke und ihre Berechnung nach Grenzkräften anstatt nach zulässigen Spannungen*, Springer-Verlag, Berlin, 1926.

[MEL 90] MELCHERS R.E., "Radial importance sampling for structural reliability", *Engineering Mechanics*, ASCE, vol. 116, no. 1, pp. 189–203, 1990.

[MEL 99] MELCHERS R.E., *Structural Reliability Analysis and Prediction*, John Wiley & Sons Chichester, 1999.

[MEL 04] MELCHER J. *et al.*, "Design characteristics of structural steels based on statistical analysis of metallurgical products", *Journal of Constructional Steel Research*, vol. 60, pp. 795–808, 2004.

[MEN 97] MENZIES J.B., "Bridge failures, hazards and societal risks", in DAS P. (ed.), *Safety of Bridges*, Thomas Telford, London, 1997.

[MIR 79a] MIRZA S.A., HATZINIKOLAS M., MCGREGOR J.G., "Statistical descriptions of strength of concrete", *ASCE Journal of the Structural Division*, vol. 105, ST6, 1979.

[MIR 79b] MIRZA S.A., MCGREGOR J.G., "Variability of mechanical properties of reinforcing bars", *ASCE Journal of the Structural Division*, vol. 105, no. 5, pp. 921–937, May 1979.

[MIR 80] MIRZA S.A., KIKUCHI D.K., MACGREGOR J.G., "Flexural strength reduction factor for bonded prestressed concrete beam", *ACI Journal*, vol. 77, pp. 237–246, 1980.

[MIR 82] MIRZA S.A., MCGREGOR J.G., "Probabilistic study of strength of reinforced concrete members", *Canadian Journal of Structural Engineering*, vol. 9, pp. 431–447, 1982.

[MOH 07] MOHAMMADKHANI-SHALI S., CREMONA C., "Damage effects on system redundancy for existing reinforced concrete bridges using response surface methods", *ESREL, Safety and Reliability Conference*, Stavanger, Norway, ESRA, 2007.

[MUR 80] MUROTSU Y., OKIDA H., NIWA K., MIWA S., "Reliability analysis of redundant truss structures", in *Reliability, Stress Analysis and Failure Prevention Methods in Mechanical Design*, ASME, New York, H00165, pp. 81–93, 1980.

[MYE 02] MYERS R.H. *et al.*, *Response Surface Methodology*, 2nd edition, John Wiley & Sons, New York, 2002.

[NAH 01] NAHMAN J.M., *Dependability of Engineering Systems: Modeling and Evaluation*, Springer-Verlag, Berlin, 2001.

[NOR 87] NORDIC COMMITTEE ON BUILDING REGULATIONS, Guidelines for loading and safety regulations for structural design, report 55, 1987.

[NOW 99] NOWAK A.S., Calibration of LRFD bridge design code, NCHRP report 368, TRB, Washington D.C., 1999.

[NOW 00] NOWAK A.S., COLLINS K.R., *Reliability of Structures*, McGraw-Hill, New York, 2000.

[NOW 03] NOWAK A.S., SZERSZEN M.M., "Calibration of design code for buildings (ACI 318)", *ACI Structural Journal*, vol. 100, pp. 377–382, 2003.

[ORC 10] ORCESI A., CREMONA C., "A bridge network maintenance framework for Pareto optimization of stakeholders/users costs", *Reliability Engineering and System Safety*, vol. 95, pp. 1230–1243, 2010.

[OTW 70] OTWAY H.J. *et al.*, "A risk analysis of the Omega West Reactor", LA4449, University of California, Los Alamos, 1970.

[PCS 02] PCSF, Precast concrete safety factors, Final Report, European Union, SMT4 CT98 2267, 2002.

[PIP 95] PIPA M.J., Ductilidade de elementos de betao armado sujeitos a acçoes ciclicas, Thesis for the Polytechnic University of Lisbon, 1995.

[PRO 36] PROT M., "Note sur la notion de coefficient de sécurité", *Annales des Ponts et Chaussées*, vol. 27, 1936.

[PRO 53] PROT M., LEVI R., "Conceptions modernes relatives à la sécurité des constructions", *Revue Générale des Chemins de Fer*, June 1953.

[PUG 73] PUGSLEY A.G., "The prediction of proneness to structural accidents", *Structural Engineers*, vol. 51, no. 6, 1973.

[RAC 78] RACKWITZ R., FIESSLER B., "Structural reliability under combined random load sequences", *Computers and Structures*, vol. 9, 1978.

[RAC 02] RACKWITZ R., "Optimization and risk acceptability based on the life quality index", *Structural Safety*, vol. 24, pp. 297–331, 2002.

[RAC 05] RACKWITZ R., LENTZ A., FABER M., "Socio-economically sustainable civil engineering infrastructures by optimization", *Structural Safety*, vol. 27, pp. 187–229, 2005.

[RAJ 93] RAJASHEKHAR M.R. *et al.*, "A new look at the response surface approach for reliability analysis", *Structural Safety*, vol. 12, pp. 205–220, 1993.

[RIC 54] RICE S.O., "Mathematical analysis of random noise", in WAX N. (ed.), *Noise and Stochastic Processes*, Dover, New York, 1954.

[ROU 04] ROUHAN A., SCHOEFS F., "Probabilistic modeling of inspection for offshore structures", *Structural Safety*, vol. 4, pp. 379–399, 2004.

[ROY 92] ROYAL SOCIETY, *Risk: Analysis, Perception and Management*, London, 1992.

[RYA 01] RYALL M.J., *Bridge Management*, Butterworth-Heinemann, London, 2001.

[SCH 81] SCHNEIDER J., "Organization and management of structural safety during design, construction and operation of structures", in MOAN T., SHONOZUKA M. (eds), *Structural Safety and Reliability*, Elsevier, Amsterdam, pp. 467–782, 1981.

[SCH 94] SCHNEIDER J., SCHLATTER H.P., *Sicherheit und Zuverlägssigkeit im Bauwesen*, Verlag de Fachvereine, Zurich, 1994.

[SCH 97] SCHNEIDER J., *Introduction to Safety and Reliability of Structures*, IABSE, Zurich, 1997.

[SEI 01] SEILER H., BIENZ A.F., "Law and technical risks: risk-based regulation – practical experience", *Safety, Risk and Reliability, IABSE Conference*, 15-21, Malta, 2001.

[SHE 97] SHETTY N.K., CHUBB M.S., HALDEN D., "An overall risk-based assessment procedure for substandard bridges", in DAS P. (ed.), *Safety of Bridges*, Thomas Telford, London, 1997.

[SHE 99] SHETTY N.K., CHUBB M.S., "Probabilistic methods for improved bridge assessment", *Current and Future Trends in Bridge Design, Construction and Maintenance*, Thomas Telford, London, pp. 649–660, 1999.

[SIL 05] SILVA R., CREMONA C., "Some considerations on the performance cycle analysis of concrete girders in France", *Structure and Infrastructure Engineering*, vol. 1, no. 3, pp. 207–220, 2005.

[SIM 98] SIMIU E., SCANLAN R.H., *Wind Effects on Structures*, 4th Edition, John Wiley & Sons, New York, 1998.

[SOB 93] SOBRINO J.A., Evaluación del comportamiento funcional y de la seguridad estructural de puentes existentes de Hormigon Armado y Pretensado, Thesis, Technical University of Catalonia, Barcelona, 1993.

[SOR 02] SORENSEN J.D., FABER M.H., "Codified risk-based inspection planning", *Structural Engineering International*, vol. 3, pp. 195–199, 2002.

[SPA 92] SPAETHE G., *Die Sicherheit Tragender Baukonstruktionen, Zweite Neubearbeite Auflage*, Springer-Verlag, Berlin, 1992.

[STR 03] STRAUSS A., Stochastische Modellierung und Zuverlässigkeit von Betonkonstruktionen, Thesis, University of Natural Resources and Life Sciences of Vienna, 2003.

[SUD 05] SUDRET B., "Analytical derivation of the outcrossing rate in time-variant reliability problems. Reliability and optimization of structural systems", *12th WG7.5 IFIP Conference*, Aalborg, Denmark, 2005.

[TAJ 60] TAJIMI H., "A statistical method of determining the maximum response of a building structure during an earthquake", *Proceedings of the Second World Conference on Earthquake Engineering*, Kyoto, Japan, vol. 2, pp. 814–91, 1960.

[THA 87] THARMABALA T., NOWAK A.S., "Mathematical models for bridge reliability", *Canadian Journal of Civil Engineering*, vol. 14, pp. 155–162, 1987.

[THE 04] THELANDERSSON S., Assessment of material property data for structural analysis of nuclear containments, report TVBK-3051, Lund University, 2004.

[THO 83] THOFT-CHRISTENSEN P., SORENSEN J.D., "Reliability analysis of elasto-plastic structures", *Proceedings of the 11th IFIP Conference on System Modeling and Optimization*, Copenhagen, Denmark, July 1983.

[THO 84] THOFT-CHRISTENSEN P., SORENSEN J.D., "Reliability analysis of elasto-plastic structures", *Comptes-rendus de la conférence IFIP'11*, Copenhagen, Denmark, 1984.

[THO 86] THOFT-CHRISTENSEN P., MORUTSU Y., *Application of Structural Systems Reliability Theory*, Springer-Verlag, Berlin, 1986.

[TIC 79a] TICHÝ M., "Dimensional variations", in *Quality Control of Concrete Structures*, RILEM, Stockholm, pp. 171–180, 1979.

[TIC 79b] TICHÝ M., "Variability of dimensions of concrete elements", in *Quality Control of Concrete Structures*, RILEM, Stockholm, pp. 225–227, 1979.

[VRI 92] VRIJLING J.K, "What is an acceptable risk?", *Annales des Ponts et Chaussées*, Paris, 1992.

[WAL 81] WALKER A.C., "Study and analysis of the first 120 failure cases, structural failures in buildings", *Institution of Structural Engineers*, pp. 15–39, 1981.

[WAS 40] WÄSTLUND G., *Säkerhetsproblemet fur praktisk-konstruktiv synpunkt*, Ingeniörs Vetenskaps Akademien, Stockholm, 1940.

[WEI 39] WEIBULL W., "A statistical theory of the strength of materials", *Proceedings from the Swedish Society of Engineers*, vol. 15, 1939.

[WIE 39] WIERZBICKI M.W., "La sécurité des constructions comme un problème de probabilité", *Annales de l'Académie Polonaise de Sciences Techniques*, vol. VII, 1939–1945.

[WIS 07] WISNIEWSKI D.F., Safety formats for the assessment of concrete bridges, Thesis, University of Porto, 2007.

[ZHO 84] ZHONG W.P., SHAH H.C., "Reliability assessment of existing buildings subjected to probabilistic earthquake loadings", *ICOSSAR'85*, Kobe, Japan, vol. 1, pp. 567–572, 1985.

Index